FOREST ECOLOGY AND CONSERVATION

FOREST ECOLOGY AND CONSERVATION

Mathew Kent
Jannet Hillborn

Kruger Brentt
Publishers
2023

Kruger Brentt Publishers UK. LTD.
Company Number 9728962

Regd. Office: 68 St Margarets Road, Edgware, Middlesex HA8 9UU

ISBN: 9781787150645

For information on all our publications visit our website at http://krugerbrentt.com/

PREFACE

The destruction of forests and wildlife is not just a biological issue. The biological loss is strongly correlated with the loss of cultural diversity. Such losses have increasingly marginalised and impoverished many indigenous and other forest-dependent communities, who directly depend on various components of the forest and wildlife for food, drink, medicine, culture, spirituality, etc. Within the poor, women are affected more than men. In many societies, women bear the major responsibility of collection of fuel, fodder, water and other basic subsistence needs. As these resources are depleted, the drudgery of women increases and sometimes they have to walk for more than 10 km to collect these resources. This causes serious health problems for women and negligence of home and children because of the increased hours of work, which often has serious social implications. The indirect impact of degradation such as severe drought or deforestation-induced floods, etc. also hits the poor the hardest. Poverty in these cases is a direct outcome of environmental destruction. Therefore, forest and wildlife, are vital to the quality of life and environment in the subcontinent. It is imperative to adapt to sound forest and wildlife conservation strategies. Forest conservation has become one of the most important environmental issues currently facing humanity, as a result of widespread deforestation and forest degradation. Pressures on remaining natural forests continue to intensify, leading to high rates of biodiversity loss. Understanding how human activities influence ecological processes within forests is essential for developing effective conservation action. Forest conservation is the practice of planning and maintaining forested areas for the benefit and sustainability of future generations. Forest conservation involves the upkeep of the natural resources within a forest that are beneficial for both humans and the ecosystem.

The present book contains thirteen chapters covering all related disciplines. These chapters include introduction, remote sensing to forest ecology and conservation, forest structure and composition, measuring trees, floristic composition, species diversity and similarity, assessing the presence, conservation and management of endangered plant species, characterizing forest disturbance regimes, seed bank in forest regeneration, ecological models for conservation, restoration, and management of forests, reproductive ecology and genetic variation, forest as habitat, towards effective forest conservation, international conventions and treaties dealing with forest and wildlife conservation. Related terminology is given at the end for ready reference. It is written in a lucid style with the most recent advances, and with theoretical discussions for easy understanding of the subject. Well labelled illustrations, diagrams, tables and figures amply support the text wherever necessary. It provides basic principles of forest ecology and conservation in a simple and easy language. The book is also supported with a comprehensive glossary and index in the end. This book would be an invaluable asset for the scientists, students, teachers and foresters working in the field of forestry, environmental sciences, plant science and botany.

We are grateful to all those persons as well as various books, manuals, periodicals, magazines, journals etc. that helped in the preparation of this book. In spite of great care taken by me and publisher team, some errors may have escaped from our notice. We shall appreciate if such errors are brought to our notice. There will be always being some scope for further improvement. Further queries, constructive suggestions and criticisms for the improvement of the book are always welcome and shall be thankfully acknowledged.

Mathew Kent

Jannet Hillborn

CONTENTS

1

INTRODUCTION

This book is dedicated to forest ecology and conservation on ecological and conservation aspects of forest. Forest ecology is the scientific study of the interrelated patterns, processes, flora, fauna and ecosystems in forests. The management of forests is known as forestry, silviculture, and forest management. A forest ecosystem is a natural woodland unit consisting of all plants, animals, and micro-organisms (Biotic components) in that area functioning together with all of the non-living physical (abiotic) factors of the environment.

Ecosystem, as defined by the Convention on Biological Diversity means a dynamic complex of plant, animal and microorganism communities and their non-living environment interacting as a functional unit. Conservation of the ecosystem entails preservation and regulated utilization of the ecosystem.

Forests occur naturally wherever environmental factors allow the growth of trees to be a successful strategy, and tend to dominate the landscape if abiotic conditions do not restrict the competitive advantages linked to being a tall plant.

Trees and other woody plants allocate more resources than do other vegetable life-forms to the production of supportive tissues like stems, a strategy which is advantageous for gaining height and a competitive edge over any smaller neighbours in capitalizing access to solar radiation for photosynthesis, but it also has the associated costs of maintenance and reproduction, a disadvantage which can limit carbon gain through primary production.

As a consequence, if environmental conditions become adverse for photosynthetic production of stems, then trees will decline in both stature and density until they are no longer able to persist as a major component of plant communities. At a large scale, unfavorable settings for trees occur mainly through the interaction of soil types with climatic variables like temperature, precipitation and moisture, but also because of variation in local-scale environmental parameters like topography, microclimates, or natural and human-caused disturbance events, which include fire regimes, as well as windstorms and irruptions of pests and pathogens. The effects of environmental variables on the survival and abundance of trees help to explain the major patterns of the biogeographical distribution of forests and other terrestrial ecosystems.

The relevance of forest ecology to understanding biological diversity lies in the fact that forests dominate the natural landscape over much of the Earth's surface and harbor a large proportion of the world's biodiversity. Undisturbed areas with moderate to high average annual precipitation tend to be covered with forests, and at least three of the eight major terrestrial biomes are forests: tropical, temperate and conifer (also known as boreal, or polar) forests.

There were an estimated 1700 million ha of tropical forests and 1600 million ha of temperate forests worldwide. However impressive, these figures represent only about 60% of the original forest cover estimated to have existed at the advent of agriculture, approximately 8000 years ago; much of the corresponding loss can be directly attributed to human impacts during the past three millennia, with increased deforestation rates in recent centuries. Depending on the sources, forest ecosystems currently cover about 32% of the Earth's land surface, and provide numerous ecologic and economical services. For example, it is known that forests perform a critical role in regulating the water cycle in nature, by mitigating floods, droughts, the erosive forces of wind and rain, and the silting of dams and irrigation canals; moreover, forests remain an important source of wood for construction, fuel, and many other uses, like pulp to make paper.

Natural habitats such as forests have long been providing local people with the means for survival, supplying food (meat, nuts, fruits, vegetables), fodder, firewood, construction materials medicinal plants and wild genes for domestic plants and animals. Forests, which are highly diverse natural ecosystems supporting this wealth of species, also provide other important ecological services such as maintaining hydrological cycles, springs, streams, regulating climate, contributing to the processes of soil formation and maturation, storing and cycling essential nutrients, absorbing and breaking down pollutants and providing sites for tourism, recreation and research.

Being among the major habitats for biological diversity of plants, animals and other organisms, it is clear that by managing the forests for sustained multiple products, one is also managing biodiversity. Forest ecosystems are many and are variable in structure and composition, and therefore management objectives differ from one forest to another. Watershed forests, also referred to as catchment forests, cover land areas that drain to some point in a stream or river.

Three major types of watersheds are recognized. First are watersheds found in forest reserves namely catchment forests; second are watersheds on deforested or degraded high plateaus or mountains; and thirdly there are wetlands that exist on gentle ground or on plateaus outside forest reserves. These wetlands are rich in diversity of water plants, birds, fish and reptiles.

The maintenance of a forest canopy reduces soil erosion while the maintenance of surface ground cover slows surface runoff and increases water infiltration in the ground. Such degradation results into loss of habitats for plants and animals and hence loss of biodiversity.

Soils support forests, animals and other organisms. Management and conservation of such watershed ecosystems have therefore, direct relationship with management and conservation of soils and biodiversity in general. However, management is guided by the policy and legislation, the latter being a tool of management.

STRUCTURE AND FUNCTIONS OF ECOSYSTEMS

Most ecosystems gain energy from the sun and materials from the air or rocks, transfer these among components within the ecosystem, then release energy and materials to the environment. The essential biological components of ecosystems are plants, animals, and decomposers. Plants capture solar energy in the process of bringing carbon into the ecosystem. A few ecosystems, such as deep-sea hydrothermal vents, have no plants but instead have bacteria that derive energy from the oxidation of hydrogen sulfide (H_2S) to produce organic matter.

Decomposer microorganisms (microbes) break down dead organic material, releasing CO_2 to the atmosphere and nutrients in forms that are available to other microbes and plants. If there were no decomposition, large accumulations of dead organic matter would sequester the nutrients required to support plant growth. Animals are critical components of ecosystems because they transfer energy and materials and strongly influence the quantity and activities of plants and soil microbes.

The essential abiotic components of an ecosystem are water; the atmosphere, which supplies carbon and nitrogen; and soil minerals, which supply other nutrients required by organisms.

An ecosystem model describes the major pools and fluxes in an ecosystem and the factors that regulate these fluxes. Nutrients, water, and energy differ from one another in the relative importance of ecosystem inputs and outputs vs. internal recycling. Plants, for example, acquire carbon primarily from the atmosphere, and most carbon released by respiration returns to the atmosphere. Carbon cycling through ecosystems is therefore quite open, with large inputs to, and losses from, the system.

There are, however, relatively large pools of carbon stored in ecosystems, so the activities of animals and microbes are somewhat buffered from variations in carbon uptake by plants. The water cycle of ecosystems is also relatively open, with water entering primarily by precipitation and leaving by evaporation, transpiration, and drainage to groundwater and streams. In contrast to carbon, most ecosystems have a limited capacity to store water in plants and soil, so the activity of organisms is closely linked to water inputs. In contrast to carbon and water, mineral elements such as nitrogen and phosphorus are recycled rather tightly within ecosystems, with annual inputs and losses that are small relative to the quantities that annually recycle within the ecosystem. These differences in the "openness" and "buffering" of the cycles fundamentally influence the controls over rates and patterns of the cycling of materials through ecosystems.

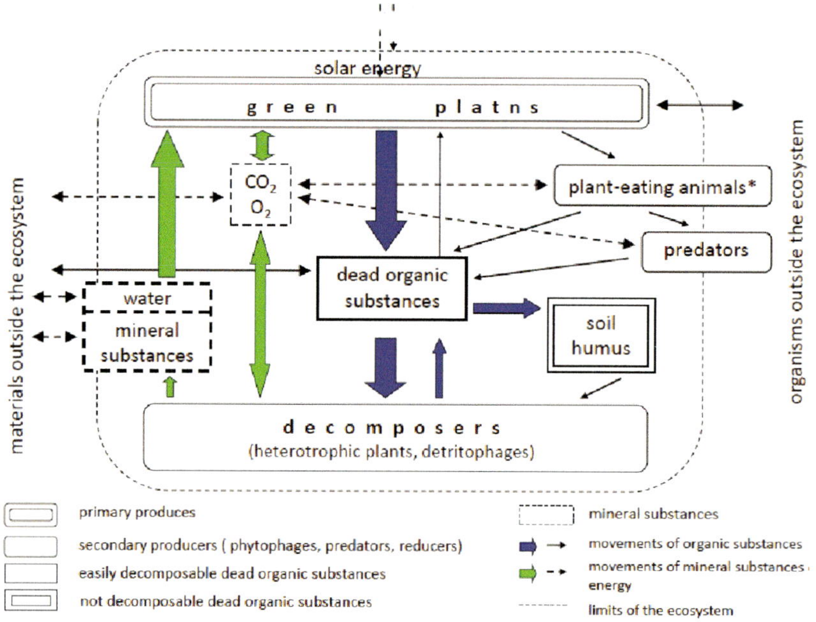

Fig. 1. Model of ecosystem structure and functionig

FORESTS AS HABITATS

Simply put, habitat is the physical area (place) where an organism lives; with some rhetoric, a habitat can be thought of as any part of the biosphere where a particular species can live, either temporarily or permanently. But all these ideas are useless as heuristic tools. In fact, the habitat concept relates the presence of a species in a site to the attributes of that place, including the various factors we commonly recognize as components of habitats: vegetation cover, quantity and availability of food, presence of water, and the like. Taking into account the diverse meanings of the term, researchers reviewed the use of the habitat concept in ornithology and proposed the following restricted definition of habitat: it is a subset of the physical environmental factors that a given species requires for its survival and reproduction. The habitat concept and the ecological niche are not mutually exclusive, since each is closely intertwined with the other, and both can be modelled using the multidimensional approach pioneered by G. Evelyn Hutchinson; actually, each habitat may provide many different ecological niches.

The more recent definition by Hall et al. can be particularly relevant to wildlifers in general, and to the students of birds in prey in forest ecosystems in particular: habitat is the total sum of the resources and conditions present in an area that produce occupancy, including survival and reproduction, by a given animal species; so, the idea of habitat is specific to a particular species, and can be much more than a single vegetation type or vegetation structure. In addition to considering habitats versus non-habitats, habitats differ in habitat quality, which is the ability of the environment to provide conditions appropriate for individual survival and reproduction, and population persistence.

The real way in which a species uses a collection of habitat elements to meet life requisites is known as habitat use, and can be broken to specific needs, like foraging habitats, or nesting habitats, and so on; finally, the behavioral responses of animals that allow them to distinguish and actively choose among various components of the environment is called habitat selection/preference, which result in the disproportional use of some conditions over another ones.

TYPES OF ECOSYSTEM

There are two main ecosystems:

1. **Terrestrial Ecosystems:** Terrestrial ecosystems are classified into various types based on temperature and annual rainfall. A few examples of terrestrial ecosystems are listed below:
   b. Grassland Ecosystems
   c. Forest Ecosystems
   d. Desert Ecosystems
   e. Cropland Ecosystems
2. **Aquatic Ecosystems:** Aquatic ecosystems are majorly two types:
   a. Freshwater/Inland Ecosystems
   b. Marine/Saltwater Ecosystems

FOREST ECOSYSTEM

A forest ecosystem is an ecosystem of forests and resources. Forests are renewable natural resources. Forests are formed by a group of plants that are structurally defined by their trees, shrubs, herbs, climbers, and ground cover. Soil, animals, insects, microorganisms, and birds are the most important interacting units of a forest ecosystem. In India, the forests occupy about 18-20% of the total land area.

1. **Abiotic Components** of the forest include inorganic and organic components present in the soil along with temperature, rainfall, light, etc.
2. **Biotic Components** are represented by producers, consumers, and decomposers.

TYPES OF FOREST ECOSYSTEMS

There are a few types of forest ecosystems listed below:

1. **Tropical Evergreen Rainforest:** Only a small percentage of tropical forests are rainforests where average rainfall is 80-400 inches in a year. This forest is characterised by deep and dense vegetation consisting of tall trees reaching different levels.

2. **Tropical Deciduous Rainforest:** The main characteristic of tropical deciduous rainforest are broad-leaved trees along with dense bushes, shrubs, etc. Two main seasons – summer and winter – are distinctly visible there. This type of forest is found in many parts of the world. A large variety of flora and fauna are found here.
3. **Temperate Evergreen Forest:** Temperate evergreen forest is a type of forest that is characterised by a smaller number of trees but an adequate number of ferns and mosses.
4. **Temperate Deciduous Forest:** Temperate deciduous forest evolves in the moist temperate region with sufficient rainfall. Here also, winter and summer are well defined, and trees shed their leaves during winter. Dominant trees are maple, oak, peach, etc.
5. **Taiga/Boreal:** Situated just south of the Tundra, Taiga is characterised by evergreen conifers. The average temperature is below the freezing point for almost half of the year.

COMPONENTS OF FOREST ECOSYSTEM

1. **Producers:** Producers can synthesise their own food by the photosynthesis process. All green plants are considered producers of the ecosystem as they convert sunlight into the chemical energy of food.
2. **Primary Consumers:** Since the consumers can not prepare their own food, they depend on producers. Herbivorous animals get their food by eating the producers (plants) directly. Examples of primary consumers are grasshoppers, deer, etc.
3. **Secondary Consumers:** Secondary consumers draw their food from primary consumers.
4. **Decomposers:** The decomposers of the forest ecosystem break down dead plants and animals, returning the nutrients to the soil so that they can be used by the producers. Apart from bacteria, ants and termites are important decomposers in the Amazon rainforest. Millipedes and earthworms also help to break down dead matter.
5. **Nutrient Cycle:** The nutrient cycle is cyclic. For the proper functioning of ecosystems, nutrients are required. Carbon, hydrogen, oxygen, and nitrogen constitute about 95%95% of the mass of living organisms. About 1515 to 2020 other elements are also needed in relatively small amounts. These are recycled repeatedly between the living and non-living components of the ecosystem.
6. **Energy Flow:** In a forest ecosystem, the grass, which draws its nutrition from sun, soil and water, is eaten by the grasshopper, which in turn is eaten by frogs, snakes, and vultures in succession (different trophic levels). In this process of eating and being eaten, nutrients are passed from one step to the next in a food chain. The flow of energy that occurs along a food chain is called energy flow. The pyramid of energy represents the total quantity of energy at each trophic level of a food chain. The flow of energy is always unidirectional.

CHARACTERISTICS OF FOREST ECOSYSTEM

1. Warm temperatures and sufficient rainfall are characteristics of forests, resulting in the formation of numerous ponds, lakes, etc.
2. The forest maintains climate and rainfall.
3. The forest supports many wild animals and protects biodiversity.
4. The soil is rich in organic matter and nutrients, which support the growth of trees.

FUNCTIONS OF FOREST ECOSYSTEM

1. **Goods Obtained from Forests:** There are various types of food products such as honey, wild meat, fruits, mushrooms, palm oil and wine, and medicinal plants obtained from forests. Other than edible parts, we can obtain timber, wood biomass, cork, etc. from forests. The fuel can be extracted from old trees that are buried under the soil.
2. **Ecological Functions:** Forests play an important role in maintaining ecological factors such as climate, carbon storage, nutrient cycling, and rainfall.
3. **Culture and Social Benefits:** The tribal people who live in the forests treat forests as nature goddesses. The traditional beliefs and spirituality save wild animals from hunters and cutting down of trees by urban people. A few modern people visit forests for recreation.

GEOGRAPHICAL DISTRIBUTION OF FOREST IN THE WORLD

Plant distributions is governed by a combination of historical factors, ecophysiology and biotic interactions. The set of species that can be present at a given site is limited by historical contingency. In order to show up, a species must either have evolved in an area or dispersed there (either naturally or through human agency), and must not have gone locally extinct. The set of species present locally is further limited to those that possess the physiological adaptations to survive the environmental conditions that exist. This group is further shaped through interactions with other species.

Plant communities are broadly distributed into biomes based on the form of the dominant plant species. For example, grasslands are dominated by grasses, while forests are dominated by trees. Biomes are determined by regional climates, mostly temperature and precipitation, and follow general latitudinal trends. Within biomes, there may be many ecological communities, which are impacted not only by climate and a variety of smaller-scale features, including soils, hydrology, and disturbance regime. Biomes also change with elevation, high elevations often resembling those found at higher latitudes.

Differences in temperature or precipitation determine the types of plants that grow in a given area. Generally speaking, height, density, and species diversity decreases from warm, wet climates to cool, dry climates. Researchers classified plant life forms based on traits that varied with climate. One such system was based on the location of the perennating organ.

Fig. 2. World biomes are based upon the type of dominant plant.

These are tissues that give rise to new growth the following season, and are therefore sensitive to climatic conditions. The relative proportions of different life forms vary with climate. In fact, life form spectra are more alike in similar climates on different continents than they are in different climates on the same continent. Regions of similar climate and dominant plant types are called biomes. This chapter describes some of the major terrestrial biomes in the world; tropical forests, savannas, deserts, temperate grasslands, temperate deciduous forests, Mediterranean scrub, coniferous forests, and tundra.

THE MAN AND THE FOREST

Soil and forest development

Much of human society today has at least some interaction with forests and the biodiversity they contain and all people benefit from the functions provided by components of this biodiversity in the carbon, water and nutrient cycles and through the links with food production. Examine the benefits that people derive from forests in terms of livelihoods, food security and human health.

Benefits to lives and livelihoods

In both low- and high-income countries and in all climatic zones, communities that live within forests rely the most directly on forest biodiversity for their lives and livelihoods, using products derived from forest resources for food, fodder, shelter, energy, medicine and income generation.

Rural people often participate in the value chains of forest biodiversity, for example by collecting wood and non-wood products from nearby forests for personal use or sale, or engaging in forest-product industries or value addition.

Non-consumptive uses of forest biodiversity, such as recreation and tourism, are also a growing part of rural cash economies. Each year an estimated 8 billion visits are made to protected areas, many of which are forest covered.

Indigenous peoples depend to a high degree on forest biodiversity for their livelihoods, although this relation is in flux as their linkages with national and global monetary economies grow. Areas managed by indigenous peoples (approximately 28 percent of the world's land surface) include some of the most ecologically intact forests and many hotspots of biodiversity.

Forests in protected areas

Globally, 18 percent of the world's forest area, or more than 700 million hectares fall within legally established protected areas such as national parks, conservation areas and game reserves (IUCN categories I-IV). The largest share of forest in protected areas is found in South America (31 percent) and the lowest in Europe (5 percent).

Forest Conservation

Ever since humans evolved, we have changed the land cover of the earth gradually. Consequently, one of the most pressing issues today is the conservation of forest. However, we need to understand the definition of the forest first:

"Forests are defined as large areas of land with trees."

There are over 800 definitions of forest that are used around the world. And it is not necessary that an area has to be populated with trees for it to be deemed as a forest. And in some cases, an area can be legally designated as a forest, regardless of the type of local vegetation. Overall, forests are the natural habitat of large scale wildlife, growth of trees, shrubs and different variety of plants which, unfortunately, are dwindling every year. Hence, conservation of forest is an important responsibility that all of us have to undertake.

Almost 31% of the earth's land surface is covered by forests. According to a 2015 report, 23% of India's land cover consists of the forest region. Trees and plants in the forest are an integral part of the ecosystem. It sustains life on the planet, provides clean air and shelter. Also, forests help conserve biodiversity.

Forests provide many resources such as food, medicine, fabric, and raw materials. Apart from keeping a check on the global temperature, forests also contribute towards preventing soil from getting eroded and shelters more than 80% of the animal species and terrestrial biodiversity. They also aid in improving the socioeconomic aspects of a country.

Wildlife refers to the animal species that are not domesticated. So, any living organism that lives in the forest region is associated with wildlife. It is found in almost all ecosystems such as rainforests, boreal forests, plains, grasslands, deserts etc. Wildlife provides great stability to **our environment** wherein they are involved in natural processes either directly or indirectly. Each living organism is placed in the food chain and is equally important, they may be a producer, a consumer or a decomposer, all these are interlinked and depend on each other for their survival.

CONSERVATION OF FOREST

Conservation of forest is the practice of planting more trees and maintaining the forested areas for the sustainability for future generations. Forests are an important natural resource and are beneficial to humans in several ways. But due to increasing deforestation activities, it has become essential to conserve forests throughout the world. Deforestation is the permanent destruction or loss of forests for the expansion of lands for agriculture, livestock, etc. The process of destructing forests for the expansion of agricultural land is referred to as shifting cultivation.

Following are some of the major steps for conservation of forest:

1. With the advent of industrialization, several trees have been cut at an alarming rate for raw materials and various other purposes. This felling of trees can be regulated by selective cutting, clear-cutting and shelterwood cutting.
2. Forest fires are one of the common causes of loss of forests. Sometimes the forest land is set on fire to make the land available for commercial purposes. Once cleared, there can be no vegetation. Natural forest fires are also responsible for the destruction of huge forest covers. Latest fire fighting techniques should be adopted to conserve the forest. However, forest fires are an important part of the ecosystem and it helps replenish nutrients in the soil from dead and decaying matter.
3. More trees should be planted to increase the forest cover. Trees should be selected according to the geographical conditions of a particular region and proper care should be taken during the growth of trees.
4. Prevention of exploitation of forestry and forest products is necessary for the conservation of forest.
5. The existing forests should be protected from diseases by spraying chemicals, antibiotics or development of pest-resistant strains of trees.

WILDLIFE CONSERVATION

Wildlife is a very important and vital aspect of our **ecosystem**. Listed below are a few of the reasons stating the importance of wildlife conservation:

- **Important for their medicinal values:** More than one-third of our pharmaceutical needs are catered by wild plants. Forests provide great scope for breakthroughs in the field of medical science and technology along with the requirements for the large-scale manufacture of antibiotics and other medicines for therapeutic uses.

- **Helps keep our environment healthy:** They are responsible for maintaining temperatures globally, thereby fighting against the greenhouse effect and in turn preventing the sea levels to rise sharply.

- **Important to maintain ecological balance:** the interdependence of plants and animals is very essential
- **Economical importance:** Fossil fuels obtained from forests help in the economic growth of the country, which in turn contributes towards a better standard of living.
- **Helps preserve biodiversity:** There are thousands of species that take shelter in these huge forests.
- Microorganisms in wildlife take part in nitrogen fixation, thus, bringing about an increase in the levels of soil fertility

Reasons Why Forests and Wildlife Matter

There is an absolute need to save our flora and fauna in the forests for the following reasons:

- It is all about breathing: Without oxygen, no living organisms can survive. Roughly 20.9% of earth's gases consist of oxygen and dependence on oxygen is higher, hence plants are essential.
- It plays a specific and significant role in the carbon cycle by absorbing carbon dioxide during photosynthesis.
- It is home for different species: Modernization has caused deforestation. This has led many animals to become homeless, leading to the death of these animals in most cases because of unfavorable conditions.
- Provides employment: Forests provide a source of livelihood to millions around the world. Raw materials that can later be sold come from these forests.
- Forests provide wood: Wood serves as an important raw material for manufacturing industries as well as commercial purposes not only animals and plants, millions of people still live in the forests and in its vicinity. They survive from the abundant resources that forests have to offer.
- Rescue during natural disasters: Vegetation in forests helps during floods as they seep in and absorb all the water, thereby preventing damage to infrastructure.
- Helps regulate climatic conditions: Forests play a key role in bringing about a balance in the atmospheric temperature.
- Biodiversity at its best: To get a closer view of life, visit a forest. Forests, as we know are home to thousands of species.
- Forests also provide great scope for exploration for science enthusiasts.
- Entertainment: Forest areas these days have been turned into adventure spots to help people get a first-hand experience of wildlife.

- Key Points on Conservation of Forest and Wildlife
- Deforestation also referred to as degradation or loss of forests needs to be controlled or hopefully stopped.
- Fuelwood and timber should be used wisely and not wasted.
- Forest fire needs to be monitored and damage control measures need to be taken at the right time. Every year, lakhs of trees are destroyed because of forest fires. 2018 saw one of the deadliest wildfires ever in the history of California causing severe damage to property and life.
- Measures and interest need to be channelled towards regrowing trees and increasing the land cover for forests. Reforestation and afforestation need to be promoted throughout the world.
- Proper forest management bodies need to be enforced to monitor wildlife actions and update on the latest.
- Care and habitation need to be provided to extinct species and efforts need to be taken to save them from becoming extinct.

ECOLOGICAL IMPORTANCE OF FORESTS

Forests are 31 % of the world's land and they collect and store water. The trees standing tall within the forest hold the soil powerfully to keep up the health of the soil. The trees facilitate the balance of greenhouse emissions by ingesting them in a cathartic atomic number 8 for the organisms.

They are an excellent supply of food for every kind of animal which has herbivores, carnivores and omnivores. The herbivores prey on the plants and also the eater species is the prey for the carnivores. Once the animals die they decompose into the soil that helps the plants to sustain and live as food for others.

Apart from the wild inhabitants, forests serve an excellent service to the social animals too, i.e. humans. Humans extremely depend on forests for their survival because it provides fuel, food, and shelter. Humans slow down trees from forests for his or her use to create homes, articles of furniture and burn down the wood for change of state.

ECONOMIC IMPORTANCE OF FORESTS

Forests contribute to quite 100 per cent of the GDP in several countries. In several developing countries, forests offer around ten million and around thirty to forty million formal and informal employment.

Forests contain several ancient herbs which will be used for medicative functions. and also the governments of the many countries export those herbs to perform trade and earn a profit. Several scientists are responsible for researching the merits and demerits of bound herbs, characteristics of their edges and positive use.

ENVIRONMENTAL IMPORTANCE OF FORESTS

Forests manage the climate and purify the atmosphere for the survival of all living beings. The trees within the forest hold the soil powerfully and maintain the health of the soil. The dense vegetation balances the temperature and weather conditions by reducing the greenhouse emission by humans. It conjointly controls sound pollution by engrossing the noise made by humans.

The forests increase precipitation by condensing low clouds and increasing precipitation to some extent. They guard the soil by holding it and supporting the basic structure of the trees, reducing erosion. They shield the watersheds and improve the water quality by filtering rain.

2

REMOTE SENSING TO FOREST ECOLOGY AND CONSERVATION

Remote sensing and GIS are complementary technologies that, when combined, enable improved monitoring, mapping, and management of forest resources. The information that supports forest management is stored primarily in the form of forest inventory databases within a GIS environment. A forest inventory is a survey of the location, composition, and distribution of forest resources. As one of the principal sources of forest management information, these databases support a wide range of management decisions from harvest plans to the development of longterm strategies.

Historically, forest management inventories were primarily for timber management and focused on capturing area and volume by species. In the past decade, forest management responsibilities have broadened. As a result, inventory data requirements have expanded to include measures of nonharvest related characteristics such as forest structure, wildlife habitat, biodiversity, and forest hydrology.

The entire forest inventory production cycle, from planning to map generation, can take several years. Except for the photo interpretation component, forest inventory production is largely a digital process. Operational level inventories, inventories, based on both aerial photo interpretation and field-sampled measurements, provide location-specific information required for harvest planning. Forest management level inventories meet longer-term forest management planning objectives. Though these levels differ in detail, they both require information fundamentally based on forest inventory data.

Remote sensing is the examination or the gathering of information about a place from a distance. Such examination can occur with devices (e.g. - cameras) based on the ground, and/or sensors or cameras based on ships, aircraft, satellites, or other spacecraft. Today, the data obtained is usually stored and manipulated using computers. The most common software used in remote sensing is ERDAS Imagine, ESRI, MapInfo, and ERMapper.

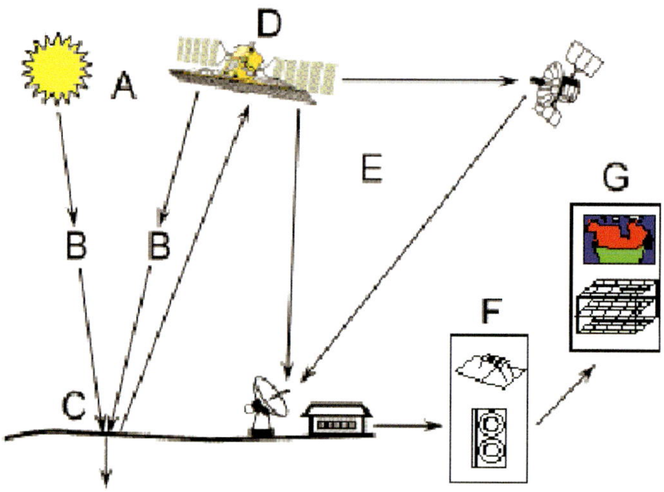

Fig. 1. Geographical information system

(A) Energy Source or Illumination, (B) Radiation and the Atmosphere, (C) Interaction with the Target, (D) Recording of Energy by the Sensor, (E) Transmission, Reception, and Processing, (F) Interpretation and Analysis and (G) Application

The expansion of GIS is Geographic Information System which consists of three words, viz. Geographic, Information and System. Here the word 'Geographic' deals with spatial objects or features which can be referenced or related to a specific location on the earth surface. The object may be physical/natural or may be cultural / man made. Likewise the word 'Information' deals with the large quantity of data about a particular object on the earth surface. The data includes a set of qualitative and quantitative aspects which the real world objects acquire. The term 'System' is used to represent systems approach where the environment consists of a large number of objects / features on the earth surface and their complex characteristics are broken down into their component parts for easy understanding and handling.

Over the past eight years, GIS technology has been widely accepted by public as well as private forestry agencies. In large part this has been a result of the benefit of using GIS technology over current forest maps. The primary management tool for timber production in America is the forest inventory. It is used to access the existing forest resource and develop harvest schedules and treatment programs to project future timber supplies and for other operational planning activities. Forest inventory data is collected using remote sensing techniques.

With GIS technology, the average age of the information in the forest data base could be reduced from 20 years to only a few weeks. The time factor alone has led to a wide acceptance and large demand for GIS applications in forestry.

THE UBIQUITY OF REMOTE SENSING: SIX KEY REASONS

Given that many forest environmental variables can be estimated directly in the field, why has remote sensing become an important data source? We note six key reasons for this situation:

First, remotely sensed imagery provides a synoptic view. The vantage provided by an Earth-observing sensor ensures that imagery captures a complete picture of the environment in its field of view. Thus, every visible feature, including its location and its location relative to that of all others in the imaged area, is captured. In short, this gives imagery a map-like format that provides a complete survey of the imaged area rather than field data, which are often based on a very limited set of samples from which inter-sample site information would have to be inferred by some form of interpolation. Because of this complete survey, remote sensing allows wall-to-wall mapping and monitoring of important ecological variables, such as land-cover change.

Second, remotely sensed data are available everywhere and often at a range of spatial and temporal scales. Key environmental remote-sensing systems, such as those carried by the Landsat satellites, have provided a constantly updateable stream of imagery for the entire planet since the 1970s. Availability can sometimes be constrained by technical problems or cloud cover, but in principle, imagery should be available everywhere irrespectively of location, enabling *inter alia* study of sites no matter how remote or hazardous they might be. Furthermore, historical remote-sensing data allow us to go back in time to look at the causes of present environmental issues.

Third, remotely sensed imagery has a high degree of homogeneity. Critically, data from key environmental remote-sensing systems are acquired under relatively fixed conditions, and the data captured relate to the way in which radiation interacts with the environment, which is constant in space and time; there are no human-induced complications, such as differences in measurement practices from one country to another.

Fourth, the imagery contains, or can easily be converted to, digital images and as such can be easily integrated with other spatial datasets in a geographical information system.

Fifth, per unit area, remote sensing is an inexpensive way to acquire data. Although the financial costs associated with remote sensing can sometimes be very large—for example, it is expensive to build, launch, and operate satellite remote-sensing systems, making some imagery expensive—much is freely available. Additionally, although commercial remote-sensing systems can appear costly, the data still provide inexpensive assessment on a unit-area basis. More critically, however, there has been an increasing trend to make key datasets for environmental science research freely and openly available. For example, the complete archive of the influential Landsat series of satellites is freely available, and recently the European Space Agency (ESA) launched a suite of new satellites and provides the data collected for free. Resources such as Google Earth Engine (GEE) also provide easy access to vast global datasets.

Sixth and finally, there has also been an increasing trend toward the provision of data products as well as the image data themselves. This reduces both the need for expert knowledge of remote sensing and image analysis and the communication gap between experts and environmental scientists, which has historically been a concern. Environmental scientists can now easily access science-quality data products obtained from remote sensing (e.g., leaf area index, land use, and land cover), although expert knowledge might still sometimes be needed.

REMOTE SENSING TECHNIQUES

Remote sensing techniques for forest and land cover mapping

Remote sensing is a powerful tool that could be used to address the problem of thematic maps which are out of date and have to be updated. The capabilities of remote sensing to map and extract information about earth resources for various applications are well documented. Among those prominently used is land cover mapping, considered as one of the most important, most direct and well established applications of remote sensing.

Forest resource mapping by remote sensing began in the first part of the twentieth century with local-scale forest mapping from aerial photography. Land cover mapping is a product of the development of remote sensing, initially through aerial photography. This is because 'viewing' large areas repeatedly is necessary for acquiring information about land cover. For the same reason, land cover mapping has been perhaps the most widely studied problem employing satellite data, beginning with Landsat 1 since July, 1972, with the launch of the first satellite sensor for monitoring Earth resources. There have been significant advances in remote sensing technology that have allowed forest resources mapping over much larger areas through the use of spaceborne sensors. Landsat 2 through 5 has provided more data about forest resource than can ever be analysed. The program is so successful that it continues even today with Landsat 7 being planned.

Remote sensing has been used in research of vegetation for many years with different perspectives such as mapping forest fires, vegetation cover and detecting changes in vegetation through different periods. Use of remote sensing for mapping vegetation is a widely used technique in ecological research since it can quickly determine the formation, distribution, and change of vegetation for very large areas. Moreover, it offers the possibility of extrapolating results of mapping to a regional extent, especially in large inaccessible areas.

Despite the different types of remote sensing data that have applied in forest and land cover map since 1972, however no kind of satellite remote sensing currently fully meets the requirements of a comprehensive forest resource assessment system. Numerous studies have been carried out regarding the relationship of forest structure and species composition to geological and topographic features. Correlations of tree species with environmental variables are used to obtain information about the vegetation distributions in different scales. Furthermore it is important to incorporate the terrain attributes (e.g. slope, elevation, local relief) into the statistical decision rules to classify ecological units using remotely sensed data. According to Boyd, D.S and F.M. Danson, "Satellite remote sensing may be used to provide three levels of forest resource information:

1. The spatial extent of forest cover, which can be used to assess the spatial dynamics of forest cover.
2. The forest type.
3. The biophysical and biochemical properties of forests.

Low resolution satellite data like NOAA (National Oceanic and Atmospheric Administration), AVHRR (Advanced Very High Resolution Radiometer) images are often used for large area forest map on a regional and global scale. This is widely used for environmental applications at regional to global scales. The first satellite-based global land cover maps have been produced. AVHRR imagery has also been used widely to estimate the extent and biophysical properties of forests.

Therefore the coarse spatial resolution remotely sensed data are the most useful for forest cover delineation for large area. However, there is one major problem that small areas may be not detected, which leads to incorrect estimation of remaining forest, this makes it difficult to observe local details.

Medium spatial resolution imagery such as the Spot HRV and Landsat MSS/TM have been used successfully to map and monitor forests. Pioneering studies illustrated the potential of these optical satellite data for the delineation of forest cover from non forest cover such as those carried on the Landsat and Spot data for estimation of forest cover at the local and national levels. The medium optical imageries of Spot and Landsat are also used in estimation of tropical forest biological and biochemical properties like leaf area index (LAI), biomass or canopy cover.

Optical remote sensing is used extensively to map forest extent, and the temporal and spatial dynamics of deforestation, re-growth, and fire. The advantages of using Landsat and Spot data in their Panchromatic tropical remote sensing survey as mentioned in the FAO's Forest resources assessment in 2000. However, the use of these optical imaging systems is that these technologies cannot penetrate clouds, which persist over many parts of the tropics during the wet season and, indeed, throughout the year in many tropical upland and montane environments.

This effectively reduces the number of Landsat passes that researchers may use to map and monitor tropical forest as months or years may transpire before cloud-free Landsat imagery becomes available for certain cloudy locations.

Another application for low and medium optical satellite data is the mapping of burned areas resulting from tropical forest fires. Therefore, remotely sensed data may be used to provide information on biomass burning, which can indicate the magnitude and temporal dynamics of forest cover change through deforestation by burning events.

High resolution satellite data like IKONOS and Quickbird represent a new generation of remote sensing satellites, the IKONOS has a spatial resolution of 1m for panchromatic imagery and 4m for multi-spectral imagery, and its future successors are reported to generate images with a spatial resolution of approximately 50 cm. The Quickbird satellite, launched in October 2001, leads the quality list of optical remote sensing with panchromatic imagery of 70 cm spatial resolution. In most aspect, the IKONOS satellite and imagery captured is similar to the Quickbird satellite. Both these satellites with its spatial resolution promise to increase the accuracy of forest resources inventories.

These imageries have been used for the validation of products derived from lower resolution data. They have also been tested for forest land use and land cover classifications as well as for the estimation of forest biomass and the detection of selective logging. However, IKONOS and Quickbird have limitations of spectral and spatial resolution.

They do not have bands in the mid and thermal infrared region of the electromagnetic spectrum and IKONOS and Quickbird data also have less bands available than Landsat which do not allow a breadth of multi-spectral analysis. This is a disadvantage for the spectral discrimination of vegetation types. Despite the differences, all data sources can be used together to create a variety of solutions.

To overcome some of the problems of frequently cloud cover in tropical areas which renders much of the optical satellite unusable and SAR data have been used. Radar remote sensing could be a better alternative for tropical region due to its ability of cloud free imaging. SAR imagery is used to produce forest cover, particularly, through the acquisition of multi-temporal datasets. In addition, the use of SAR data for mapping forest stand properties such as the classification of forest cover types from SAR imagery exploits difference in the macro structure between stands of different species, age or density.

However, despite the potential of SAR shown for provision of forest and land cover information, currently it is the optical remote sensing approach that is more useful as SAR data are not available in tropical regions and are a relatively expensive option.

The importance of remote sensing to generate information for effective forest protection and management has been widely recognized as mentioned above. It is the only way to acquire repetitive biophysical data for large geographic area at reasonable cost, accuracy and effort. Different remote sensing data have been successfully used to map different forest types and land cover. Therefore, there is good possibility of using remote sensing to provide forest and non-forest, which would be needed for assessment of forestry's land potential, for sustainable forest management.

On the whole, in many tropical developing countries like Vietnam, the application of high spatial resolution satellite data so far has rarely been used for forest types and land cover mapping. Among various satellite images used in studies of tropical forest, Landsat TM (Thematic Mapper) and ETM+ (Enhanced Thematic Mapper Plus) are the sensors most commonly used in tropical forest applications and remain the standard to which other optical satellite data are usually compared. The spatial resolution of Landsat TM and ETM+ (30 m for 6 optical bands and 15 m for Panchromatic band) appropriate for the detection of change in canopy condition, extent as well as land use around forested areas. On the other hand, Landsat TM and ETM+ generally provide a clear delineation forest and non-forest cover types and allow both manual. In addition Landsat scene covers approximately 170 m x 170 m km is suitable for the study area.

Classification issues

Digital image classification is the process of assigning pixels to class. Usually each pixel is treated as an individual unit composed of values in several spectral bands. Unsupervised classification method is used to identify natural groups, or structure, within multispectral data. Only afterwards information labels are assigned to the resulting groups. The disadvantage and limitation of these methods primarily arise from a reliance upon "natural" grouping and difficulties in matching these groups to the informational categories that are of interest to the interpreter. In addition, the interpreter limit controls over the menu of classes and their specific identities.

By contrast, supervised classification method is the process of using samples of known identity (training area or training field) and extends it to the entire image. Each primitive image is characterized by n observations (the values in n data channels). The training samples are vectors in an n-dimensional space (the feature space). A supervised classifier uses the distribution of the training samples for each class to estimate density functions in the feature space and to divide the space into class regions.

This method has many disadvantages. First, the analyst, in effect, imposes a classification structure upon the data. Second, training data are often primarily defined with the reference to informational categories and only secondarily with reference to spectral properties. A training area that is "100 % forest" may be accurate with respect to the "forest" designation, but may still be very diverse with respect to density, age, shadowing, and the like, and therefore form a poor training area. Third, the training area selected by the analyst may not be representative of conditions encountered throughout the image. Fourth, conscientious selection of training data can be a time consuming, expensive, and tedious undertaking, even if ample resource are at hand.

Finally, supervised classification may not be able to recognize and present special or unique categories not represented in the training data. Therefore, the use of these methods in separate classification may be unsatisfactory.

Considerations for classification accuracy assessment

Accuracy assessment is an important step in the process of analyzing remote sensing data. There are a number of reasons why this assessment is so important including:

1. To let the end user need to know how good the data is and to learn from the mistakes.
2. The ability to use the information resulting from the spatial data analysis in some decisionmaking process.

Thematic maps derived from remotely sensed data are used in a lot of applications, but especially applied in forestry, these maps should be accompanied by information about their accuracy. Meaningful and consistent measures of thematic map reliability are necessary for the map user to assess the appropriateness of the map data for forestry application; additionally, the accuracy of the thematic map may significantly affect the outcome of forestry application.

Map accuracy measures are equally important for the producer of a thematic map to analyze sources of error and weaknesses of a particular classification strategy. Yet, documenting map accuracy is not a straightforward task. While individual measures of map accuracy are well established in the literature, considerable ambiguity remains concerning the implementation and interpretation of thematic map accuracy assessment. Uncertainties include the selection of which accuracy measures to report, how to interpret them, and the nature and quality of reference samples. As a result, map quality remains "a difficult variable to consider objectively.

Accuracy assessment determines the quality of the information derived from remotely sensed data. Accuracy assessment can be qualitative or quantitative, expensive or inexpensive, quick or time-consuming, well-designed and efficient or haphazard. The purpose of quantitative accuracy assessment is the identification and measurement of map errors. Therefore, quantitative accuracy measures are necessary to compare the relative accuracies resulting from different data processing methods.

Typically, most quantitative methods to assess classification accuracy involve an error matrix built from the two data sets (i.e., remotely sensed map classification and the reference data). The error matrices represent the number of correctly mapped pixels by comparing information from reference sites to information of a map for a number of samples areas. The overriding assumption in the entire accuracy assessment procedure is that the error matrix must be representative for the entire area mapped from the remotely sensed data. Overall map accuracy is calculated by dividing the correctly classified pixels/samples (sum of the value in the major diagonal) by the total number of pixels/samples checked. When the total number of correct pixels in a category is divided by the total number of pixels of that as derived from the reference data, the result is an accuracy measure called "Producer's accuracy". On the other hand, when the total number of correct pixels in a category is divided by the total number of pixels that were classified in that data), this measure is called "User's accuracy." Or the reliability is indicative of the probability that a pixel classified on the map/image actually represents that category on the ground.

Considerations about Kappa Statistics or K (KHAT)

Another discrete multivariate technique of use in accuracy assessment is called KAPPA and has been utilized for land cover and land use accuracy assessment derived from remotely sensed data. The result of performing a KAPPA analysis is a KHAT statistic (an estimate of KAPPA), which is another measure of agreement or accuracy. The values can range from +1 to −1. A Kappa of zero occurs when the agreement between classified data and verification data equals chance agreement. The values of Kappa greater than 0.80 (80%) represent a strong agreement; a value between 0.40 and 0.80 (40– 80%) represents moderate agreement; and a value below 0.40 (40%) represents poor agreement.

The Kappa coefficient was introduced to the remote sensing community in the early 1980s and has become a widely used measure for classification accuracy. It was recommended as a standard by Rosenfield and Fitzpatrick-Lins. The Kappa coefficient is a powerful tool because of its ability to provide information about a single matrix and as a means to statistically compare matrices. He also stated that the power of the Kappa analysis is that it provides two statistical tests of significance. Using this technique, it is possible to test weather an individual land cover map generated from remotely sensed data is significantly better than that of the map that had been generated by randomly assigning labels to areas. The second test allows for the comparison of any two matrices to see if they are statistically significantly different. In this way, it is possible to determine that one algorithm is different than another and, based on a chosen accuracy measure (e.g. overall accuracy), to conclude which is better. Therefore, Kappa analysis must still be considered a vital accuracy assessment measure. Although some authors have criticized its characterization of actual map accuracy.

Sampling design considerations

The overriding assumption in the entire classification accuracy assessment procedure is that the error matrix must be representative of the entire area mapped from the remotely sensed data. Using the wrong sampling design can be costly and yield poor results. Researchers stated that "Designing a poor sampling scheme can easily result in significant biases being introduced into the error matrix, which will then affect the classification accuracy". For this reason, a proper sampling approach must be chosen. Therefore, the sample design is a critical part of the accuracy assessment of image classification.

The sampling design includes a number of considerations about the relations between study area and sampling such as:

1. The spatial distribution of samples within a study area,
2. The size of the individual sample,
3. The number of samples to be collected within a study area,
4. The choice of the appropriate sampling scheme.

Spatial distribution of samples

The spatial distribution of samples depends on the classification. The reference data need to be collected and labelled using the same classification scheme that was used to generate the map. Important considerations are the discrete nature of map information and the spatial interrelationship of that information. Therefore the distribution of map categories will affect both, how to select accuracy assessment and the outcome of analysis.

Size of each sample

Researchers stated that a point unit is, in practice, also an area unit because a point can not be accurately verified. Sampling units should be at least the size of one pixel but, in order to take into consideration of geometric distortions also, these sampling units should be more than one pixel. Researchers recommended, for a multistage approach, a 3 x 3 pixel environment as the sampling unit.

Number of samples

Depending on the goal of the accuracy assessment the number of samples can be calculated with different methods. Researchers indicate that the number of sample sites within a study area is dependent upon the size of the training data set and the number of classes in the final analysis. They conclude that a larger sample size than 30 should be collected in remote sensing studies if sampling errors are to be kept low. A balance between what is statistically sound and what is practically attainable is to be reached. Typical remote sensing analysis calls for approximately 50 samples per class as an appropriate sample size. The more the better, this is obviously a huge and money commitment. The larger the area being mapped, the more samples should be collected. Important categories can be sampled more than less important categories.

Choice of the appropriate sampling scheme

The choice of sampling scheme is a part of any accuracy assessment. The selection of the appropriate scheme is critical to generate an error matrix that is representative of the entire map. There are five common sampling schemes that have been applied for collecting reference data: Simple random sampling, systematic sampling, stratified random sampling, cluster sampling and stratified systematic unaligned sampling:

- Simple random sampling is a method of selecting the sample units out of the population, such that each element in the population has an equal chance of being selected. Therefore, sampling design should be random in order to meet the statistical requirements of the Kappa analysis, among other statistical techniques that are based on the error matrix. Simple random sampling can be appropriate if the sample size is large enough to ensure that all classes are adequately represented. The adoption of a simple sampling design is also valuable in helping to meet the requirements of a broad range of users) although the objectives of all users cannot be anticipated. However the simple random sampling yields too many samples in the large areas and too few samples in small areas. This can lead to classes of small extent failing to be represented in the sample.

- Stratified random sampling is similar to simple random sampling, however, some prior knowledge about the study area is used to divide the area into groups or strata and then each strata is randomly sampled. The major advantage of stratified random sampling is that all strata (i.e., land cover types), no matter how small, will be included in the sample, so that stratified random sampling is frequently used in order to ensure that samples from all classes are included in the accuracy assessment. On the other hand, simple random sampling may lead to problems if the resulting samples are located in inaccessible areas where the true class cannot be determined. Because of this, some authors use cluster sampling.
- Cluster sampling is a technique of sampling in which units are not single pixels but groups of pixels, the cluster sampling is also frequently used in accuracy assessment of image classification, especially to collect information on many samples quickly. However, cluster sampling must be intelligent. Simply using very large clusters is not a valid method of collecting data, because each pixel is not independent of the other and adds very little information to the cluster.
- Systematic sampling is a method in which the sample units are selected at some equal interval over the study area. The advantage of systematic sampling is the convenience of obtaining the sample and the uniform spread of the sampled observations over the entire population. An obvious problem in systematic sampled population is the bias that exists if the population shows some kind of spatial autocorrelation. If the presence, absence, or degree of certain characteristics affects those in neighbouring units, then the phenomenon is said to exhibit spatial autocorrelation.
- The cluster sampling, and also systematic sampling do not ensure that every individual in the population has an equal chance of being included in the sample, thus violating the requirements for inferential statistics.
- Stratified systematic sampling combines the advantage of randomization and stratification with the useful aspects of systematic sampling, while avoiding the possibilities of bias due to the presence of periodicity. When dealing with very large areas where cost and time are of great concern sampling from a car can be a necessary choice. Some consider this sampling technique as a mixture of random and a systematic sampling. Nevertheless, disadvantages are limitation of the sampling to passable roads. Extensive classes that are not covered by roads, e.g. mountainous terrain, may then be under- or unsample. Land use and other human activities may also not be representative for the whole area.

The concept of randomness is a central issue when perform almost any statistical because a random sample is one in which each member of population has an equal and independent chance of being selected. For the purpose of using the Kappa coefficient in the analysis of the confusion matrix, a random sample sampling methods should be chosen.

GIS TECHNIQUES

Geographic Information System (GIS) is a powerful tool for collecting, storing, retrieving, transforming and displaying spatial data from the real world.

GIS can be used in concrete applications ranging from resource assessment to land evaluation and land use planning. It provides a powerful set of tools for data visualization, data analysis and evaluation of scenarios. GIS is an integrating technology that allows, encourages and expects users to bring together data from many different sources through the unifying medium of geography.

The typical application of GIS is in a kind of spatial modelling which can be described simply as combining information from several different images to produce a single output image for land suitability assessment. One major part of a GIS is the ability to overlay various layers of spatially referenced data, this allows the user to determine graphically and analytically, how structures and objects, interact with each other.

Land evaluation by map analysis techniques may be accomplished with any geographical information system (GIS). GIS technology continues to be used for making forest and land suitability maps. The use of this technology leads to the rapid generation of thematic maps and area estimates, and enables many of the analytical and visualization operations to be carried out in a spatial format, by combining different sets of information in various ways to produce overlays and interpreted maps. On the other hand, the modelling capability of the GIS is being used for the site selection, and for natural resources decisions, GIS can be a powerful tool base for evaluation of choice alternatives based on spatially related criteria. GIS have been recognized as the most promising tool capable of providing reliable information for both planning and decisionmaking for the sustainable management of resources. Furthermore, GIS have been widely applied in the land suitability analysis, according to Guillermo A. Mendoza "An integrated GIS-based multi-criteria approach to land suitability analysis and allocation offers significant advantages.

The GIS environment enables the spatially explicit evaluation of site suitability and the assignment of various measures of suitability to specific sites or geographic areas. The integration also allows area allocations at specific spatial or geographic locations. Hence, the integrated GIS-based model combines the spatial capabilities of GIS, with the analytical power of multi-criteria analyses.

It means that the GIS-based integrated model permits both analytical planning and optimization of land use decisions at different levels, namely:

1. Site suitability assessments based on different factors and specific land uses;
2. Generation of suitability indices based on combinations of different factors (i.E. Composite index/measure of site suitability); and
3. Generation of an optimal land use plan that simultaneously considers the individual site suitabilities, and the optimal allocation to the most suitable land use i.e., mix of land uses that yields the highest overall cumulative suitability".

GIS methods are combined with production modelling to support for land suitability assessment. GIS and expert knowledge has been used for land suitability assessment for Cherimoya in southern Ecuador.

The combination of GIS technology with expert system to inform a land suitability evaluation model. Therefore, GIS technology is already a prerequisite for managing the massive datasets required for spatial land evaluation application - a simple map subsystem (e.g. ArcView).

Data structure

There are two types of spatial data: geometric data and descriptive data. Geometric data - points, lines and polygons, express location or shape of entities. Descriptive data, on the other hand, relate to geometric entities.

Layers

A GIS has the capability to manage many layers of data for the same points in space in order to perform analyses. Layers represent different sets of geographic information. Layers can provide information on geologies, soil, land use, species distribution, vegetation, etc.

GIS applications involve everything from basic to complex functions. These applications include the analysis of spatial reference data, queries on point or polygons, data classification, map production, area calculations, the creation of buffer as well as elevation models.

LAND SUITABILITY MODEL

A model is a representation of an object or a phenomenon, a design to be copied, an ideal or standard situation, or a perfect thing to be imitated. In the sense in which it is used here the term model means a simplified representation of reality, where we can compute outcomes without having to perform actual experiments. The model is designed here to facilitate visualization, prediction and calculation, and which can also be expressed mathematical or computable form.

In land evaluation, Ramond Caldwell notes that "models are computer programs that predict the performance of a land use on a land area, when given information on that area's land characteristics ". Therefore, many land assessment computer models were developed by various scientists.

Modelling is the process of designing and using these models. Spatial modelling are the analytical procedures applied with GIS and GIS modelling is the process of creating new GIS products from existing products. There is a range of methodologies for environmental modelling. Some of these are based on an assessment of factors and combinations of factors, and others primarily on a process modeling approach. In spatial modelling there are three different classes of models, e.g. logical, empirical and conceptual:

- A logical model simulates a new attribute by applying simple rules.
- Models that are set out principally to describe or mimic the observed relationships between variables are called empirical models.
- A conceptual model describes essential features of a meteorological phenomenon and identifies the principal processes taking place.

LAND SUITABILITY EVALUATION METHOD BASED ON GIS TECHNOLOGY

The land suitability model is a set of procedures by which the suitability of a cell for a new land use is quantitatively assessed on the basis of selected environmental variables.

Land suitability evaluation is to determine the most suitable land use spatial model according to the needs, desire and expectations for future activities of mankind. The purpose is to serve the land use planning and kinds of land use decision-making. Evaluation method is the hotspot research area in land suitability. In late 19th and early 20th century, some landscape architects began to evaluate land suitability with the methods of manual drawing and superimposing maps in the United States, which mean a new land suitability evaluation method emerged, and this method had a similar philosophy with the one based on GIS technology. With the rapid development of computer technology and the 3S (GPS, GIS and RS) technology, GIS-based land suitability analysis had gradually become the mainstream of this research field, and a series of evaluation methods emerged. Harvard laboratory evaluated land suitability with the technology of computer-aided superimposed mapping, which had obvious advantages in dealing with massive data than traditional method of superimposing maps. Traditional method of overlaying maps was widely used in the field of Land suitability evaluation by integrating multi-criteria decision-making methods and GIS technology, which could be considered as the decision-making process of getting the expected results through integrating spatial and non-spatial data, and judging multi-index comprehensively. Artificial intelligence methods played a catalytic role in the development of land suitability evaluation, and could solve the problem of multi-index decision-making method that different multi-index analysis may yield substantially different evaluation results.

According to current research results of land suitability evaluation, two aspects of shortage can be found as the follows: First, the evaluation method is not yet mature. Because of the inherent complexity of the land system and evaluation methods, evaluation results are difficult to fully and accurately reflect the true suitability of land. Even to the same object, different methods may lead to different evaluation results. Secondly, it is necessary to address how to express the results of land suitability evaluation. In the past time, most of researchers took the suitability classification for a particular purpose (suitability evaluation map) as the evaluation results. Actually, this kind of evaluation result was not detailed enough and some useful information was missed and ignored, which limited the service value of land suitability evaluation to land use decisions. Based on the above analysis, taking pan agriculture mixed-use suitability evaluation as target, this paper discuss the program, method and result-expression of land suitability evaluation in order to improve the objectivity of evaluation methods and practicality of evaluation results.

Land suitability analysis

As mentioned above, land suitability is fitness of a given type of land for a specified kind of land use. The land may be considered in its present condition or after improvements. The process of land suitability classification is the appraisal and grouping of specific areas of land in terms of their suitability for defined uses. The main objective of the land assessment is the prediction of the inherent capacity of a land unit to support a specific land use for a long period of time without deterioration, in order to minimize the socio-economic and environmental costs. Land suitability analysis/assessment of suitability is made by comparison between land use and land quality, couple with analysis in environmental, economic and social terms (land evaluation for forestry). Researchers noted "land suitability analysis incorporates human use information, such as demographic data, distance measures, and economic factors in addition to the physical data used in land capability analysis". Therefore, land suitability analysis is an interdisciplinary approach by including the information from different domains like soil science, crop science, meteorology, social science, economics and management.

Analysis of land suitability is an evaluation problem involving several factors. The principal problem of land suitability analysis is to measure both the individual and cumulative effects of the different factors. In the other hand, analysis of land suitability generally involves determining an appropriate approach to combine these factors. Some approaches to combine these factors include:

1. The weighted linear combination method which used GIS based decision rules. In addition, this analysis method is often applied in site selection, analysis of land suitability and problems of resources evaluation.
2. The heuristic rules of combination expressed through verbal logic rather than in numeric terms,
3. The various forms of multi-criteria approaches such as compromise programming.
4. Composite rating consists of weighted composite rating. These approaches use a weighting scheme to combine criteria into a single measure of utility.

On the whole, the weighted linear combination method is one of the most widely used in the land suitability analysis. The reason for selecting this method is that the method is very easy to implement within the GIS environment using map algebra operations and cartographic modelling, and the method is also easy to understand and intuitively appealing to decision makers.

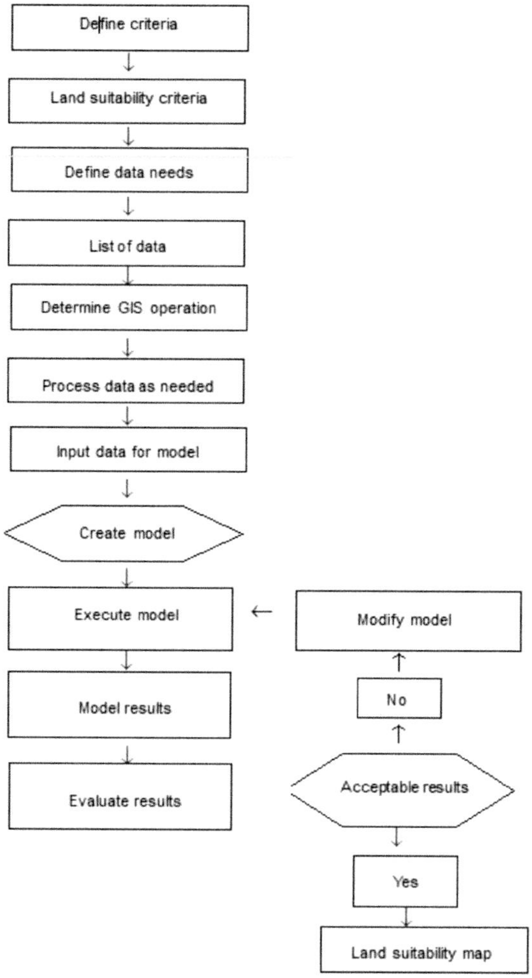

Fig. 2. Illustrates the steps necessary for performing a land suitability model

INTEGRATION OF REMOTE SENSING AND GIS FOR LAND RESOURCES ANALYSIS

Remote Sensing and GIS technology are both tolls for managing spatially distributed information in large quantities and at a variety of scale, both provide a systematic or synthetic view of spatial information, both increase the capabilities of human decision makers and planners to gap relations at larger scales and in more complex settings than has been possible before.

According to researchers, both Remote Sensing and GIS play an important role in assessing natural resources. They are a tool for resource survey and inventory that provides a primary source for up-to-date geographic data. These technologies are complementary to each other and it has been seen that the analysis of remotely sensed data is a complex subject to many difficulties especially with regard to classification and generalization. However, pre-existing GIS data can have impact roles both on the performance of classification and also in spatial operations in the image domain. The GIS data have reported on the use for image stratification using Boolean operations between data layer, to incorporate GIS data in natural resource network classifiers, and in evidential reasoning with geographical context.

Integration of remote sensing and GIS will be a powerful tool to interpret real word situation in most realistic and transparent way. Researchers has shown that "integrated Remote Sensing and GIS technology can be threefold. Firstly, GIS can be used to provide image analysis algorithms with a priori information, which is used, e.g. to restrict the search space or impose constraints. Secondly, image analysis and processing methods can be used within a GIS for data analysis, especially for raster data (e.g. buffering of a corridor by using mathematical morphology), visualisation, content-based image retrieval etc. The third case is when image analysis and GIS fully interact, e.g. GIS information is used to guide image analysis, which extracts more complete and accurate information, which is in turn used to update the GIS database".

GIS information can be used in image analysis for image classification purpose to produce forest and land cover map, e.g. in supervised classification to automatically select training areas, and in unsupervised ones to automatically assign detected information classes to thematic classes. In contrast, the use of remote sensing, especially image classification will provide a more rapid updating of GIS database.

WILDLIFE HABITAT AND CORRIDOR MONITORING

The Northeast region of India is rich biodiversity and highly significant as it lies at the confluence of the Indo-Malayan, Indo-Chinese and Indian biogeographically realm and it supports diverse floral and faunal diversity. The region is also home to a large number of rare, endemic and endangered wild animals. However, the habitats of the wildlife are under severe threats due to depletion, degradation and fragmentation of forest cover and loss of corridors for the movement of the wildlife. The fragmentation of habitat and loss of animal corridors have consequently led to an escalation in animal–human conflict.

Towards management of wildlife habitat in NER region, studies on mapping the suitable habitat of elephant and tigers (Dampa Tiger Reserve, Mizoram) have been carried out. The elephant habitat study had been undertaken in collaboration of the Environment and Forest Dept., Govt. of Assam. Habitat suitability for tiger and elephants were mapped and potential corridor for the movement of elephant has also been identified. These projects were taken up as part of R&D projects of NESAC. Apart from this the wildlife habitat mapping was done for Keibul Lamjao National Park (KLNP), Manipur and grassland burnt area of the floating meadows in KLNP was also done on the request from the state forest department and Wildlife Institute of India.

ANALYSIS OF VEGETATION PHENOLOGY USING TIME SERIES SATELLITE DATA

A pilot study was carried out in R&D mode to analyse the distribution pattern and variation in vegetation phenology for several natural vegetation types in North East India using time series satellite data. This will help in understanding the vegetation composition pattern and aid in classification of forest types. MODIS's 16 days' composite NDVI time series data for a period of 18 years has been analysed for the NER.

After smoothing the data, phenological parameters were computed by defining the threshold for the start and end of seasons. Seasonality parameters such as start of the season, the end of the season, length of the season, base level, mid of the season, maximum NDVI value and seasonal amplitude have been calculated. Satellite based derived vegetation phenology has been analysed and correlated with the major forest types found in the region.

REMOTE-SENSING PLATFORMS AND SENSORS

Remote-sensing systems are available as a diverse array of sensors and platforms. Sensors can be divided into passive and active sensors, whereas platforms range from Earth observation satellites, planes, and helicopters to unmanned aerial vehicles (UAVs) with fixed wings and rotaries.

The most common sensor used in remote sensing is an optical imaging system, which is similar in design and application to a standard digital camera except that it can acquire data beyond the visible wavelengths (i.e., infrared and thermal wavelengths) across the electromagnetic spectrum. Materials reflect and absorb at different wavelengths, and through these differences, land covers (i.e., forest and canopy cover) can be identified.

Optical sensors vary in terms of the number of bands (and the widths of those bands) from which image data are captured. Multispectral sensors have a limited number of bands, whereas hyperspectral sensors have thousands of much narrower bands. Optical systems (and thermal systems) are passive sensors, which rely on reflected sunlight or emitted thermal energy, and consequently cannot penetrate clouds or smoke, are affected by haze from clouds, and cannot be used at night.

Active sensors include light detecting and ranging (LiDAR) and synthetic aperture radar (SAR) systems. These sensors emit a pulse and measure the backscatter reflecting back to the sensor. A key advantage of such sensors is their ability to penetrate clouds and smoke and operate at night. SAR sensors can differentiate land-cover features according to their surface roughness, the 3D structure of the targets, and water content.

Depending on the wavelength of the sensor (e.g., X-band, L-band, or C-band), the signal can penetrate vegetation, canopy, and soil. Conversely, LiDAR systems emit pulse from lasers and measure distance to a target and the reflected light. Differences in laser return times and wavelengths can then be used for making digital 3D representations of the target.

In recent years, as a result of developments in sensor technology, all passive and active sensors have versions that can be mounted on all platforms, although larger platforms such as satellites and larger planes can carry heavier payloads, allowing for larger sensor systems that are of higher quality and accuracy. However, because of sensor miniaturization, rarely found combinations, such as SAR mounted on a UAV, are starting to be more common and affordable.

Spaceborne sensors take consistent measurements at specific time intervals according to the time it takes for the sensor to revisit the same location (e.g., Landsat sensors revisit the same location on Earth every 16 days). However, many commercial satellites, such as WorldView series, can be tasked for specific locations, whereby the sensor head turns to acquire data on an angle to increase revisit time. In addition, the latest satellites are commonly developed as part of a constellation of multiple satellites to increase revisit time (i.e., Sentinel). Conversely, airborne platforms have the advantage of being able to be flown in response to specific events, such as fire, and can also fly under clouds (especially UAVs), addressing this key limitation of Earth observation satellites. Airborne platforms such as UAVs can acquire very-high-centimeter spatial-resolution data.

APPLICATIONS AND CLASSIFICATION METHODS

Within forest ecology and management, there is a diverse range of applications for remote sensing, including the measurement of cover, vegetation structure, vegetation chemistry and moisture, biodiversity, and soil characteristics. These variables are critical for understanding forest ecosystem functions and processes, as well as classifying forests into specific communities, ecosystems, and biomes.

For forestry applications, remote-sensing measurements can be used for producing forest inventories for calculating the number of trees per acre, the basal area, and the value of timber. For forest monitoring, measuring change in these variables is important for understanding ecosystem dynamics and anthropogenic impacts in both the short term (i.e., fire) and long term (i.e., climate change). From the day-to-day management perspective, monitoring forest change is critical for determining potential risks such as fire hazard due to fuel loads and overall forest health.

There are a range of ways in which remote sensing is used to represent different forest variables. Both optical and SAR data are provided in a (flat) raster format (i.e., as a grid of values), whereas LiDAR data are represented by 3D point clouds. These data are then classified into either categorical or continuous outputs.

For example, land use and land cover are categorical, whereas foliage projective cover is continuous. However, depending on the resolution, the same variable can be represented as continuous or categorical. For example, medium-resolution Landsat can be used for classifying pixels according to the percentage of foliage projective cover, whereas high-resolution 5-cm data derived from a UAV can characterize the actual canopy extent.

These two perspectives for representing forest variables determine the general types of analyses conducted with remote-sensing data. For continuous biophysical measurements (e.g., the fraction of absorbed photosynthetically active radiation and biomass), the correlation between field measurements and vegetation indices, such as the normalized difference vegetation index (NDVI), is most common. However, for categorical mapping, classification algorithms such as the maximum likelihood classifier and machine-learning approaches such as Random Forests are "supervised" with training data.

For high-spatial-resolution data, pixels can be aggregated first to form objects that represent natural spatial units of relevance (e.g., individual trees) on the basis of their similar spectral and textural properties. Rather than being classified as pixels, these image objects are classified according to a method known as geographic object-based image analysis.

Where remote sensing can really demonstrate its great potential is in measuring forest variables for multiple time periods or between multiple locations. Although any remote-sensing application requires the acquisition of high-quality cloud-free data, for applications where more than a single scene is analyzed, this is even more important. A key decision point that will depend on the type of analysis is whether to pre-process data to reduce atmospheric effects (i.e., illumination and cloud haze). For certain applications, radiometric correction is necessary for converting the raw remote-sensing data (digital numbers) into surface reflectance, which represents the fraction of incoming solar radiation that is reflected from Earth's surface.

AERIAL PHOTOGRAPHY

Introduction

Across the global south, charcoal is frequently produced from clear cutting or selective logging of woody plants that resprout. The sustainable management of resprouters for charcoal production requires data on their structure and growth capacity; which in turn are affected by the species' sprouting ability (strong versus weak), forest management practices (clear cutting versus thinning; short versus long rotations), and local biophysical conditions. In other words, good quality data on the resprouters' structure and growth are site-specific and should be generated *in situ* if adequate sustainable forest management plans are to be designed and implemented.

In order to design sustainable charcoal production programs in data-poor landscapes, cost-effective methods are needed for measuring the aboveground woody biomass productivity of resprouters. One way of attaining this is to estimate the biomass stocks between multiple time intervals, so as to model regrowth rates. However, while past field measurements of biomass stocks are missing for most charcoal production areas of the world, historical aerial photographs could be available.

The most accurate methods for estimating aboveground woody biomass are those involving direct measurement of individual trees by destructive sampling. Common field-measured predictive variables for the development of allometric models are diameter at breast height (DBH) and total tree height. However, these labor-intensive methods alone do not provide information about the continuous spatial distribution of woody biomass. This can be overcome by integrating field-based allometric equations with remote sensing and spatial analysis, a technique that has proved useful even for regional and global wall-to-wall aboveground biomass (AGB) estimates.

Methods less dependent on field data rely on the availability of high spatial resolution products (< 1 m per pixel) such as high-resolution digital aerial photographs, multi- and hyperspectral sensor imagery as well as light detection and ranging (LiDAR) systems coupled with modern computation and data management capabilities. These allow us to infer tree descriptive parameters such as crown area, biomass volume, tree density, tree height and species composition. However, these tools are often not feasible to implement due to budget constraints. But most important, the newer the sensor, the less the availability of past images.

When dealing with high-resolution biomass estimations (such as those needed for local woodland management plans), allometric models have been calibrated by relating field-based DBH measurements with AGB determination in order to develop empiric equations to estimate biomass.

The estimation of tree biomass from aerial imagery has been around for at least two decades. Earlier approaches were based on either automated or manual analysis of digitized analog aerial photographs and photogrammetrical software. These techniques gave a good overall agreement between image-based, ground-based crown diameter and total stem volumes, but rather limited accuracy regarding individual tree height in dense forest.

But the level of agreement depends on adequate single-tree crown delineation, which is affected by forest density, structure, species composition as well as tree position within the photograph. Researchers reports that the best delineation occurred in trees close to the photograph nadir. More recent approaches include digital aerial photographs, multispectral sensors and LiDAR sensors.

None of these approaches deal specifically with young sprouts typically exploited for charcoal production, which have quite a different architecture to non-managed stands. Researchers, for example, report a canopy to the biomass model that could be useful for bioenergy purposes, but oaks in their study area were never cleared cut, but '...pruned with a periodicity of approximately 5 years in order to enhance fruit production and the stands are predominantly mature'.

Furthermore, none of the previous studies dealt specifically with wood suitable for charcoal (WSC), as the portion of AGB actually useful for charcoal making. Oak forests and rangelands (*Quercus spp.*) in many developing countries across the tropics represent a major woodfuel source, and are intensively coppiced for charcoal on rotations of varying duration, usually by clear cutting in order to save time and effort carrying trunks and thick branches to the kilns.

Oaklands traditionally managed for charcoal are no exception regarding the lack of legal permits for exploitation and commercialization. It has been reported that many charcoal producing areas across developing countries have been traditionally managed without any kind of permits. Barring a few exceptions, charcoal producing areas have been neglected by foresters and ecologists due to their low economic potential as timber sources, and high level of anthropization, respectively.

The present study was designed and performed to calibrate allometric models relating the crown area of young oak sprouts measured in aerial photographs with AGB and WSC at tree and stand level. The underlying objective was to show that high-resolution estimates of AGB and WSC, aimed as a source of information for sustainable charcoal production, can be produced in a simple and relatively cost-effective way with little previous information on the characteristics of the managed woodland. Therefore, this framework could be implemented where applicable in similar settings.

Methods

Photogrammetric measurements of the tree crown area were obtained from very high-resolution (30 cm) small format digital aerial photographs to derive indirect allometric equations and to estimate the basal area (BA), AGB and WSC. To account for the existing variation, the estimates of BA, AGB and WSC were obtained at two different levels of analysis: (1) individual trees, and (2) stand level (aggregation of individual trees). The stand level will be referred to as 'grouped trees' from now on.

Crown area estimation using aerial photographs

Individual tree crowns were delineated using a mirror stereoscope and high-contrast color prints of the aerial photographs. The tree crown limits were on-screen digitized using GRASS GIS v6.4.3. Individual tree measurements were aggregated to obtain grouped tree areas. Field verification and delineation of interpreted features occurred within the first month after the acquisition of the image.

Biomass estimation using photo-derived crown measurements

To estimate the total AGB and WSC using the crown area measured in aerial photographs (CA_{ap}) as the predictor variable instead of DBH, first we evaluated the existence of an allometric relationship between the field-measured DBH (expressed as BA) of the oaks' coppice-shoots and CA_{ap}.

The data gathered during the two field surveys were divided into two sets as follows: (1) field measurements of the first field survey (2014) were used to evaluate the BA versus CA_{ap}, and (2) field measurements for the second field campaign were used to estimate the biomass and generate the allometric equations between CA_{ap} versus AGB and CA_{ap} versus WSC.

We calculated the BA from the DBH field measurements for *Q. castanea* and *Q. spp.* coppice shoots. Since typical sprouting oak trees or stools managed for charcoal making exhibit several coppice shoots, as opposed to unharvested oaks trees, it was necessary to sum up the BA of the coppice shoots belonging to the same tree. In this way it was possible to establish a 1:1 relationship between the total individual tree BA and individual tree photo-measured crown area (CA_{ap}). We followed the same approach to evaluate the allometric relationship between the BA for grouped trees (*Q. spp*) and the CA_{ap} corresponding to those grouped trees, i.e. we sum up the total stand BA and total stand CA_{ap}.

Image acquisition

If financial resources are available, it may be possible to commission an organization or company to provide the photographs required. Specialist companies are now available in many areas that will acquire specific imagery on request. However, it is often the case that aerial photographs are already available for the area of interest.

National forest services or conservation agencies may possess extensive archives of air photo imagery, although it is important to remember that interest in such photographs goes well beyond the forestry and conservation sectors. Land planning and rural development agencies, local or regional government administrations, hydrological surveys, and agricultural departments may all have commissioned aerial surveys at various times. Failing that, national military or defence organizations can usually be relied upon to possess comprehensive air photo coverage, which in some cases is made available to the public sector.

Some companies, such as those involved in mining or the construction of oil, gas, or water pipelines, also invest in developing extensive archives of air photos. Increasingly, comprehensive coverage is offered by specialist companies who can provide specific images to order, from archives that they have already developed.

Whatever the source of the imagery, a key consideration is scale, which determines the area on the ground that the photograph can detect. The *spatial resolution* or resolving power of a photograph describes the degree of detail on the ground that can be observed, and is influenced by the properties of the camera lens and film used in taking the photograph, as well as the characteristics of the object itself, such as its degree of contrast with its surroundings.

The *spectral sensitivity* of an aerial photograph refers to the sensitivity of the film to different parts of the electromagnetic spectrum, which depends on the type of film used.

Radiometric resolution refers to the image contrast or density on the film, and is influenced by the dyes and metallic silver used in the manufacture of the film, and its degree of exposure. Films also differ in terms of their types of emulsion and how they are processed, which can affect the graininess of the image, and therefore its spatial resolution.

The characteristics of films used in aerial photography can most readily be obtained by reference to the websites of the relevant manufacturers. The quality of the image produced can also be influenced by the type of paper used in printing the image from the negative.

It is important to check the date when the photograph was taken. The time of year influences the solar angle and therefore affects how objects within the photograph are illuminated by the sun, and the size of shadows that are cast. The phenology of the vegetation changes through the growing season, even in evergreen forests, and this influences the characteristics of the vegetation on the photograph. Details of the flightpath may also be useful as an aid to interpretation. It is also important to note that aerial photographs vary in terms of the angle above the ground at which they are taken.

Whereas vertical photographs are taken with the axis of the camera arranged vertically, oblique photographs will result if the camera axis is tilted. In forest ecological work, vertical air photographs are almost always preferred. If obliques are available, they may be of some value in terms of interpreting the characteristics of a forest area, but are likely to be of limited value with respect to development of forest maps unless they can be adequately converted through some form of digital processing (orthorectification).

Image processing

Much useful information can be gained by examining aerial photographs visually, for example by using a hand lens or binocular microscope. Stereoscopes can be used to view stereo pairs of photographs, enabling a three-dimensional image to be viewed. However, at least for most ecological applications, photographs are generally viewed on a computer screen following digitization (it should be noted that instruments for on-screen stereoscopic examination of digitized images are now available).

Once in the digital domain, the photograph will often need to be rectified if it is to be used as a basis for spatial analysis or mapping. *Image rectification* refers to the process of producing an image that is geometrically corrected, removing any distortions introduced during the photographic process. The process of correcting for distortion caused by variation in topography is called *orthorectification*, and photographs that have been processed in this way are often referred to as *orthophotos*. Orthorectification is almost always necessary, unless photographs are obtained that have already been processed in this way.

There are two main methods by which an aerial photograph can be orthorectified. First, *ground control points* (GCPs) are taken at selected locations within a landscape; these may be obtained from field surveys, perhaps by using a *global positioning system* (GPS), or directly from a published map. These points are then located on the image and their coordinates entered. At least 3–5 GCPs must be established in this way, but this is a minimum, and more accurate results are obtained if a larger number of GCPs is used. An alternative method is to use *digital elevation models* (DEMs), which may be derived from digital maps, some remote sensing data (such as lidar) or from stereoscopic models by photogrammetric methods.

As with GCPs, a relationship is determined between the map coordinates in the real world and locations on the digitized aerial photograph, and the digital image is then resampled to create the rectified image. This resampling involves warping the image so that distance and area measurements made on the image are closely related to those in the real world. Orthorectification is generally done by using specialist software.

Whether or not a photograph is rectified, it will certainly need to be referenced if it is to be used as a basis for producing maps. The process of *georeferencing* (sometimes called ground registration) involves processing an image so that it is aligned according to a ground coordinate system. Although the process shares some similarities with orthorectification, it is important not to confuse the two processes: whereas orthorectification corrects for distortion in the image, georeferencing enables the image to be related to existing map coordinate systems. Again, GCPs are selected in the image for which coordinates on the ground are available. These are then entered, usually by clicking on the appropriate points on a computer screen using a mouse. Most GIS software packages have the capability to georeference images in this way, although it should be noted that their ability to warp images (as required during orthorectification) is often limited.

As a result, a GIS package cannot be relied upon to provide all of the tools that may be required to process a photograph to the degree necessary for its use in mapping activities; additional image-processing software may also be needed. Once an image has been georeferenced it can be combined with other spatial data in a GIS with the same coordinate system. Measurements of variables such as distance and area can now be made, enabling different forms of spatial analysis to be carried out. The accuracy of these measurements depends on the accuracy of the georeferencing process, as well as the characteristics (such as spatial resolution) of the image, so care should be taken throughout the processing procedure, and attention paid to the levels of accuracy achieved.

Image interpretation

Aerial photographs are usually interpreted visually, a skill that can be developed through training and practice. Guidance on the interpretation of air photographs is provided by researchers. Areas or objects can be differentiated by inspection of characteristics such as tone, texture, pattern, size, shadows, shapes or associations. These may be defined as:

- Image tone or colour. Many objects have a characteristic colour or tone, depending on the specific signatures of electromagnetic radiation that are reflected or emitted. Different types of vegetation may therefore vary in how they appear on either black-and-white or colour images, according to their species composition, phenological state and canopy characteristics. Coniferous tree canopies, for example, often appear darker than those of broadleaved tree species. Usually, similar objects emit or reflect similar wavelengths of radiation. The types of camera and film used can also influence how objects appear on an image. For example, on colour infrared images forest canopies tend to appear pink or red rather than the usual tones of green.
- Texture. Vegetation canopies differ in their surface texture, or whether they appear rough or smooth. Texture can readily be used to differentiate between different types of land cover, such as forest and cropland; for example, agricultural crops often appear to have a smoother, more homogeneous texture than most natural forest canopies. Forest canopies with many strata or canopy gaps can appear rougher than even-aged stands. Texture, as with the size of the objects being observed, is related to the scale of the image.
- Pattern. How objects are arranged within an image can help aid their identification. Plantation forests or orchards, for example, tend to be characterized by a more regular pattern of tree distribution than natural forests.

LANDSCAPE ECOLOGY, HABITAT, AND BIODIVERSITY

Sustainable forest management requires that landscape ecological characteristics related to habitat and biodiversity be included in forest inventory and certification procedures. The characteristics of interest are:

1. Spatial patterns within the landscape,
2. Specific habitat-related forest conditions, and
3. The ecological processes that link spatial pattern, habitat, and ecosystem functioning.

Land-cover information is one example of spatial patterns readily obtainable by classifying remotely sensed data. Other useful datasets include forest canopy information (e.g., crown closure or leaf area estimates), understory information, and measures of the distribution and boundaries of landscape units such as forest fragmentation. Remote sensing can provide repeatable and consistent methods to develop these data layers such that changes over time can be monitored and habitat models can be developed and validated for individual species.

Habitat assessment is typically GIS-based; it involves selecting data layers likely to be of value in developing predictive models for the occurrence and distribution of individual species or species assemblages, as well as the identification of species useful as indicators of ecological condition. The use of remotely sensed data together with other spatial datasets integrated within a GIS environment has greatly enhanced the habitat assessment process.

FUTURE DIRECTIONS OF REMOTE SENSING IN FORESTRY

A key development in remote sensing has been the increased availability of high-spatial- and high-spectral-resolution remotely sensed data from a wide range of sensors and platforms including photographic and digital cameras, video capture, and airborne and spaceborne multispectral sensors.

Hyperspectral imagery promises to provide improved discrimination of forest cover and physiological attributes. Radar applications are being developed that penetrate the forest canopy to reveal characteristics of the forest floor. New technologies such as LiDAR can provide estimates of forest biomass, height, and the vertical distribution of forest structure with unprecedented accuracy. The use of advanced digital analysis methods and selective use of complementary data have provided more detailed information about forest structure, function, and ecosystem processes than ever before.

3

FOREST STRUCTURE AND COMPOSITION

Forest structure is the three-dimensional arrangement of trees and other plants, in combination with nonliving spatial elements such as soils, slopes, and hydrology. In short, structure is the physical geography of the forest, considered at a range of spatial scales. Structure includes such characteristics as canopy and understory geometry, continuity or fragmentation of canopy cover, homogeneity or patchiness of species distribution through the landscape, soil characteristics, and the species composition and age structure of stands. Most of these elements may be considered at micro, local, or landscape scales. Forest structure influences forest biodiversity directly through the formation of microhabitats as well as the determination of larger-scale habitat characteristics, but the relationship between structure and species diversity is complex and not well understood.

In recent years, an increasing number of biodiversity experiments have been designed for forests. However, trees examined in these experiments have often been very young. As an alternative method, inventories of natural and managed forests have been analysed.

Several field studies of forests have revealed positive effects of biodiversity on productivity, but others have found invariant or even negative relationships. A meta-analysis of published field measurements reveals equal abundances of species and a broad variety of light strategies among species as important prerequisites for positive biodiversity-productivity relationships.

In addition to field studies and experiments, forest models serve as a means of investigating biodiversity-productivity relationships. They allow the analysis of a variety of species mixtures over long periods. Researchers used a well-established forest gap model to show that tree species richness promotes productivity (defined as biomass increments per year) in mature European temperate forests. In addition, the authors show that competition for one resource (here light) is sufficient to generate an increase in forest productivity with species richness.

Despite the widely recognized positive trends found in diversity–productivity relationships in forests, it remains unclear why different relationships can be found. Besides the influence of species diversity it is well known that basal area also influences productivity. Recent studies also highlight the influence of tree size heterogeneity on productivity. For instance, some studies found a negative effect of size heterogeneity in several monocultures, whereas researchers revealed positive effects of size heterogeneity (and species diversity) in mixed stands.

CHANGE IN FOREST COMPOSITION AND STRUCTURE

Natural Succession

Succession is a natural process that alters a forest's composition and structure over time. Through succession, ecosystems constantly undergo change. One plant community replaces another plant community as species present decline and die.

This progressive replacement of one type of plant community with another is called succession. Disturbance plays a role in driving succession. Succession goes through a series of stages. Early successional stages usually consist of faster-growing, shorter-lived trees like aspen. Later stages are made up of slower growing trees that live longer lifespans, such as oak, sugar maple, or hemlock.

Natural Disturbance

Disturbance is considered anything that alters the current composition of the forest. Wind throw, wildfire, frost, insects, herbivores, and disease are considered natural disturbances. Disturbances can be small to large scale events. The composition of the forest is altered by disturbance. Wind throw may remove several trees from the forest. Some species of trees are more susceptible to wind throw, like aspen.

Intense wildfires are a disturbance that can eliminate all trees in an area. Dry, hot weather, a buildup of dead plant material, and lightning start many natural wildfires. Wildfires are part of the natural succession process of forests.

Fires kill weaker trees, open areas to sunlight, preserve fire-adapted species and ensure a healthier forest in the long run. Fires sometimes create ideal conditions for some species not normally found in an area to come in and colonize. The openings created by fires and wind throw create spaces for sun-loving species to establish.

Other climatic disturbances also change forests. A winter with temperatures below zero for several consecutive days weakens some trees, making them more susceptible to diseases. At the same time, a hot dry summer in Wisconsin causes water stress in some trees. While this may not kill plants and trees right away, it will weaken them. This makes it harder to store food for the winter, or to fight off disease and pests.

Insects can alter the composition and structure of forests by removing entire species from an area. Herbivores such as deer browse tree bark, branches, and young seedlings.

This browsing severely affects the tree's growth potential and limits the ability of the next generation to regenerate.

Browsing by insects and herbivores weakens and exposes trees to natural and exotic diseases. Although natural disturbance causes dramatic changes in forests, these events are important for forest health. These changes provide opportunities for the species to change over time.

Human Disturbance:

Forests are changed by human disturbance. For over 10,000 years, humans have influenced the composition of forests by both the introduction and removal of species. When humans remove trees, this opens the over story to suppressed trees, leading to quick changes in structure and function.

Human change often mimics natural forest changes but at a much faster pace. The use of many management techniques are often meant to speed along the natural succession process of forests and alter the composition and structure. Examples of human disturbances are forest logging and thinning, fires, and the introduction of exotic species.

FOREST INVENTORY

Brief History

The history of conducting forest inventory in India goes back to eighteen centuries. Assessment of the Forest Resource on a relatively large area basis (catchment basis) using statistically robust approach and aerial photographs began in 1965 when the Pre investment Survey of Forest Resources (PISFR) was launched in the country with FAO/ UNDP assistance. Besides, trained and committed team from India, foreign experts were also involved in designing the methodology of forest inventory and data processing. This was also a beginning of era where assessment of the forest resources was linked to the requirement of wood based industries. The forest inventory was continued in different parts of the country with varying sampling design till 1981 when PISFR was reorganised as Forest Survey of India (FSI). Inventory remained as one of the important activity of FSI even after creation of FSI with a uniform design.

Forest Inventory conducted by FSI during 1965 to 2002

During 1965 to 1980 generally, systematic cluster sampling was used in study area. The study area was divided into suitable grid sizes (5' × 5' or 2½' × 2½') depending upon the optimum sample size (which defines the sampling intensity) and within the selected grid a cluster of 3 to 8 sub-plots was considered for recording the data on different parameters of inventory. Whenever the information on stratification variable was available (generally, pre-stratification on the basis of aerial photographs), stratified random sampling was used, otherwise the collected data was post-stratified to increase the precision of estimates.

The late 1970s and early 1980s, were very important in international as well as national forest scenario and influenced a paradigm shift in realization of role of forest resources. The forests which initially seemed inexhaustible resource under the pressure of human and cattle population were rapidly depleting and leading to overall deterioration in the environment, threatening the very existence of civilization. Therefore, strategies were framed to divert production forestry to conservation forestry. New programs like social and agro-forestry came into being. To formulate suitable strategies for the new scenario there was a need to have information at national level.

During the last decade of 20th century, the role of forests was again redefined by including the additional parameters like carbon sequestration in plant community and in forest soil, regeneration status, naturalness etc. Taking it as opportunity to fulfill its dream of conducting national level forest inventory (NFI), FSI conducted a series of workshops and meetings with experts and came out with an integrated approach in 2001-02 in which forest inventory with provision of capturing above mentioned parameters be conducted in a cycle of two years. This approach was followed and the estimates were being improved cycle after cycle till 2016.

INITIATION OF NATIONAL FOREST INVENTORY

As the forest inventories carried out in different parts of the country since 1965 were in a different time frame, it was not possible to generate national level estimates on growing stock, area statistics and other parameters with reference to one point of time. At the beginning of the 10th five-year plan in 2002, considering the resource limitation, a new sampling design was adopted to generate national level estimates of various parameters including those of growing stock from forests which is described in the following paragraphs.

Sampling Design for NFI

A two-stage sampling design was adopted for national forest inventory. In the first stage the country was stratified into homogeneous strata, based on physiography, climate and vegetation called the 'physiographic zones'. The civil districts within the physiographic zones were taken as sampling units. The 14 physiographic zones were identified in the country; Western Himalayas, Eastern Himalayas, North East, Northern Plains, Eastern Plains, Western Plains, Central Highlands, North Deccan, East Deccan South Deccan, Western Ghats, East Ghats, West coast and East coast.

A sample of 10 percent districts (approx. 60 districts in the country) distributed over all the physiographic zones in proportion to their size were selected randomly for detailed inventory of forest and trees outside forests. In the second stage, separate sampling designs were followed for detailed inventory of forest and TOF.

Second stage sampling design for forest area

The 60 districts selected in the first stage were taken for detailed inventory of forests. The Survey of India (SOI) topographic sheets on 1:50,000 scale (size 15'x15' i.e. 15 minutes latitudes and 15 minutes longitudes) of the district were divided into 36 grids 2½ × 2½ which were further divided into sub-grids of 1¼ × 1¼ forming the basic sampling frame. Two of these 1¼ × 1¼ sub-grids were then randomly selected to lay out the sample plots. Other forested sub-grids in the districts were selected systematically taking first two sub-grids as random start. If the centre of the selected, 1¼ × 1¼ sub-grid does not fall in the forest area then it was rejected.

If it fell in the forest area, it was taken up for inventory. The intersections of diagonals of such sub-grids were marked as the centre of the plot at which a square sample plot of 0.1 ha area was laid out to record the measurements on tree diameter at breast height (dbh) and height of sampled trees species. In addition, within this 0.1 ha plot, sub-plots of 1mx1m were laid out at north-east and south-west corner for collecting data on soil and forest floor (humus and litter) etc.

The data regarding herbs, shrubs and climbers was collected from four square plots of 1m × 1m and 3m × 3m in all four directions along the diagonal, 30m away from centre of 0.1 ha plot.

Besides forest inventory, TOF inventory was also started. This design was continued till 2016. However, as per the national and international requirement, FSI again modified its sampling design.

SAMPLING SCHEME FOR NFI

Under the new design, a nationwide uniform grids of 5 Km × 5 Km have been created. A forest layer has been developed by combining the digital layer of RFA boundaries/ greenwash. Overlaying the forest layer on grid layer, the intersection of these two layers provided the general sampling frame for NFI. On the basis of forest layer, every grid was identified as either forested grid or non-forested grid (TOF grid). All the grids have been numbered from 1 to 5. Choosing one of these numbers as random start and thereafter all grids containing that number marked as forested grid, were chosen for data collection. Within the selected forested grid denoting 25 km^2 forest area, a random point using GIS gives the centre of the sample point from where the detailed data is collected as per sample design. The diagram shows, the cluster of four circular sub-plots each with radius of 8m in a fixed pattern. Here, the central subplot-1 is laid out at the point with assigned lat-long. Other three subplots-2, 3 and 4 are to be located at a distance 40m in north, 120° and 240° from north respectively. It is also envisaged that out of these plots, around 10% plots will be measured for biodiversity indicators, health indicators, climate change indicators and soil characteristics.

Scope

The NFI is designed to include all forest area of the country. The forest area is notified by the Government. The digital boundaries of forest area are not available for entire country. This is available for 12 states and for remaining states green-wash area as depicted in

Survey of India topographic sheet will be taken as proxy of RFA (Survey of India shows some area as green in their topographic sheets usually referred as 'green-wash'. This green-wash depicts: recorded forest areas and other traditional forest areas at the time of survey for topographic sheets preparation).

Sample size

The NFI envisages measurement of about 35,000 sample points across the country in the five years. Every year 20% sample points will be covered. In other words, it will measure a fixed proportion of the plots in each State, each year. Around 10% of the identified sample plots will be used for special studies related to biodiversity and soils.

Precision

The sample size is optimized so that enough plots are measured to satisfy precision standards for area and volume estimates. There is provision that if States wants more precise estimates then they may choose to increase the sample size by installing additional plots.

Layout of the sample plot

- Cluster of four sub plot of 8 metre radius each from the centre of sub-plot-1 at Azimuth at 360°, 120° and 240° at a distance of 40 metre.
- Three 1 m × 1m plot for soil and forest floor with centre at a distance of 20 metre from the centre of sub-plot -1 in the direction of sub-plot 2, 3 and 4.
- Two circular plots for NTFPs, for Herbs (0.6 m radius), Shrub, Climber and Regeneration and woody litter (1.7 m radius) in all four sub plots at a distance of 5 metre from the centre towards east.
- A circular plot of 2.8 metre for stump & dead wood collection will be laid out in all four sub plots at a distance of 5 metre from the centre towards east.
- Set of Herbs, Shrubs-climbers-regeneration-woody litter and stump, dead wood circular plots is concentric.

TYPES OF INVENTORIES

National forest inventories (NFIs) have been implemented in many countries, particularly those with extensive forest resources, to provide information on forests at the national scale in support of forest planning and policy development. NFIs are usually established according to a systematic grid or network of plots across the whole country, with field plots arranged in clusters to make the process of data collection more efficient.

Many developing countries still lack comprehensive NFIs. This is something currently being addressed by the FAO through provision of capacity-building and support. According to the methods developed by FAO, the country is first stratified into relatively homogeneous regions, then tracts are established within which field sample plots are located.

The number of tracts may vary between 50 and 500, depending on country size and homogeneity. Four field plots are established per tract, which are marked permanently on the ground. For example, in Guatemala, the country was divided into three strata according to national ecological zones, namely lowland, southern coastal plain, and central high mountain areas. One hundred and eight tracts (1×1 km) were systematically distributed throughout the country, at higher density in the heterogeneous central stratum and at lower density in the relatively homogeneous northern and southern strata. In each tract four sample plots (25 × 20 m^2) were established, oriented south–north, west–east, north–south, and east–west. Some variables were measured in subplots located within these plots, to save effort.

It is unlikely that any of the readers of this book will ever be in the position of needing to design an NFI - if so, useful guidance is provided by de Vries (1986) and Shiver and Borders (1996). However, NFIs can be of great value as a source of information for forest ecology and conservation. For example, Researchers used the Indonesian NFI to estimate the population size of *Aquilaria* spp. (gaharu), threatened tree species of great commercial importance for which limited information on conservation status was previously available. Field visits were undertaken to check the accuracy of the NFI data used in these analyses, and a number of sources of error were discovered, including inaccurate descriptions of plot locations and taxonomic confusion caused by the use of local, rather than scientific, names of tree species.

Despite such problems, this study demonstrated that NFIs can provide information of value to conservation that is difficult to obtain in any other way, particularly when the species concerned are widespread and occur in inaccessible areas. Another example is provided by researchers, who used NFI data in Guyana for development of a national protected area strategy. However, the limitations of NFIs should also be noted: they often lack information of great interest to conservationists, such as production of non-timber forest products.

Sampling Designs

Inventories are used to make management or policy decisions. For that purpose the most common variable used is the average value of whatever is being measured. Whether this average value is representative and whether it is a reliable estimate of the true value of what has been measured, depends very much on the sampling design used. Complex or specific situations require complex or specific sampling designs. The reader is referred to the specialist literature for these designs. In this section, only some of the basic aspects of forest inventory and sampling design are dealt with.

To assure that the information is *representative*, it is important to try to include as much as possible of a population's variety within the sample. This can be achieved by choosing populations as homogeneous as possible, by making sure that the sample units are well distributed over the whole population, by increasing the size of the sample or by adjusting size and shape of the sampling unit.

Optimum sizes and shapes need to be determined for each different type of forests, while statistical calculations help determine the optimum sample size. In even-aged single species forests that are classified according to their site productivity, few sample units of small size in each productivity class may be very representative. In the natural wet tropical forests local diversity in structure and floristic composition may require large sample units, while high diversity between units may require a large number of sample units in order to capture the whole variety of the forest.

The *reliability* of the data obtained is important to be able to estimate the probability that the averages or totals estimated for the forest are different from the real values in a particular sample of that forest. A company that for example wants to invest in a sawmill will want to know the risk of not being able to harvest the quantity of harvestable timber estimated for its forest. This risk can be expressed in terms of the sampling error, a statistical measure of the reliability of the data. If this error is high, the risk is high, and the company may not want to invest. Thus, it is important to design forest inventories in such a way that the risk, i.e. the sampling error, is reduced to acceptable levels. One can, for example, decrease the variability between sample units by increasing their individual size. This decreases the sampling error. Increasing the number of units has similar effects.

It is rarely possible to measure all of the trees in a particular forest. Therefore, a *sample* must be taken from the complete *population* of all sample units. The most important basic principle is to ensure that the sample is representative. Otherwise, the information obtained will be biased in some way, and the inferences drawn from the data are likely to be invalid. The first step is to divide the forest area to be surveyed into sampling units, for example by dividing it into grid squares or forest stands or patches. The next step is to decide which sample units to include in the survey, according to an appropriate design. Four basic sampling designs are used in forest surveys: *simple random*, *stratified random*, *systematic sampling* and *cluster sampling* (Fig. 1).

Whichever method is used, it is important that the survey is as accurate as possible. If a sample is taken from a population, the values of measurements obtained from the sample may differ from those of the population. This difference is referred to as the *sampling error*, and is expressed as the standard error of the mean. Sampling error can be reduced by increasing the sample size. Other (nonsampling) errors may arise as a result of mistakes or inaccuracies in data collection, bias in the estimates because of a lack of independence in the sample units, inaccurate production of maps and calculation of areas, and poor data processing or management. Attempts should be made to document and minimize such sources of error.

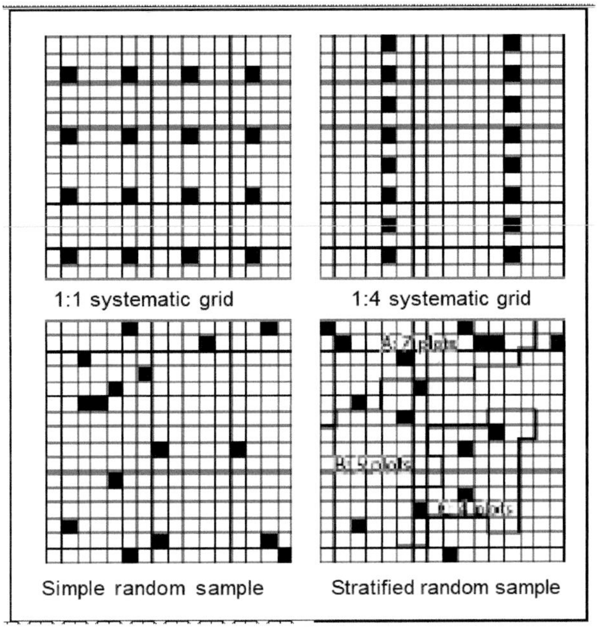

Fig. 1 Different sampling designs, based on four possible arrangements of 16 samples in a population composed of 256 square plots.

Distribution of Sample Units

In this approach, the forest to be surveyed is first divided into relatively homogeneous areas (or strata). Sample units are then randomly selected from each stratum (usually at least two from each), using the same approach as for simple random sampling. There are a number of key advantages to this approach. Forests are usually spatially heterogeneous, as a result of variation in environmental variables such as topography, soils, aspect, and altitude, as well as in patterns of natural disturbance and previous management. This variation can often be detected on aerial photographs or during preliminary field surveys. Strata can therefore be defined on the basis of such information, in a way that is relevant to the objectives of the survey. For example, it may be known that in different parts of the forest the stands are dominated by different tree species because of variation in soil conditions and drainage. In this case, if the objective is to assess stand structure in each of the different kinds of forest stand present, stratified approaches offer an advantage by enabling each type of forest stand to be adequately represented within the sample.

How should the number of sample units in each stratum be determined? There are two main options. Samples can be allocated among strata in proportion to their relative areas (*proportional allocation*); in other words, more samples are allocated to larger strata than smaller strata. Alternatively, sample units can be allocated to strata taking into account both the size and expected variance of the strata (*optimal* or *Neyman allocation*). In other words, more variable strata are sampled more intensively than less variable strata of the same size. The latter method results in a more precise estimate of the population mean, but requires prior estimates of the sample variance within the individual strata. However, optimal allocation is generally preferred if stratum areas and variances can be reliably determined. Both of these methods provide separate estimates of mean values for each of the strata, which might be different forest types or administrative units.

Different methods exist to distribute the sample over an area. Statistically, the most reliable distribution is obtained through random selection of the sampling units, where each unit of a population has equal probability to be selected. While such a distribution statistically is the more correct one, it has as a disadvantage that sample units may be grouped in certain areas, with other parts of the forest not being sampled. This may result in statistically reliable but little representative data.

This form of distribution also reduces the use of the information for area calculations or for the analysis of forest types or species and tree distribution patterns. The random distribution is useful in some types of research and for exploratory inventories over very large areas, where sample units exists of clusters or groups of plots. In that case, the clusters can be located randomly within the different forest areas (stratification). Random selection, however, is little practical for the distribution of sample plots in most forest inventories.

The most common form of distribution of sample plots is that obtained through systematic sampling: of the sampling frame of N units, the first sample unit to be measured is selected randomly, after which every xth sample unit is selected in such a way that N/x is approximately n, the sample size. As a result one obtains a grid of sample units at equal distances between units within sample lines, and equal distances between the lines. While statistically this form of distribution gives some problems, if applied correctly, it usually results in a better distribution of the sample units and allows better planning and easier access to each unit.

Another form of distribution is that through expert selection: the forester or researcher selects its sample sites as "representative" for the forest area to be sampled. This is acceptable for certain types of research, but in general will give biased information in the case of forest inventories: most people tend to select well-stocked areas as representative, leaving areas with clearings or affected by other natural phenomena out of the sample, thus creating overestimates for the averages of most stand variables (e.g. volume per hectare data).

Systematic sampling

In this method, the first sampling unit is selected randomly or arbitrarily located on the ground, and locations to be sampled are thereafter spaced at uniform intervals throughout the area to be surveyed. For example, if 10% of the area were to be sampled, every tenth sampling unit would be selected. Typically, sample units are established on a grid.

This method has been widely used in forest survey because the sampling units are easy to locate on the ground, and because the samples are distributed over the entire area, giving the impression that a representative sample has been obtained. The main problem is that it is less statistically powerful than random sampling methods. Specifically, it is not possible to obtain a genuinely valid estimate of the sample variance, because the sample units are not truly independent. This makes it difficult to estimate the precision of the measurements taken. This is particularly the case when there is a regular spatial pattern in the forest being surveyed, which may be caused by regular variation in soils, topography, or hydrology. The effects of topography can be minimized by referring to soil maps or by orientating grid lines to be surveyed up and down slope, rather than along the contours. However, in situations where estimations of precision are not required, systematic sampling may be preferred, primarily because it may be more efficient (in terms of information gained per unit effort expended) than random sampling approaches.

Cluster sampling

A cluster is a group of smaller units (subplots) that taken together make up the sampling unit. Clusters can be arranged in many different ways, depending on the number of subplots included, the distance between them, and their spatial relations. As noted earlier, clusters are often used in NFIs, particularly in areas that are remote or difficult to access. This is because time and resources can be saved if information is collected from a number of locations within a particular area. However, cluster sampling is also sometimes used in regeneration surveys at a local scale.

Clusters are randomly selected from the population. However, they are not stratified, as in stratified random sampling. Cluster sampling may also be divided into a number of stages. For example, in two-stage clustering, clusters are randomly sampled and then, instead of each subplot within the cluster being surveyed, a subsample of these is randomly selected for survey. When implementing cluster sampling, the first step is to specify appropriate clusters. Ideally, the number of subplots within a cluster should be relatively small relative to population size, and the number of clusters should be relatively large. Cluster sampling will be more precise than simple random sampling if variation at the local scale is high relative to variation at the scale of the entire population. It should be remembered, though, that the subplots within a cluster are not independent of each other, and therefore the independent sample unit is the cluster rather than the subplot.

Permanent Sample Plots

Forest inventories in general are designed to describe a static situation: those aspects of the forest that are of interest in a particular point in time. The number of trees per hectare, volume per hectare, basal area per hectare, relative importance of a species, biodiversity indices are measured at a specific point in time. In order to get an idea of changes over time, be these natural or caused by human activities, it will be necessary to repeat these measurements over time. These should preferably be done within the same (permanent) sampling units. These plots need to be carefully laid out, well-marked in the field and on maps, and should be established in areas where it is likely that they will not be removed by natural or human activities.

Locating sampling units

Traditionally forests were mainly used and managed for timber production. However, social expectations regarding forests have notably changed in the last few decades. These changes are reflected in the initiatives of the pan-European FOREST EUROPE process (formerly known as Ministerial Conferences on the Protection of Forests in Europe – MCPFE), as well as in the statements on the role of forests in modern societies presented in the EU Forest Action Plan (EC 2006), and the New EU Forest Strategy. In addition to the traditional production function (timber), the other – often conflicting – functions that forests should fulfil include protective functions (e.g., of site, drinking water basis), biodiversity and ecosystem functions, social and cultural functions (e.g., health promotion, education, recreation) and economic functions (i.e., sustainable use of forest related goods and services).

One possible way to achieve these goals is to spatially separate the different functions by designating nature reserves, protective forests (soil, landslide), recreational areas and forests serving primarily wood production. However, in most parts of the world – especially in Europe with a long history of intensive land-use – the magnitude of landscape transformation (habitat destruction, fragmentation and degradation) has exceeded the threshold of habitat loss where this approach is feasible. Even in other, less deteriorated regions, because of social and economic hindrance, there is a shortage of land available for reserve allocation, and proper reserve management is often compromised.

In order to be able to plan forest management and conservation actions in an integrative and efficient way, higher data quality must be achieved both in terms of thematic richness and applicability over various spatial scales. Otherwise, we are not able to set quantitative let alone qualitative goals for actions favouring forest-dwelling organisms. We need data complementing existing sources of information. New data should fulfil the requirement of serving both management and conservation aspects and need to serve as the basis for complex management action plans.

GENERAL CRITERIA AND SAMPLING SCHEME

The forest state assessment protocol was developed within the framework of the project "Multi-purpose assessment serving forest biodiversity conservation in the Carpathian region of Hungary". The main aim of the project is to collect data that complement available information in forestry and conservation databases and integrate all for the enhancement of conservation of these habitats. The project includes the development and implementation of forest state assessment and zoological surveys on forest-dwelling bats and birds and xylophagous insects. All work has been carried out in the Northern part of Hungary, in the Börzsöny, Mátra and Aggtelek Mountains.

The main goal of the forest state assessment was to collect structural and compositional data that were missing and that could be used by forest management companies and conservation agencies (national parks) for strategic as well as daily planning of forestry and conservation activities. The assessment was also planned to support the analysis of forest naturalness and to enable us to find potential hotspots of biodiversity. Methods to be chosen had to meet the following criteria:

1. Simple and fast measurements or estimates without the need for special equipment,
2. Reproducible methods to allow for many field workers,
3. Data comparable to existing databases,
4. Appropriate data for supporting forest management planning and conservation planning. It was also required to produce thematic maps at the scale of 1:10 000 for direct use in conservation management plans and produce traditional maps such as habitat type maps.

From the very beginning we decided to use systematic point sampling in order to achieve spatially explicit data based on which maps with several distinct thematic contents could be produced. Since many national forest inventories and other investigations use 500 m^2 circular plots, this size was chosen as the main sampling area at each sampling point.

VARIABLE SELECTION AND CHOICE OF MEASUREMENT METHODS

Potential variables were selected from existing forest inventories, forest naturalness surveys, forest reserve programs, Natura 2000 habitat monitoring, specific conservation projects and international projects e.g., ForestBIOTA. A list of variables considered for our multipurpose forest state survey was put together. The extended variable list contained over one hundred items and for a number of variables several optional definitions and measurement methods were identified.

Several stakeholders (national park experts, representatives of local state forest enterprise and state forestry authority) were involved in the selection of relevant features and in defining measurement scales that could be applied in the field. Where alternative methods were available, they were tested in the field and evaluated for reproducibility, simplicity of method and time consumption as the main criteria. An example of such a pilot study is given below in detail for canopy composition.

Existing data on canopy composition include admixing ratio, height, closure and volume of commercially important tree species. Therefore, our complementary data had to provide better estimates for species richness by being sensitive for species with low cover and for structural diversity by differentiating between age classes and vertical complexity

To develop the method for describing canopy composition, 30 plots located in tree stands of different ages and composition have been sampled. In our novel method we planned to record canopy cover estimates for each tree species in several diameter classes. Canopy cover was estimated by using broad cover categories: 0-5%, 6-20%, 21-50%, 51-100%. Two independent crew recorded canopy composition using two DBH category scales:

Method A (five categories): 0-8 cm, 9-20 cm, 21-35 cm, 36-50 cm, >50 cm;

Method B (six categories): 0-8 cm, 9-15 cm, 16-25 cm, 26-35 cm, 36-50 cm, >50 cm.

The extent of the 500 m^2 circular plot was visually estimated. As a reference (Method C, complete survey), following estimation according to A and B, the extent of the plot was measured then species name and DBH were recorded for each tree individual in the plot. Methods A and B were compared against the reference (C) based on species admixing ratio and relative cover in diameter classes.

For methods A and B admixing ratio of a species (its relative importance in the canopy) was calculated as its cover divided by total cover. Relative cover in a diameter class was derived similarly but using only data of the respective diameter class. For these calculations we used the middle values of cover categories (e.g., 75% for 51-100%). For method C basal area (cross-sectional area of a tree species at 130 cm height expressed in m^2/ha, derived from DBH) was used for calculations.

Distributions of relative cover in diameter classes were compared between A-C and B-C using Chi-square test (test of homogeneity) plot-wise. The Chi-square test showed that method A was better than method B in describing diameter class distribution, while correlation between admixing ratios showed that both estimation techniques performed well ($R^2 = 0.84$, $p < 0.05$). As a conclusion, taking into account time consumption (method A was 30 seconds faster on average than method B), method A was finally chosen for the description of canopy composition and structure.

THE NOVEL METHODOLOGY AND DATA COLLECTION SYSTEM

The multipurpose forest state survey is based on a systematic grid of field points. The intended average density of the grid is two points per hectare. Three different and joint grid densities were established that could be used for stands with varying structural complexity. For the most heterogeneous stands (e.g., forest reserves), a dense grid of 50 m × 50 m was laid out. By selecting every second or fourth point in more homogeneous stands, a network of sampling plots in every 70 m or 100 m was set up (Fig. 2).

From the latter grid sizes the denser, 70 m × 70 m grid was chosen during the first year of sampling.

We use three distinct sampling units. At every grid point two concentric sampling plots are surveyed. Most data are collected for the 500 m^2 circular plot (radius: 12.62 m), while shrub and regeneration data are gathered in the 30 m^2 subplot (radius: 3.09 m).

Between plots, variables occurring at coarse spatial scales and attributes carrying particular interest even if found only outside the plots are recorded on the route. The route is defined as the visible area from the grid line including the plot of arrival.

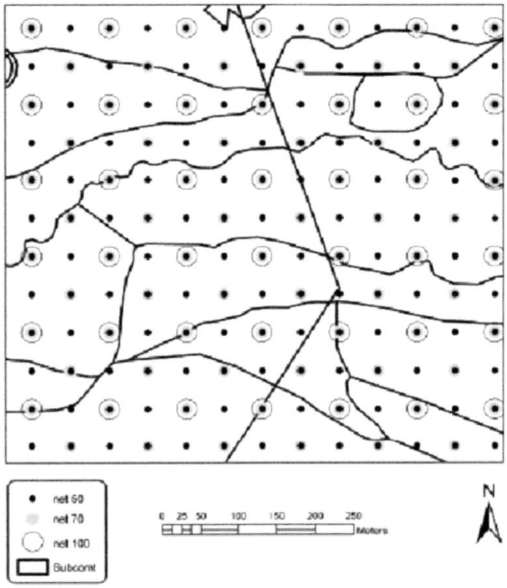

Fig. 2. Systematic sampling grid of various densities used during our survey. In the basic grid, points have a distance of 50 m (black filled circles)

A digital data collection system has been developed for the purposes of the forest state survey. Forms are filled in using the ForestDataCollect (FDC) Android application developed by our team. EVOLVEO StrongPhone Q4 is used to run the app, while GPS coordinates are recorded using Garmin GPSMap 64 equipment. The empty forms are downloaded safely for fieldwork in predefined packages containing approx. 30 grid points using the FDC app.

There are several data input widgets including table question (Fig. 3a) and search list question (Fig. 3b) that were not found in the tested available data collection apps. With the table question (Fig. 3a), several attributes can be collected for one entity (in our case tree species), and since species lists are long, the search list widget provides easy access to species through favourites and a search field. As an example of other widgets, picture select question has been used to aid visual estimation of lying deadwood (Fig. 3c). After completing the data recording, the finished packages are uploaded and integrated into the PostGIS database together with the photographs and GPS coordinates. Processes are semi-automated using Python scripts.

A crucial prerequisite of guaranteeing high quality data is the training of field crew. All crew members had to take part in a series of indoor and outdoor training lessons and had to pass exams. Field training takes place at the beginning of every field season. Quality assurance is also implemented during field work, data import and data analysis. The smartphone application excludes misspelling of species names and incoherent data through constraints and within-form rules. Only consistent data are imported into the database and a record-based manual quality control ensures credible data. Differences between pairs of surveyors is checked by analysing data that were collected in packages set out in a chequerboard design (Fig. 4).

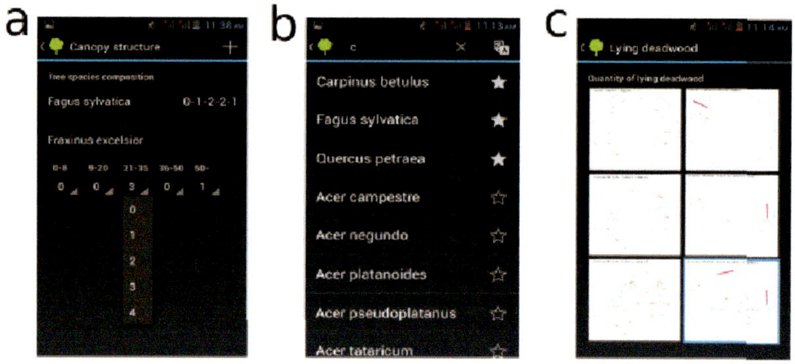

Fig. 3. ForestDataCollect android data collection application screenshots.

a) Table widget used to collect cover categories in each diameter class and species. b) Search list widget used to select species from long lists aided by dynamic selection of favourites moved to the top of the list and search field that narrows down the list to matching entries. (c) Image select widget showing 500 m² circular plots with varying amount and diameter distribution of lying deadwood.

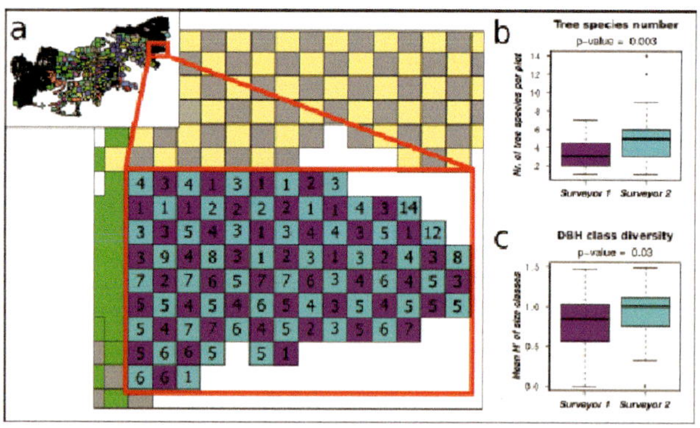

Fig. 4. Graphical illustration of data quality assurance based on chequerboard-design and two examples for testing surveyors' data quality.

a) One of the chequerboard subsets is shown from Aggtelek Mts. (overview map). Pixels are derived from the 100 m × 100 m grid. Points belonging to a given field surveyor are represented by a specific colour. On the map, species number observed by the surveyors are marked within the pixels. b) Comparison of the two surveyors based on tree species number using t-test. c)

Illustrative example applications

The whole forest state survey system is designed to enable the collection of large numbers of samples. During the lifetime of the project several tens of thousands of plots have to be sampled. The case studies shown below are based on data collected in roughly 12,000 plots during the first field season in 2014.

An important application of the forest state survey is that distribution maps can be drawn for tree species that can serve as hosts for specialist species. The location of such rare or economically unimportant species is often unknown, since they are not surveyed. During the first field season of our survey, more than 20 000 new records on the occurrence of 36 native tree species were gathered in the 11,194 sampling points that contained canopy trees. Compared with data taken from the National Forestry Database, new tree species were recorded in 97% of the sampled forest subcompartments. Since several specialist birds, bats, insects or fungi colonize their host tree individuals if they are above a certain age or size (translated to diameter at breast height (DBH), e.g., > 50 cm DBH), tree species distribution maps can be refined by diameter classes (Fig. 5). Such maps can be used to specifically search for new occurrences of specialist animal species.

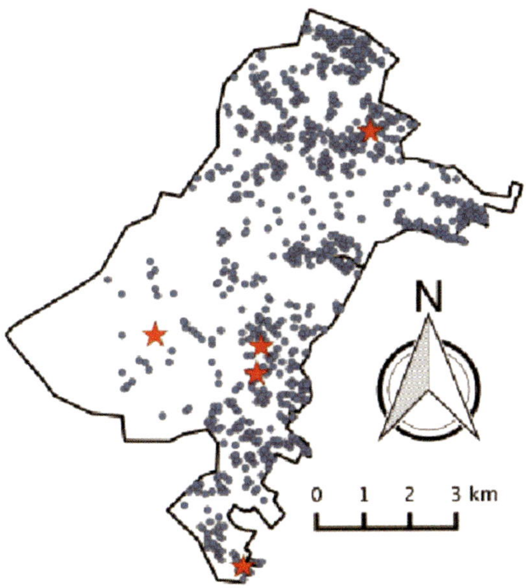

Fig. 5. Distribution of Cerasus avium (L.) Moench in the mapped area of the Börzsöny Mts., Northern Hungary. Dots show occurrences of the species in any diameter class, while large (> 50 cm DBH) trees are indicated by stars.

Tree species number varied between 1 and 12 with an average of about 4 species/plot. The diameter distribution of trees within plots seemed rather diverse. All diameter classes appeared in 15% of the plots, while only 9% contained just one diameter class. As Fig. 6 shows, there is a general positive – though not necessarily significant – relationship between species number and diameter class diversity (DCD, calculated as Shannon-diversity based on relative cover of diameter classes), however relatively high DCD values occurred even if tree species number is low, and vice versa. Using both variables, we can identify forests that have the most diverse canopy (see Fig. 6).

USING A COMPASS AND MEASURING DISTANCE

Finding Distance With A Compass

An interesting compass application is to be able to measure the distance across a lake, or large area with a good degree of accuracy. There is an easy way to do this with your compass.

d = (Tan (90 - (A -B))) × Ref

d = Distance (to be calculated)

Tan = Tangent value of the resultant angle

A = Greater value of the two measured bearing angles

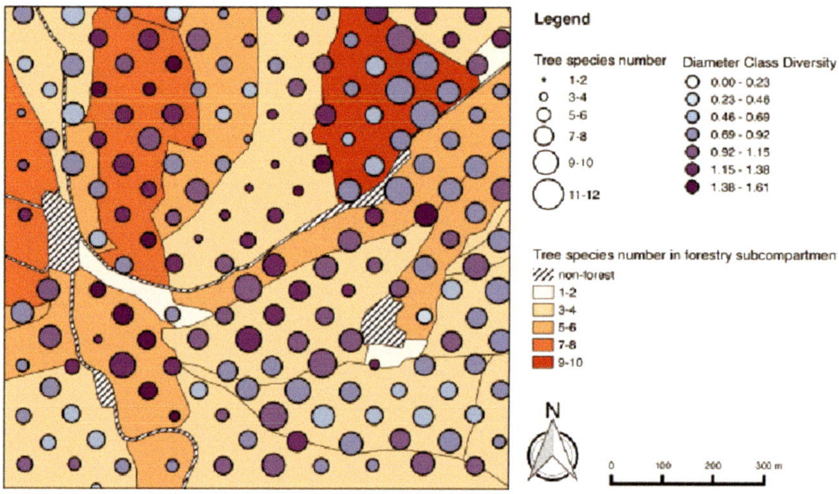

Fig. 6. Tree species richness and diameter class diversity.

Sizes of circles represent species richness, while their colour represents diameter class diversity. The applied sampling density is 70 m × 70 m. Some "holes" can be seen in the grid: those are either meadows or roads that could not be sampled. Species richness and diameter class diversity (DCD) was calculated for single plots. For DCD, middle cover values were summed in each diameter class and Shannon diversity index was calculated for the classes to account for evenness. The polygons of forest subcompartments are shown displaying species number recorded in the National Forestry Database (NFD).

B = Lower value of the two measured bearing angles

Ref = Measured reference distance

The method used here is to find a fixed point at the spot you would like to measure the distance from, to where you are. You must sight two reference bearings, one on each end of a measured reference (Ref) distance perpendicular to the spot to be measured. In practice, the more accurate you measure this reference distance, and the greater this distance is up to a 45° total offset, the more exact your results will be.

Sight a bearing to the spot that you will measure the distance to, from the exact location where you are standing. This will be bearing or angle (A).

Walk to a measured perpendicular reference point (Ref) to take another reading to the same spot that will be measured. This will be bearing (B).

If bearing (B) is greater than bearing (A) then swap the (A) and (B) values. You want to always subtract the smaller number from the greater number.

Now perform (A) - (B).

Take this difference and subtract it from 90. Now you have 90-(A-B).

The next step is to find the Tangent of this final angle or bearing differential. Now you have the Tan of (90-(A-B))

The last step is to multiply this value by the measured perpendicular reference distance. You now completed the function (Tan(90-(A-B))) × Ref.

Note that if the first bearing reading is less than 360N (or 0)°, and the second bearing reading is past the 360N (or 0) degree value, then you must assume the compass card to exceed the 360N value. For example, if you read 358N° for the first value, and 2E° for the second value, then the result would be equal to 362°. The differential value would be 4° as required.

For example, if you are looking at a tree across a lake and you want to find its distance from you, sight the first bearing to it. Let us say that it is 210°. Now walk exactly 10 meters perpendicular in relation to the tree (As perpendicular as possible). Next, sight the second bearing to it. Let us say that it is now reading 214°. Now calculate 90 - (214-210). You should get 86. Now find the Tangent of 86. This will give you 14.3 . Multiply this value by the perpendicular reference (Ref) distance (10 meters). This would be (14.3 × 10)=143. The distance (d) to the tree across the lake is 143 meters.

Engineer's Lensatic Compass

There are specialized compasses for specific applications. Some lensatic compass applications are used in industry, the military, and exploration.

The Lensatic Compass Scale was designed to provide a quick way to measure the distance of an object or landmark. It has been worked out to be scaled in 64 units representing the full 360°. However, it offers a slight compromise in accuracy compared to using detailed trigonometry functions. Each division of lensatic angle is equivelent to 5.625° of compass azmuth. The accuracy of distance measurement using the lensatic scale should be within 1% under most applications.

It has been worked out that if a sighted object has a perpendicular width of 1 meter, and measures 1 lensatic division (unit) on the compass scale, it is 10 meters from you.

The formula is: (S / Le) × 10

S = Size of target (perpendicular)

Le = Lensatic divisions, or number of lensatic units

For example, if you have sighted an automobile facing you that you know measures 1.8 meters wide (perpendicular measurement facing you), and the lensatic divisions are 1.6, the distance from you would be: (1.8 / 1.6) × 10 = 11.25 meters from you.

The reading of 1.6 divisions lensatic is equal to 9° of azmuth. Using some basic trigonometry reveals the car to be 11.36 meters distance. The error in this case is about 0.9%. For most applications, and considering possible parallax errors when using a compass, this is considered acceptable accuracy.

Using The Compass Clinometer Scale To Find Height

Here is a simplified way to find the height of a structure or tree. This method applies to compasses that have a clinometer scale. If your compass clinometer scale is in° only, you must convert this to the percent of angle (% Grade).

- Measure the total Vertical Angle in percent (% Grade) from the base of the structure to its top. If your compass does not have the % Grade scale, use°. You will have to look up the Tangent Value, and convert it to % Grade.
- Measure the Distance from your position to the base of the structure.
- Multiply the Distance by the % Grade.
- Divide this result by 100.

Height = (d) × (%G) / 100

d = Distance

%G = Angular Percentage (% Grade)

To find Percent Grade (%G) from the Tangent Value, multiply the Tangent Value by 100.

%G = Tan × 100

%G = The Percent Grade or Angular Percentage

Tan = The Tangent Value

For example, say the distance to the tree is 100 feet and the %Grade is 30%. The tree height is:

height = 100 feet × 30% Grade / 100 = 30 feet tall ((100 × 30) / 100) = 30

(The measured angle was 17°. Tan 17 = 0.30).

If the Tangent is at or exceeds 100% (45°), then use the direct Tangent value. For example if the angle is 60°, the Tangent value is 1.73. You would use 1.73 directly. In this case the formula is: Height = (d) × Tan.

Using a GPS device

A GPS (Global Positioning System) device is a hand-held receiver that picks up signals from satellites circling the earth, enabling you to locate your current position to an accuracy of 10-20m. Most modern GPS devices can then show that position on a map and, if you have stored a pre-programmed route, help you navigate along it.

GPS devices are similar to the SatNav devices now used by many drivers. However, while SatNavs use information about the road network to navigate step by step towards your destination, there is no equivalent "dataset" of information about off-road paths and access areas for walkers. So although the GPS can help you find your way, it won't give step by step instructions like SatNavs.

Many walkers find GPS devices a useful gadget but they aren't a substitute for traditional maps and compasses in outdoor navigation. This is because:

- To use a GPS successfully you will already need to have a reasonable understanding and experience of map reading and navigation.
- GPS can't be relied upon, especially in remote areas where accurate navigation is safety critical. The signal may fail in particular conditions, the device may fail, or the batteries may run flat, particularly if you're unexpectedly delayed.

Some things to bear in mind when buying a GPS

- The more channels a receiver has the more chance the unit will find enough satellites to pinpoint your location accurately.
- An external power socket is a useful feature, for example if you want to save power when using your GPS in a car connected to a mobile charger or cigarette lighter, or use it with a solar powered charger.
- Good battery life is essential – and make sure you pack some spares.
- A good screen that displays quality colour will make the map more readable. Large screens are easier to read but are likely to use more battery power.
- Will the buttons work if you're wearing gloves? Will they be pushed accidentally when you put the unit in your pocket?
- If you want to store lots of routes then you will need plenty of memory space.
- Compatibility. Will your GPS connect to your computer (most will by using USB connections), and will it integrate with any electronic maps and mapping packages you want to use? Will it accept routes from your favourite online sites?

- Is it also equipped with a magnetic compass? This is a useful feature to have as a failsafe it connection with the satellite fails.

Advanced GPS Features

All handheld GPS models do basic navigation. What you pick—and pay—largely depends on which and how many of the following features a unit has:

Barometer/altimeter: While all GPS receivers can tell you altitude based on satellite data, a unit with the barometer/altimeter feature provides more accurate elevation readings by also considering barometric data. This feature also lets you gauge weather trends.

Electronic compass: All GPS receivers can tell you what direction you're going while you're moving. If you get a unit with an electronic compass, though, it will also tell you what direction you're facing while you're standing still. That's a big convenience as it lets you orient yourself and plan your route at any rest stop.

Wireless data transfer: This allows your GPS to communicate with compatible (same-brand) GPS units. It provides a quick and easy way to share data like tracks, waypoints and routes with others.

Fig. 7

Preloaded maps: All GPS units come with a simple base map. Some units also include full-fledged topo maps. Most units also accept separately purchased topo maps that are downloaded or provided on a CD or a microSD card. You can manage all your maps and plan trips by connecting to your PC or Mac, then using the program that the maker of your GPS unit provides.

Third-party maps and software: Because maps from GPS makers can be pricey, you can search online and find an ever-changing array of sites that have inexpensive (or free) maps. If you aren't a tech wiz or prefer guaranteed products, then stick with maps and software from your GPS manufacturer. Shop REI's selection of mapping software.

Memory and waypoint totals: The more you have, the more map and navigation data you can store. In addition, most units also let you expand your memory, usually via a microSD card.

Unless you're doing a thru-hike, almost any unit you get will have enough internal memory (and waypoint capacity) to store detailed maps and key location data for a single trip. And it's easy to delete unneeded maps and waypoints as you go to free up memory for another trip or another section of a thru-hike.

Geocaching features: Many units have geocaching functions that allow for paperless geocaching and managing your found caches. To learn more about the activity, read Geocaching: How to Get Started.

(Note: You don't need to buy a GPS receiver to try geocaching; phone apps work fine for that. If you're already a geocacher and want a GPS unit for backcountry navigation, then geocaching features are handy to have.)

Digital camera: Offering comparable quality to a basic phone camera, it can also geotag photos so you can see where on the map you snapped each shot.

LINE INTERCEPT TECHNIQUES

Basics for Line Measurements

A common and objective way to assess cover is by laying down a transect and tallying up how much of a plant intersects the transect. This is called the Line Intercept Method first described by Canfield in 1941.

A full description of how to conduct a line intercept in the field is written in the Sampling Vegetation Handbook

The line intercept method is fairly objective. All the observer must decide is what the plant species is and where the edge begins and ends.

The beginning and end point for each time a plant intercepts the line can be recorded and the total distance of intercept can be calculated back in the lab.

If sampling objectives can be met without keeping track of the exact place where a plant intercepted the line, then all that needs to be recorded is the distance of intercept along the line.

Lines should be as thin as possible " in theory, a line has no width at all. When reading distances along a transect created with a measuring tape, read only along one edge of the tape. It is also important to ensure that the tape is not inadvertently moved to include or exclude certain plants.

Usually lines are used to estimate basal and canopy cover, but can be used for foliar cover for species that have large leaves and dense canopies.

Notice that distances are measured and recorded in 10ths of feet. This may seem a bit odd because there are 12 inches in a foot. It therefore, takes a special transect tape measure to do this " one that divides feet into 10 segments. But, it makes sense for calculating distances. As in this example, when distances (recorded in 10ths of feet) are totaled for a 100 foot tape, the total distance = % cover.

It turns out that a 100 foot (or 30 meter) distance is often used in grassland and shrubland studies. This distance is about right " generally long enough to get a good representative sample, but short enough that several lines per site, pasture, or landscape can be read.

Issues of Overlap and Gaps

The line intercept method is simple, just measure the distance of first contact to the last contact of the species along the line. But, what should be done when plants overlap on the line? And what if there is a gap in intercept for a single plant?

A few rules to deal with issues of overlap and gaps in the field:

- Ignore overlap of individuals of same species.
- Disregard overlap of differing species. In other words, count the total distance of a species regardless if it overlaps with another species.
- It may be difficult to tell where the canopy starts or ends. Create a set of rules to use in the field to clearly describe how you determined canopy dimensions.
- Decide what to do with a broken canopy within a shrub or tree. For example, if there is a gap less than 2 inches you could ignore it but count gaps greater than 2 inches. In other words, don't count the distance of the gap (if > 2 inches) in total distance of intercept by the plant you are measuring.

In the line intercept method, the line is the experimental unit. Therefore, determine % cover for each line then average lines together to estimate cover for the total landscape, site, or pasture.

Remember that because each plant species is examined separately and overlap between species is allowed, the total cover along a transect determined by distances of vegetation cover can exceed 100%.

It might seem intuitive that to get total cover for plant type (i.e., grasses, forbs, shrubs, or trees) along a transect you could simply add the cover for all plants of each plant type on a line to get the total % cover of by type. For example, you could add all distances covered by a grass to get total grass cover. But, **this is not appropriate**. Think about how overlapping vegetation is handled along the line. If there was a lot of overlap between species of the same plant type, then the total cover by plant type would be inflated.

It is possible to use the line intercept method to ascertain the total cover by grasses, forbs, shrubs, or trees, but these distances must be determined in the field.

If the beginning end ending point of vegetation intercept is recorded for each species, it would be possible to look at data back in the lab and determine the total distance of intercept by plant type. It would be a bit tedious, but measures of overlap between different species within a plant type could be determined and removed from the total distance calculation.

TREE TAXONOMY, IDENTIFICATION, AND MEASUREMENT

There are many ways of identifying trees. This chapter describes dichotomous keys, the taxonomy of trees, and tree families. Dichotomous keys are a scientific method where the user is led through a series of choices that ends with the selection of a specific tree. Trees may be understood in terms of where they are located in the hierarchical categorization of living things (taxonomy). Tree families share certain common characteristics: understanding those characteristics enables the user to more readily identify specific trees. As later discussed in Chapters Eight and Ten, tree identification is particularly important for Tree Stewards because it is an essential element of selecting the right tree for the right place and a critical aspect of determining tree disorders, associated causal agents, and appropriate courses of action.

Dichotomous Keys

A dichotomous key is a scientific tool for identifying trees. Dichotomous keys also exist for identifying other items in the natural world such as various types of living things or even rocks. Any dichotomous key leads the user through a series of steps where at each step the user must select among two choices. A common first step for trees is:

- Leaves are needle or scale-like, go to x
- Leaves are broad and flat, go to y

The selection at each step leads to a designated next step ultimately resulting in the identification of the tree.

There is no single dichotomous key for trees. Keys may be simple or complicated. Some keys, such as the one in *Tree Finder: A Manual for the Identification of Trees by Their Leaves*, are based solely on characteristics of the leaves. Others may include traits such as bark, twigs, leaf scars, buds, and seed production (size and shape of cones, acorns, etc.). The power of any key is only as good as the set of plants that it includes. However, in general, the more trees included in the key, the more complicated the key.

The ability to correctly identify a tree based on a dichotomous key is limited by the accuracy of the user's observations, the ability of the user to understand the choices posed by the key, the season of the year (a key based on leaves is not very useful for identification in winter), and the scope of the key. The tree of interest may not be included. For example, the dichotomous key in Common Native Trees of Virginia will not help identify a zelkova. While zelkovas are increasingly planted in Virginia, they are native to East Asia and thus not included in a book on native Virginia trees.

Taxonomy of Trees

The various types of trees may be understood in terms of the related groupings (taxa) to which they belong and the similar characteristics that are shared within their group. All living things are classified into hierarchical categories based on shared characteristics. The contemporary method for classifying living things into taxonomies was developed by Carl Linnaeus in the 1700s. At that time, living things were grouped based on observable forms and structures. Over time classifications have changed as knowledge and scientific tools change. DNA analysis, chromosome numbers, isozymes (proteins produced by particular genes), and nucleic acids sequences are examples of new techniques used in more recent years. Different taxonomists may draw different conclusions. Taxonomies from different sources may not completely agree in their terminology. Taxonomy information in this section is based on the classification scheme for plants as used by the US Department of Agriculture (USDA) Plant Database.

Table 1 Taxonomy schema for plants

Classification Division	Distinguishing Characteristics
Kingdom	Structural Organization (Single cell/multi-cell, whether nucleus is enclosed in a membrane)Method of Nutrition (Absorb, photosynthesize, or ingest food)
Subkingdom	Whether plants have a vascular system or not
Superdivision	Whether plants produce seeds or spores
Division	Whether seeds are naked (non-flowering) or encased (flowering)
Class	Areas of commonality based on important, more detailed, similarities within subsets of each division.
Subclass	Further sub setting into groups with like characteristics.
Order	Further sub setting into groups with like characteristics. All names for orders end in "ales."
Family	A group of individuals directly related by descent from at least one common ancestor. (Dictionary of Forestry)
	Based on commonality of distinguishing characteristics especially those inherent in their reproductive structures (flowers, fruit, seed).
	All names for plant families end in "aceae."
Genus	Groupings based on commonality in fundamental traits such as flowers, fruits and sometimes, roots, stems, buds and leaves. Members of a given genus have more characteristics in common with each other than other groupings within the same family.
Species	A group of similar interbreeding individuals sharing a common morphology (external and internal structure of plants), physiology and reproductive process. There is generally a sterility barrier between species or at least reduced fertility in interspecies hybrids. (Dictionary of Forestry)

A note about the terms subkingdom and superdivision in Table 1: A common taxonomy categorization hierarchy is kingdom, phylum, division. Instead of phylum this chapter uses subkingdom and superdivision because it is at those levels that basic plant characteristics diverge.

The end result of the taxonomy of living things is a two-part scientific name composed of the genus and species and called the "binominal nomenclature." The taxonomical classification of a given plant and even its binominal nomenclature may change as more is understood about a given plant.

Further detail within species is provided by variety, cultivar, and hybrid, as defined in Table 2. Note that while the terms in Table 1 are hierarchical, that relationship does not hold for these terms. A cultivar is a type of variety, not a subset. The hybrid could result from mating different varieties, cultivars, or species.

The utility of the various taxonomy levels varies with the kingdom. For example, when studying insects significant characteristics are captured in orders.

Academic researchers deal with division, class, and order. These classification levels are most frequently impacted by the use of current scientific methods. For practical applications the focus tends to be on the family level and below. In contrast to division, class, and order, plant families are quite stable. Families are important in gardening because plants within a given family generally share comparable cultural requirements and similar insect and disease problems. Pest management and cultural techniques are frequently discussed at the family level. Plants within some families may share common characteristics in appearance, leaf shape, seed form and structure, and growth habit. Other families may exhibit much more diversity.

Table 2 Sub classifications below species

Classification Division	Distinguishing Characteristics
Subspecies	A geographically isolated variant within a species.
Variety	A grouping within a species, usually not geographically isolated, having distinct, though often inconspicuous differences and breeding true to those differences. A variety may be naturally occurring, managed, or a hybrid.
	A category based on fewer correlated characters than are used to differentiate species or subspecies, given a Latin name preceded by "var."
Cultivar	A linage in a species that has been selected for a specific attribute and is distinct, stable, and uniform when propagated. Cultivars are usually propagated by asexual means, such as grafting, to preserve genetic makeup.
Hybrid	A cross between two species that results in a viable new plant. May occur in nature or artificially.

An important grouping for trees occurs at the division level. At that point, plants are separated into gymnosperms (non-flowering plants that produce naked seeds not enclosed in an ovary) and angiosperms (flowering plants with seeds enclosed in an ovary).

Gymnosperms include the Conifers, Cycads, Ginkgo, and Gnetophytes Divisions. The only Cycad native to eastern North America is a shrub native to Florida and SE Georgia. The only Gnetophyte in eastern North America is a shrub that grows in Texas and Oklahoma. Since our focus is trees, these two divisions are not addressed further. The taxonomy of Conifers and Gingko gymnosperms from division to family is provided in Fig 9.

The Pinales Order contains two additional families that are not addressed here because they do not include trees relevant in our area. It is interesting to note that the ginkgo tree is the only living species within the Ginkgophyta Division. The ginkgo has existed for over 200 million years and is considered a living fossil.

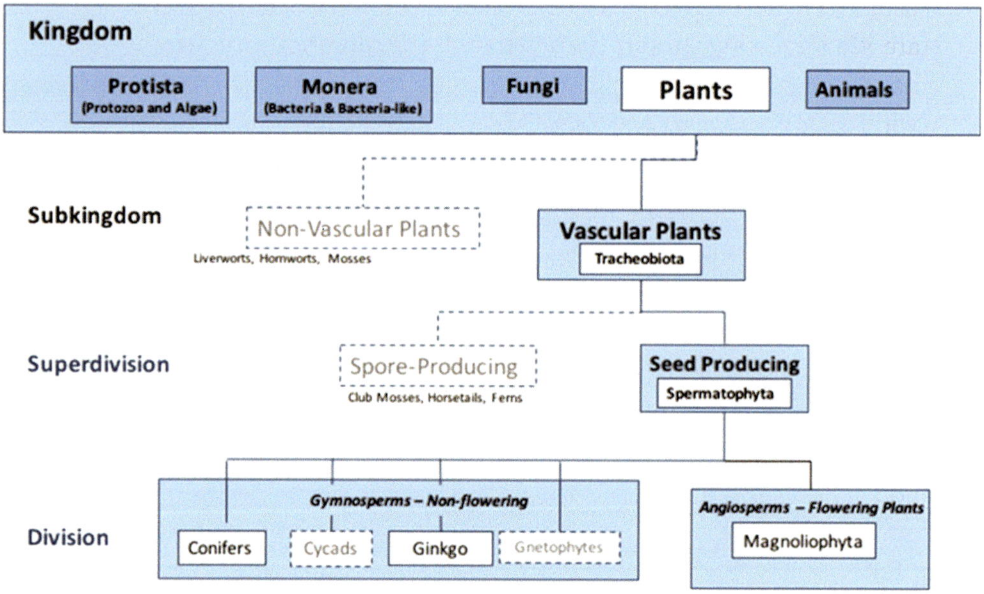

Fig. 8 Taxonomy of Trees through the Division Level

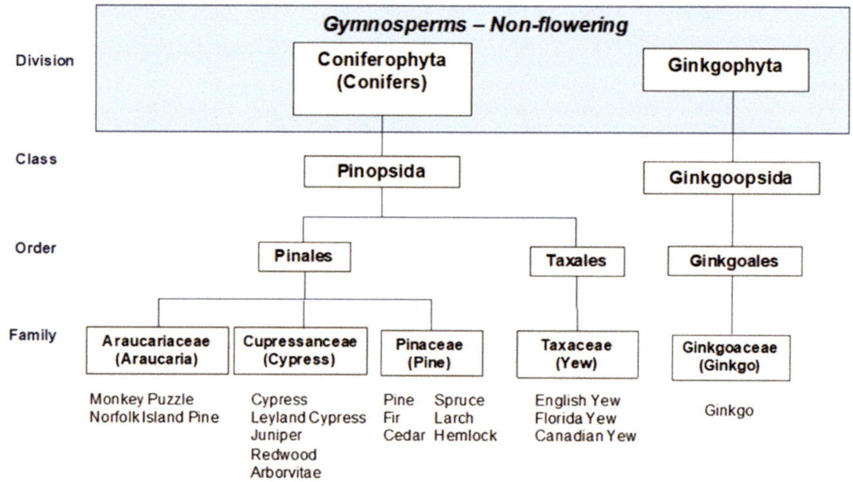

Fig. 9. Taxonomy of Gymnosperms

Several types of trees within the Pine Family have multiple cotyledons within their seeds and may be referred to as polycots.

All angiosperms are contained in the Magnoliophyta Division and have traditionally been subdivided into monocots and dicots. This has been a morphological classification, that is, the classification is based on the form and structure of the plants, such as the usual number of flower parts, the number of cotyledon (embryonic first leaves of a seedling) inside the seed, and additional related characteristics.

- Monocots have a single seed leaf within the seed. Most monocot flowers have flower parts in sets of three, so that there may be three or six petals. Leaf veins are usually arranged in parallel. The vascular bundles of monocots vary in size and are scattered throughout the stem and, therefore, cannot form annual rings of hardened tissue, i.e. wood. This limits the strength of their stems. The only trees within the Monocot Class are palms (Arecaceae Family).

- Dicots, as traditionally defined, have two seed leaves with flower parts that usually occur in multiples of 4 or 5. Leaf veins are netlike. Vascular bundles are uniform in size and arranged in a ring around the stem of the plant. This enables dicot trees to form wood from annual rings of hardened tissue.

The scientific consideration of dicots has become more complicated due to recent advances, especially in the ability to evaluate plants based on DNA and also to study plants at a molecular level: this is called phylogenetic classification. This more recent research has modified the traditional monocot/dicot grouping for taxonomical purposes into monocot, eudicot, and basal angiosperms to better reflect the evolution of the plants. A distinguishing structural characteristic of eudicots is that their pollen grains have three grooves. The pollen of both monocots and basal angiosperms have only one groove.

Scientists now believe that the initial primitive angiosperms were dicots; monocots evolved out of that group. Monocots now account for about 22% of all plant species. Palms are the only major type of tree that is a monocot. Most of the remaining dicots evolved into eudicots.

Eudicots make up 75% of flowering species and contain the largest proportion of tree families. The remaining 3% of flowering species are considered basal angiosperms, a group of the most primitive flowering plants. Magnolias, tulip trees, paw paws, and sassafras trees are basal angiosperms.

An illustrative taxonomy of angiosperm trees from division to family is provided in Fig. 10. Note that the taxonomy of angiosperms includes an additional level of classification "subclass" not used in the gymnosperm taxonomy. The subclasses, orders, and families with relevant trees tend to be broader and more numerous than those levels in the gymnosperm taxonomy. Only one categorization line is shown for each relevant subclass.

Tree Families

Tree families and trees discussed below were selected based on their prevalence in this area, economic/historical/ornamental value, environmental importance, and likelihood of problems. Given the large number of trees that are either native or have been planted in this region, the selection below is necessarily subject to modification based on one's local experience.

The intent is to provide information on observable characteristics to help Tree Stewards in identifying trees and providing right plant/right place tree selection and care. The discussion focuses on observable and distinguishing, non-technical characteristics associated with a family, genus, and species. In general, the discussion does not go into scientific aspects that are not readily observable. One of the easiest tree guides to use (for native trees) is VDOF's Common Native Trees of Virginia.

Fig. 10. Atlantic White-cedar

Except as noted, the black and white illustrations accompanying the family paragraphs are copied from this book, with kind permission from the Virginia Department of Forestry. They are the work of Juliette Watts of the US Forest Service.

The native growing range for trees indigenous to Virginia can be found in Common Native Trees of Virginia. For trees indigenous to North America, the native growing range is provided at the USDA Plant Database website.[9] Trees of Eastern North America also provides ranges. The Digital Atlas of Virginia Flora website notes Virginia counties in which a given species has been observed.

The stated number of genera and species in a family can vary based on the source of the information. This document uses the numbers provided in Trees of Eastern North America.

Also, stated expected heights and diameters of trees vary based on the source of the information.

Gymnosperms

Cypress Family (Cupressaceae)

The Cypress Family is composed of cone-bearing plants with male (pollen) and female (seed) cones either on separate plants (dioecious) or on the same plant (monoecious). In most Cypress species, the seed cone is small and woody but some have small fleshy cones that look like berries. While some species may be deciduous, most members of the Cypress Family are evergreen with leaves that are either scale like or needle like. When needle like, the needles may cover the twig or be flattened and two-ranked (leaves arranged in two vertical columns on opposite sides of the stem). The bark is fibrous and furrowed and may be peeled off in long, thin strips. Generally, the wood contains resin, is aromatic, and is rot-resistant. Given its resistance to decay, the wood is frequently used for construction and fence posts.

There are some genera within the Cypress Family. Key representative genus and trees of interest in this area are discussed below:

Genus *Chamaecyparis Spach* – False Cedar

The trees in this genus are sometimes referred to as false cedars or new world cedars to distinguish them from the true cedars in the Pine Family, discussed below. Common characteristics are:

- Tiny, scale-like leaves that overlap like shingles and form flat sprays like a fern.
- Distinctive, small cones that remain on the tree long after their seeds are gone.
- Aromatic wood.

The **Atlantic White-cedar (*Chamaecyparis thyoides* (L.) B.S.P.)** is found in southeast Virginia and favors swampy habitats. It is generally of medium height with a 1 to 2 feet diameter trunk. It looks very similar to the more common eastern redcedar but the needles are blue-green with small, round (¼ "diameter) woody cones with a waxy grayish coating.

Genus *Juniperius* L. – Junipers

Junipers are evergreen, have scale-like leaves that are pointed and prickly, and have fleshy cones that look like berries. The genus contains many shrubs and some trees.

The most common juniper tree in our area is the **Eastern Redcedar (*Juniperus virginiana* L.)** (Fig. 11). With a medium height and a trunk of 1 to 2 feet in diameter, it is the largest and most tree-like of the junipers. The eastern redcedar grows throughout most of the eastern United States on a variety of soils and is able to thrive where few other trees are found. The fleshy, berry-like cones appear only on female trees. Cones are small, ¼ to 1/3 inch in diameter. Cones start out as green and turn blue when they mature. It is one of the longest-lived of North American conifers and has been known to live almost 800 years. Its aromatic wood is used to line closets and chests.

Genus *Taxodium* Rich.

This genus contains three species of large trees that are extremely flood tolerant. The needle-like leaves, 0.2 to 0.8-inch-long, are borne spirally on the shoots, twisted at the base so as to appear in two flat rows on either side of the shoot. The seed cones are shaped like globes, 0.8 to 1.4 inches diameter. The male (pollen) cones are produced in pendulous racemes, and shed their pollen in early spring.

Bald Cypress (***Taxodium distichum***) (Fig. 12) is the most familiar species within this genus. It is a tall, deciduous conifer with a trunk that can be as large as 3 to 6 feet in diameter. Needles occur flattened and two ranked on its twigs.

The tree frequently naturally occurs near water and is considered the characteristic tree of southern swamps. Its unique characteristic is knobby wooden knees. The purpose of the knees is not known but are thought to provide additional oxygen to the plant. The tree generally does not produce knees in a non-wet environment.

Fig. 11. Eastern Redcedar

Fig. 12 Bald Cypress

Other

Other well know trees within the Cypress Family that Tree Stewards likely will want to be knowledgable of are:

- Arborvitae
- Leyland Cypress
- Cryptomeria
- Dawn Redwood
- Giant Sequoia

Pine Family (Pinaceae)

All members of the Pine Family are monoecious (both male and female parts on a given plant). The bark varies across the species. Leaves are either simple needles or needles occurring in bundles of 2-5. Cones are usually woody with many scales.

Key representative genus and trees of interest in this area are discussed below.

1. *Genus Abies* Mill. – Fir. Firs have needlelike leaves, arranged in a spiral but often appearing 2-ranked. The needles are generally soft to the touch. Firs live in regions that are covered in snow many months of the year, generally at elevations from sea-level to 6500 feet.

Fraser Fir (*Abies fraseri*) is the only native fir in the Appalachian mountains. It is a medium height tree with a trunk around 20 inches in diameter. The fraser fir is usually found in stands with red spruces in small areas in the south Appalachians that include areas in Tennessee, North Carolina, and Virginia. The twig is covered by red hairs with needles that are approximately ½ .to 1 inch long. Cones may be 1.5 to 2.5 inches long.

2. *Genus Picea* A. Dietr. – Spruce. Spruces look very similar to firs but have stiff, usually sharp-pointed needles. Their native habitat is mostly in the mountains at high altitudes.

Red Spruce (*Picea rubens* Sarg.) A tall tree with a trunk that could be 3 ft in diameter, the red spruce is found in small scattered areas at elevations above 4000 feet in the west of Virginia. Needles are ½ to 5/8 inches long, pointed and attached to the twig on tiny raised pegs. Cones are 1 ¼ to 2 inches long, light reddish-brown, and shiny with smooth edged scales. Cones drop their first winter. The wood is light, somewhat soft and elastic. Example uses are lumber, pulpwood, and musical instruments.

Fig. 13. Fraser Fir

3. Genus *Pinus* L. – Pine. Pines are evergreen trees with needle shaped leaves occurring in bundles of 2-5. Representative pines of interest are discussed below.

Fig. 14 Red Spruce

Loblolly Pine (*Pinus taeda*) Tall and straight trees with trunks of 2 to 3 feet in diameter. Needles are 6-9 inches long, three needles per cluster. Lower branches self-prune as the tree grows and ages. Cones are narrowly oblong, 2 to 6 inches long, and stay on the tree for a year after maturing. Loblollies are native to the eastern part of Virginia and are the most abundant tree in the Virginia coastal plain. They are also the most important commercial timber tree in the state. The wood is used for lumber, paper pulp, plywood, poles, pilings and fuel.

Virginia Pine (*Pinus virginiana* Mill.) A tall tree, somewhat smaller than the loblolly, with trunks of 12 to 14 inches in diameter. Needles are 1 ½ to 3 inches long, two needles in a bundle,. Needles are usually twisted. Cones are egg shaped, 1 ½ to 2 ¾

Fig. 15 Loblolly Pine

inches long. Cones mature in two years but stay on the tree several years before falling. The wood warps easily and is used for rough construction and paperpulp.

Longleaf Pine (*Pinus palustris* Mill.) Tall with trunks of 2 to 2.5 feet. This tree has the longest needles and largest cones of any pine in the eastern part of North America. Needles are 8 to 15 inches long with three in a cluster. Cones are cylindrical, 6 to 10 inches long. Cones mature in two years and drop from the tree after releasing their seeds in the fall.

Fig. 16 Virginia Pine

The longleaf seedling appears to be a clump of grass and only develops a stem after several years. The tree prefers to grow in acidic, infertile soils and likes sandy soils. It is very adapted to fire, generating seedlings best when fire burns off the fallen needles, clearing the understory but not harming the mature trees.

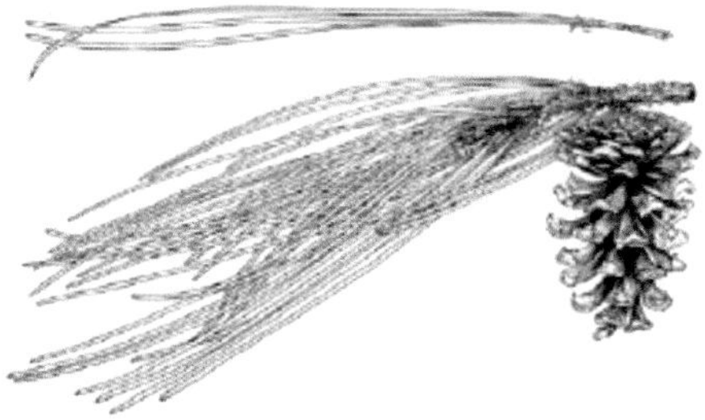

Fig. 17 Longleaf Pine

Before European settlement, longleaf pines dominated the coastal plain from Virginia to Texas. Today they are found only in isolated sections. Heavy cutting of the trees for commercial use, combined with fire suppression, has lead to significant reduction of the species. Restoration efforts are ongoing in Virginia. American's Longleaf Restoration Initiative (ALRI) was formed in 2007 to restore longleaf pine ecosystems. It is lead by the USDA Forest Service, Department of Defense, and U.S. Fish and Wildlife Service with participation from various public and private organizations across nine southern states.

Other well known trees within the Pine Family that Tree Stewards likely will want to be knowledgeable of are:

- Shortleaf Pine
- Eastern White Pine

4. Genus *Tsuga* – Hemlock. Hemlock needles are arranged in a spiral on the twig but appear to be 2-ranked and flattened. There are only 7 species worldwide with only 2 species, **Eastern Hemlock (*Tsuga canadensis*)** and **Carolina Hemlock (*Tsuga caroliniana*),** in the Virginia/North Carolina area. When crushed the foliage smells a bit like the European native poisonous herb called hemlock and thus the name.

Eastern Hemlock (*Tsuga canadensis*) A tall tree with a trunk 2 to 4 feet in diameter, the eastern hemlock is found primarily in the western portion of Virginia. Needles are short, 1/3 to 2/3 inch long, with two pale lines on the lower surface. The ¾ cones grow on the tip of branchlets.

5. Genus *Cedrus* – True Cedar. True cedars are native to the mountains of the western Himalayas and the Mediterranean region and are adapted to mountain climates. They have rather short evergreen needles occurring primarily in dense clusters at the end of short woody pegs. This tufted arrangement of needles is only seen in true cedars and larches. Large, barrel-shaped cones sit on the top of branches and have thin scales that fall apart when mature. They are generally very large (tall and wide) trees and are planted in parks and public areas in the United States and especially in the Pacific Northwest. Most true, old-world cedars seen in North America are ornamentals.

Fig. 18. Eastern Hemlock

Deodar Cedar (*Cedrus deodara*) is frequently seen in this area and is rumored to be naturalized in North Carolina, South Carolina, and Georgia. Its needles are 1-2 inches long, grouped in clusters with individual needles on the branch tips. There is a noticeably droop in the branch tips and leader.

Other true cedars include the **Atlas Cedar (*Cedrus atlantica*)** and **Cedar of Lebanon (*Cedrus libani*).** Both trees have needles about 1inch long with a stiff branching habit.

Fig. 19 Deodar Cedar

Angiosperms

While all angiosperms flower, not all flowers are notable. This is especially true for trees that are angiosperms.

Birch Family (Betulaceae)

The Betulaceae family includes birches (*Betula*), alders (*Alnus*), hornbeams (*Carpinus*), and hazels (*Corylus*). The trees are deciduous and have simple, alternate leaves that are generally thin with serrated edges (saw-like teeth) and pinnate veins (major veins branching off from a central vein). The trees are monoecious with male and female catkins forming on the same plant. A catkin is an elongated cluster of single-sex flowers bearing scaly bracts, usually lacking petals.

Key representative genus and trees of interest in this area are discussed below.

1. Genus *Betula* – Birch. Birches usually have smooth bark, dark brown to silvery white, that often peels into plates or papery layers. Trees often have several trunks.

Fig. 20. Hophornbeam Leaf

Fig. 21 Filbert Catkins

River Birch (*Betula nigra* L.) is the only birch native to the Coastal Plain in the southeastern United States. It can be a tall tree and is often divided into several trunks. The trunks may be 1 to 3 feet in diameter. The bark peels back in papery layers. The leaves are 1½ to 3 inches long with an oval to triangular shape and doubly serrated edges. The wood is quite hard and close-grained but the tree is seldom harvested. It is commonly planted for erosion control situation such as stream bank restoration and is frequently used as an ornamental tree in home yards.

2. Genus *Carpinus* – Hornbeams and **Genus *Ostrya* – Hophornbeam.**

The **American Hornbeam (*Carpinus caroliniana* Walt.)** and **Eastern Hophornbeam (*Ostrya virginiana* (P. Mill) K. Koch.)** are both native in this area. While they are very similar and both belong to the Birch Family, they belong to different genera. Both are small trees growing 20 to 30 feet and can have a single trunk or multiple trunks. Both trees have leaves that are alternate, simple, 2 to 4 inches long with doubly toothed edges. The leaves of the American hornbeam are oval and long-pointed while the leaves of the eastern hophornbeam are oblong with narrow tips. The trunk of the American hornbeam is smooth and rippled, resembling well-defined muscles. This leads to alternate common names of ironwood or musclewood. The fruits of the American hornbeam are 4 to 6 inch hanging clusters of slightly folded, 1 inch, 3-lobed leafy bracts; each bract contains ribbed nutlet. These clusters somewhat resemble the samaras (helicopter seeds) of maples. However, the hophornbeam has fruits hanging in clusters of leafy, oval, papery

Fig. 22. River Birch

sacs 1½ to 2½ inches long, with each sac containing a ¼ inch nutlet. These clusters of papery sacs resemble hops and thus the name Hophornbeam.

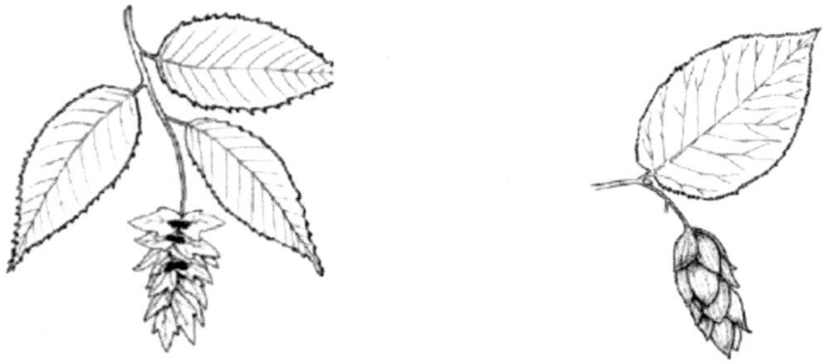

Fig. 23. American Hornbeam and Eastern Hornbeam

The wood of both trees is strong and durable but is seldom harvested.

Bean or Pea Family (Fabaceae)

The Fabaceae Family is the third largest plant family and contains approximately 730 genera. More than 45 genera of trees exist north of Mexico across North America. All members have compound leaves, mostly pinnately or bipinnately compound. However, the leaves of some species, such as the redbud, are unifoliate, that is they are theoretically compound but have only one leaflet and therefore appear to have a simple leaf. The unifying feature of the family is that the fruit is a legume, that is multiple seeds reside in a pod that opens along two sides. Roots often have nitrogen-fixing bacterial nodules. This document addresses one genus and two species in other genera within the *Fabaceae* Family.

1. Genus *Gleditsia* – Locusts. Locusts are medium height, single trunk trees with pinnate or bi-pinnate compound leaves. The compound leaves alternate and within the compound leaf, leaflets also alternate. A distinguishing characteristic is thorns that occur on the trunk and/or branches. Fruits are flattened legume pods with one to multiple seeds. Two species are native to North America.

Fig. 24 Honey Locust

Honey Locust (*Gleditsia Triacanthos L.*) The leaves are pinnately compound, 5 to 8 inches long, with 15 to 30 leaflets. However, the leaves can also be bipinnately compound with 4 to 7 pairs of minor leaflets. Leaflets are elliptical and ½ to 1 ½ inches long. Trunk, large branches and twigs have branched thorns that are strong and sharp. Small greenish-yellow flowers are fragrant and appear on 2 to 3 inch hanging clusters in late spring. The seed pod is 6 to 8 inches long and flattened. The pod matures in late summer and early fall, becoming dry and twisted. Seeds are oval, dark brown and 1/3-inch-long with multiple seeds in each pod. Because the wood is hard, strong, and moderately resistant to decay, it is sometimes used for fence posts and crossties. The tree may be planted for erosion control and windbreaks. Thornless varieties are available and are commonly chosen for planting in urban landscapes.

Other Genera

Eastern Redbud (*Cercis canadensis* L.*)* is a short tree, frequently multi-stemmed, that grows as an understory tree generally at the edge of a forest. It is one of the most common ornamental trees. In spring, it is covered with very small (½ inch) pink to purple flowers that resemble the flowers of a pea. The flowers occur in clusters along the twigs and small branches and appear before the leaves. Leaves are alternate, heart shaped, and smooth edged, 3 to 5 inches long and wide. While the leaves appear to be simple, they are an example of the infamous unifoliate. Although the wood is heavy and hard, it is not strong and has little commercial value.

Mimosa (*Albiziz julibrissin* Durazz) is a lovely small tree with smooth gray-brown bark. It was introduced from Asia in the late 18th century. The leaves are alternate, bi-pinnately compound, 10 to 20 inches long with a feathery appearance reminiscent of ferns. Fragrant flowers appear in the form of showy pink fluffy heads in mid to late summer. The fruit is a flat pod, 5 to 6 inches long, and turns a light brown. The single trunk is short, sending out branches at a low height and spreading into a wide V-shape crown that makes it an excellent climbing tree for children. Unfortunately, the seeds germinate easily across a wide range of conditions making the mimosa far too good at reproducing itself. It is treated as invasive and commonly considered a trash tree. So sad.

Other trees in the Fabaceae Family that the Tree Steward may find interesting include:

- Yellowwood
- Black Locust
- Kentucky Coffeetree
- Golden Chain Tree

Beech or Oak Family (Fagaceae)

The Fagaceae Family contains deciduous and evergreen woody plants with simple leaves arranged alternately. The leaves may have entire, toothed, or deeply lobed leaf margins. Flowers are typically unisexual with male and female flowers appearing on the same plant (monoecious) in catkins (elongated cluster of highly reduced unisexual flowers). Fruits are nuts within a scaly or spiny cap.

There are 9 genera in the Beech/Oak Family. Key representative genus and trees of interest are discussed below

1. Genus *Fagus* – Beech. Beeches have smooth bark that is either silvery gray or blue-gray, though some species develop shallow furrows and ridges. Leaves are simple, elliptic or ovate, with margins that are bluntly or sharply toothed. The leaves have lateral veins that are parallel and terminate in the tip of a marginal tooth. The fruit is two nuts in a four-valved prickly capsule. Beeches are tardily deciduous trees. Many have a tendency to retain their leaves in a dry, lifeless state through the winter, especially when young.

Fig. 25. Mimosa Leaves, Blooms

American Beech (*Fagus grandifolia*) is a common understory tree in the Coastal Plain. It can grow to be fairly tall with a trunk of 2 to 3 feet. It can be easily identified by its smooth silvery bark. The leaves are coarsely toothed with 9-14 pairs of lateral veins. Buds are distinctly elongated and torpedo-shaped.

2. Genus *Quercus* – Oak. Though the Mid-Atlantic representatives of the genus are typically deciduous, many evergreen species occur throughout the world's warmer temperate and tropical regions. There are more local species, at least 16, of oak than any other genus.

The fruit of all oaks is the familiar acorn which is botanically identified as a nut covered by a cap of scales. Any plant that produces an acorn is an oak. The size and shape of the acorn and its cap is specific to the species of oak.

The leaves are arranged alternately and vary in shape. Many are conspicuously lobed, but others have simple leaves without lobes. The bud scales overlap and most oaks feature a cluster of buds at the terminal end of each stem.

Oak wood is heavy, strong, hard, close-grained and durable. It is used for lumber, furniture, and flooring.

The majority of oak species are placed in either the Red Oak group or the White Oak group. Red Oaks have a bristle on the tip of their lobes and their acorns require two years to ripen. White Oaks do not have bristles. Their acorns ripen in a single season and are sweeter (have less tannin) than acorns from trees in the Red Oak group. White Oaks generally are slower growers than Red Oaks. The Sawtooth Oak is neither in the Red or White group but is considered an intermediary between the two groups.

White Oak (*Quercus alba* L.) is the most common oak in Virginia forests. It is a tall tree with a trunk that can be 3 to 4 feet in diameter. Leaves are alternate and simple, 4 to 7 inches long, with 7 to 10 rounded lobes. The sinuses between the lobes may be shallow to deep, almost reaching the midrib; this can give the leaf a very different appearance even among trees of the same species. The acorn is light chestnut brown, ¾ inch long, with a bowl-shaped cap that covers ¼ of the acorn. The cap detaches at maturity. The wood is close-grained, strong, and durable. Uses include furniture, flooring, and fuel. The wood is well suited for whiskey and wine barrels because it is highly water-tight due to the presence of a substance called tyloses in the wood vessels.

Willow Oak (*Quercus phellos* L.) is long lived, fast growing, and frequently planted in public spaces. The tree has an erect trunk that is medium to tall in height with a diameter of 1 to 2.5 feet, sometimes considerably wider in the taller specimens. Leaves are alternative, simple, and narrow oval (lanceolate) with a bristle tip, making them members of the Red Oak group. The acorn is small (¼ to ½ inches) and tan. The cap covers ¼ of the acorn nut and is flat, thin, and scaly. The wood is heavy and strong, but coarse-grained. It is used for rough construction and pulpwood. There are many oak species. Examples include:

- Southern Red Oak
- Northern Red Oak
- Swamp Chestnut Oak
- Live Oak
- Sawtooth Oak
- Water Oak
- Overcup Oak
- Scarlet Oak
- Pin Oak

3. Genus *Castanea* – Chestnut. Chestnuts have alternate leaves with bristle-tipped marginal teeth and conspicuous parallel lateral leaf veins. The trees produce large, sweet, edible fruit. The nuts are enclosed in a spiny, prickly capsule that may contain up to three nuts. There are 4 native species in North America. Of particular note is the greatly missed American chestnut.

American Chestnut *(Castanea dentate)* used to be a tall tree that was a dominant species in forests in much of Virginia. Trunks could be quite large with diameters up to 4 feet. In the early 1900s chestnut blight was introduced into eastern North America. By 1940 the chestnut had disappeared as a major tree in American forests. Today the plant can be found as sprouts from the stumps of dead American chestnut trees. The sprouts may get up to 20 feet in height and may live as long as 40 years. They do not produce fruit.

Another common tree in the Chestnut Genus is Allegheny Chinkapin, *Castanea pumila.*

4

MEASURING TREES

The size of the tree is best described with the diameter of the tree. The diameter of the tree provides a measure of tree performance and is a useful starting point for estimating tree volume. It is the linear measurement, the main objective of which is to estimate the volume of the trees. The volume of a tree is dependent on diameter or girth at breast-height, total height and form factor.

It is not only necessary for calculation of volume of logs, but also necessary for making inventory of growing stock as well as to correlate height, volume, age, increment of trees.

In forestry, diameter should always be used for the description of the tree. In the past, girth has been used but this is no longer recommended.

A carefully defined point should be used for measuring the diameter of the tree. In many countries, especially India, this will be at 1.37 m above the base of the tree where the tree meets the ground. By convention, the diameter of forest trees is measured in centimeters at 1.37 m above the ground and is termed the "Diameter at Breast Height" or DBH. Because trees are measured with the bark on, this is also called the Diameter at Breast Height Over Bark (DBHOB).

When measuring live trees most information is presented as over bark dimensions. Diameter at Breast Height (DBH) is the most common parameter used in the measurement of standing timber.

Generally DBH is not measured on dead trees or on those of less than 7 cm DBH. Universally adopted standard height for measuring girth, diameters and basal area of standing trees is 1.37 m and being practiced in India, Burma, America, Union of South Africa and other British Colonies.

In Europe, Australia and UK, DBH is taken as 1.3 m which is recommended by FAO as standard. In New Zealand, DBH is measured at 1.4 m.

AGE

Calculating Tree Age Using DBH (Diameter at Breast Height)

Now, when we talk about the age of a live tree, you can estimate the age of a living tree, without knowing when the tree was planted, by the following method.

In the example below, we are measuring an oak tree.

Step – 1: Measure the circumference (c) of the tree trunk using a measuring tape that measures in feet and inches. This should be done at 4.5 feet above the ground which is the Diameter at Breast Height (dbh). Here, the oak tree measures 12 feet, 10 inches in circumference.

c = 12 feet, 10 inches = 154 inches

(multiply 12 feet x 12 inches per foot, then add 10 inches)

Step – 2: Calculate the diameter (d) and radius (r) of the trunk in inches.

Divide the circumference by "pi" 3.14.

c/3.14 = d; 154 inches/3.14 = 49 inches

Divide the diameter by 2 to get the radius (r).

d/2 = r; 49 inches/2 = 24.5 inches= r

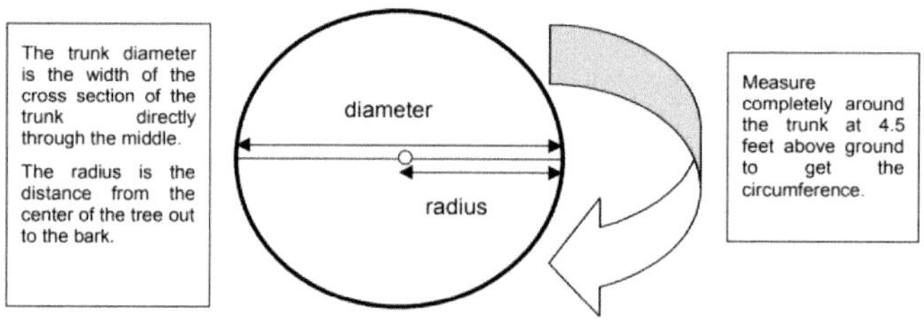

Fig. 1

NOTE: You should deduct at least .5 inch from the radius for the width of the tree's bark, and more for trees with very thick bark (1.0 inch), less for trees with very thin bark (.25 inch). In our example, we will deduct .5 inch, for a radius of 24 inches.

Step – 3: Determine the average width (w) of an annual ring of wood for the species of tree you have measured. In ring-porous species of trees (distinct annual rings), the annual ring will contain both early or spring wood (large vessels) and late or summer wood (small vessels). In diffuse porous woods, it is much more difficult to see where the rings begin and end. A hand lens or magnifying glass is useful.

- The width of an annual ring of wood for a particular species depends upon many factors, including:
  - Genetic differences
  - Tree age

- Height above ground
- Local site and climatic conditions
- Competition from other trees
- Water availability
- Nutrient availability
- Presence of insects or diseases
- General tree health

You can also calculate the average width of an annual ring using cross sections of wood from the same species. Measure the radius from just inside the bark to the center of the trunk, count the number of rings, and divide the radius by the number of rings. The result will be the average width per ring.

10 inches/50 rings = .2 inches/ring

Conversely, if we divide the number of rings by the number of inches, the result will be the average number of rings per inch.

50 rings/10 inches = 5 ring/inch

If the absence of actual data for white oak we will use in our example an assumed average of .2 inches per ring, or 5 rings per inch.

Step – 4: Divide the radius (r) by the average width (w) of one annual growth ring to get the approximate age of the tree.

r/w = approximate age

24 inches/.2 inches per year = 120 years!

STEM DIAMETER

Measurements of stem diameter are widely used in both forest ecology and management to characterize the size distribution of forest stands and to estimate timber volumes. Diameter measurements are usually taken at a standard height, the diameter at breast height (dbh), which is defined as 1.3 m above ground level (or 4.5 ft in the USA).

Measurements are complicated by the fact that tree stems are often not circular in cross-section, and may be leaning or surrounded by prominent buttresses, making them difficult to measure. The following standard procedure is recommended:

- Meters separately, and produce an overall figure by calculating the mean of the two values.

The two most commonly used instruments for measuring tree stem diameters are calipers and diameter tapes. Calipers are usually used when the trees are less than 60 cm dbh; although larger calipers are available, they can be difficult or unwieldy to use in the field. Calipers are usually constructed out of metal, wood, or plastic, and enable the diameter to be read directly off a scale when the arms of the caliper are placed around the tree stem. The caliper arms should be pressed firmly against the tree stem with the main beam of the caliper placed perpendicular to the axis of the tree stem, and the arms parallel and perpendicular to the beam.

Standard measurement tapes can be used to measure stem diameter by placing the tape around the circumference of the tree at breast height. Measurements of circumference taken in this way can be converted to diameters (assuming a circular cross-section) by dividing the values by w. Diameter tapes can be obtained, however, that are graduated at intervals of w units (in cm or inches) enabling diameter to be measured directly. Care should be taken to ensure that the tape is positioned correctly: it should be in a plane perpendicular to the trunk of the tree, and pulled taut around the trunk so that accurate measurements are obtained. Although more accurate results can be obtained with calipers, measurements with tape tend to be more consistent if repeated measurements are made, because caliper measurements are more sensitive to the positioning of the instrument.

Bark thickness can be determined by using a bark gauge, which consists of a steel shaft that is pushed through the bark. The thickness of the bark can be read directly off a scale with the instrument in place. A minimum of two readings should be taken. When measurements of diameter are made, whether or not the bark was included in the measurement should be recorded.

A number of instruments are available for obtaining measurements of upper stem diameters, which can be useful for assessing the form of the tree or extent of stem taper. Options include optical forks, optical calipers, and fixed-based or fixed-angle rangefinders.

Many of these instruments are expensive and are prone to inaccuracy. Most commonly, a relas-cope is used, which is a form of optical rangefinder. However, calipers or diameter tapes provide more accurate measurements of upper stem diameters, if the upper parts of the tree can be accessed through use of climbing ropes or ladders.

HEIGHT

Height is the vertical distance from the base of the tree to the highest sprig at the top of the tree. The base point is the point at which the tree germinated. If the tree is on a slope, the base point is the ground level halfway between the upper and lower sides of the tree.

Girth is a measurement of the circumference of the tree measured at 4.5 feet above ground level. If there are significant branches below this height, then the girth should be measured at the narrowest point below the lowest branch and that height noted. If there is a burl or protuberance at the measurement height, then the girth should be measured immediately above the protuberance or at the narrowest point of the trunk below the protuberance and that height noted. If there are multiple trunks, measure with a tape measure around all trunks at the 4 ½ feet height.

Crown spread measures the footprint of the crown of the tree expressed as a diameter. The simplest approach is to take measurement along the longest axis of the crown from one edge to the opposite edge. A second measurement is take perpendicular to the first lie through the central mass of the crown. The two measurements are then averaged to reach a number for crown spread.

Total height of the tree is the straight line distance from the tip of the leading shoot (or from the highest point of the crown where there is no leader) to the ground level, usually measured on slopes from the uphill side of the tree. It is the straight-line distance from the highest point of the crown to the ground level. It is used to find out tree volume.

It is also essential to read volume tables, form factor table, yield tables, etc. Tree height is normally used to find out productive capacity of the site and site quality of a locality.

There are different measurements of height on trees:

Bole Height

It is the distance from the ground level to the position of the first crown forming living or dead branch.

Commercial Bole Height

The height of bole that is usually fit for utilization as timber is called commercial bole height.

Height of Standard Timber Bole

It is the height of the bole from the ground level up to the point where average diameter over bark is 20cm.

Timber Height

It is the vertical distance from the base of the tree to the point on the main stem where the diameter is 7 cm for conifers. For hardwoods, timber height is the vertical distance from the base of the tree to the point on the main stem where the diameter is 7cm or where the main stem becomes the crown, whichever is the lower.

Stump Height

It is the height of the top of the stump above ground and it is normally 20-30 cm.

Crown Length

It is the vertical measurement of the crown of a tree from the tip to the point half way between the lowest green branches forming green crown all round and the lowest green branch on the bole.

Crown Height

It is height from the ground level to the point half way between the lowest green branches forming green crown all round and the lowest green branch on the bole.

Breast Height

If refers to the usual point of measurement of standing tree or stem diameter (1.37 m or 1.3 m) above ground on the uphill side of the tree.

Mean Crop Height

The mean height of all the trees growing in forest.

Top Height

Mean height of the trees with the largest dbh in a stand.

Height Measurement Principles

Trigonometric Principle

Trigonometry is a branch of mathematics dealing with measurements of the angles and sides of triangles and functions based on these measurements.

Tangent Law

The height of the tree is calculated with the help of the tangents of the angles to the top and base of the tree and the distance of the observer from the tree. Simple example is of standing tree in plane ground.

Tree Height (BD) = AE + ED Tan α

Height sticks are used to measure the height of small trees. Hypsometers, altimeters and clinometers are used to measure height of tall trees.

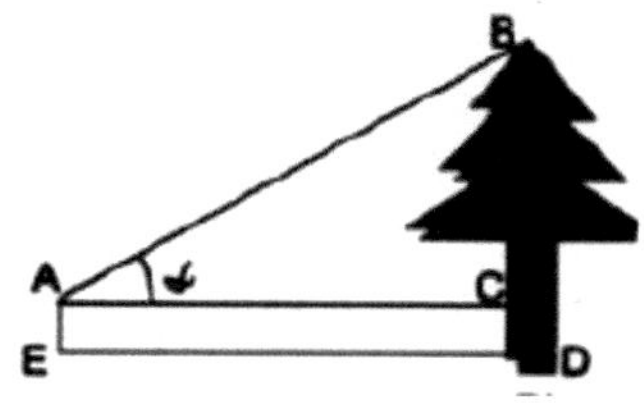

Fig. 2

i. Hypsometers

Used for determining the height of standing tree from observations taken at some distance from the tree.

ii. Altimeters

Generally altitude measuring instruments, which can be devised to determine heights of tree.

iii. Clinometers

It measures angle of slope. Any instrument which measures angles of slope can be used for height of tree by trigonometrical methods.

iv. Laser Dendrometers

They basically rely on the same geometric principles as the hypsometers.

Some clinometers designed for this purpose called hypsometers are based on geometrical principles of similar triangles or based on relations between the sides of right angled triangles. Christen Hypsometer, Improved Callipers, Smythies' Hypsometer, Improved Smythies' Hypsometer are the instruments based on similar triangles.

MEASURING CANOPY CLOSURE

Canopy closure is the proportion of the sky hemisphere obscured by vegetation when viewed from a single point. Note the difference between this term and canopy cover, which refers to the proportion of the forest floor covered by the vertical projection of the tree crowns (Fig. 3).

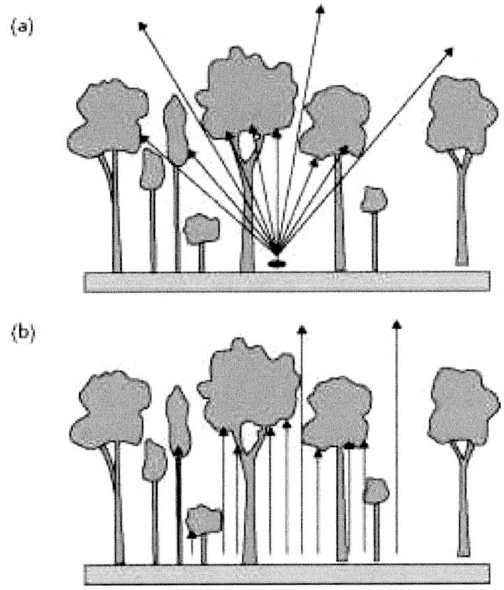

Fig. 1 The difference between canopy closure (a) and canopy cover (b).

Importance of Canopy Cover Measurement

Canopy cover is measured as the proportion of a fixed area of the ground covered by tree crowns. The canopy cover will be determined by the tree species, as they have different crown sizes, shapes, and heights.

Tree canopy measurements are important for various reasons:

- The canopy can be affected by the health of a tree, which in turn is affected by nutrition, water access, disease, pest infestations, and stress. Thus, canopy cover acts as an indicator of these factors for both natural as well as cultivated groves and forests.
- The efficiency of fertilization, irrigation, and pruning methods in agriculture, horticulture, and floriculture also affect the canopy cover, so its measurement sheds light on these factors as well.
- In forests, canopy cover is also used to study the effects of fire, pollution hazards, and microclimate. The impact of forest use, in terms of deforestation, degradation, thinning or afforestation, can also be evaluated by canopy cover.
- The canopies of forests and other natural vegetation are important in watershed areas for replenishing groundwater and hydrology, so they need to be monitored.

Measuring Canopy Cover

There are many methods to measure canopy cover, and, likewise, there are several commercial tools available on the market. Some of the more important ones are described below:

- **The convex spherical densiometer** is a pocket-sized instrument that uses a convex mirror with a grid of 24 squares. Canopy cover is calculated from the number of squares on the mirror filled with vegetation.
- **The angular densiometer** has a convex mirror, a compass, a bubble level, and a fixed eyesight. It produces a clearer image and a more exact estimation of canopy cover than the convex spherical densiometer.
- **The moosehorn** uses 25 dots on a 5 x 5 grid on a transparent screen; the dots covered by canopy are counted through a side aperture to give an estimate of the vertical projection on the canopy.
- **The canopy-scope** is an improvement on the moosehorn and uses the same principle of 25 dots on a 5 x 5 grid. This instrument is affordable, compact, and accurate. It has the advantage of being less prone to user error compared to other tools.
- **The vertical tube** is a 20cm-long brass tube mounted in such a way that it hangs vertically. It is used to measure crown projection but requires many more points of observation within a given area compared to other methods.
- **Single-point and handheld light sensors** estimate the light that comes through the canopy cover. They do, however, need a reference measurement under an open and clear sky. The disadvantage of this method is that it measures only light transmittance.
- **Hemispherical photography (HP)** and digital analysis uses the fisheye lens approach and produces the most accurate measurements. It can be used for estimating canopy cover, light transmittance, leaf area index and photosynthetically active radiation. It is suitable for tree level measurements of crown architecture and dimensions, and density. These estimates are scalable, as they are related to stand level canopy cover.

In all the above methods, except HP, the resolution decreases with canopy cover over 30%.

Our in-house example of hemispherical photography measurement is the CI-110 Plant Canopy Imager which calculates leaf area index (LAI) and photosynthetically active radiation (PAR) levels. Using these two indices, the canopy cover is calculated.

The LAI is estimated by a 150 fisheye image of the canopy. The non-destructive method Gap Fraction Method measures the area of sky not visible due to canopy cover utilizing a classification system of 0-1. Here, 0 means the canopy cover has no gaps, while 1 means there is no canopy above, and the sky is fully visible.

Photosynthetically Active Radiation (PAR), which is the part of the visible light used by plants for photosynthesis, lies in the wavelength range of 400-700 nm. The CI-110 finds out the wavelength of the light filtering through the canopy.

The CI-110 is versatile, as it needs no above canopy reference measurements and is suitable for trees, shrubs and other low plants. A GPS connected to four satellites and an internal compass allow time series measurements at a single location.

The Best Time for Measurements

The best time and weather conditions for canopy cover estimation will depend on the objective of the study, as well as the instrument being used.

- Hemispherical photography should not be used midday or in bright sunshine. Cloudy skies give better results with HP.
- Single-point and handheld light sensors require a reference measurement of light transmittance under clear skies.
- Photosynthetic active radiation (PAR) is recorded at midday.

Understanding Forest Canopy Structure

A forest or any biome is made up of several trees that belong to many species. A tree has a stem and a crown of branches that bear leaves, flowers, and fruits. Though forest canopy structure can mean different things, it basically refers to the arrangement of tree crowns; it has two components: the vertical structure and horizontal structure.

- **Horizontal structure** is created by the placement of trees and how close their crowns are to each other. A collection of these crowns creates the canopy (See Fig. 4).
    - **Vertical structure** is formed because trees and shrubs in a forest will have different heights. So, two or more layers of tree crowns can be found in forests. These layers are called forest storeys or strata. The top-most stratum is made of emergent trees; below them are the canopy trees, which are usually present as a continuous layer. The tree strata below them are called the sub-canopy and the understorey. Next, close to the ground could be shrubs, herbs, and grasses (See Fig. 5).

Complicating this concept of forest structure is the issue of spatial scale, i.e., space or area that is considered for estimating the canopy structure.

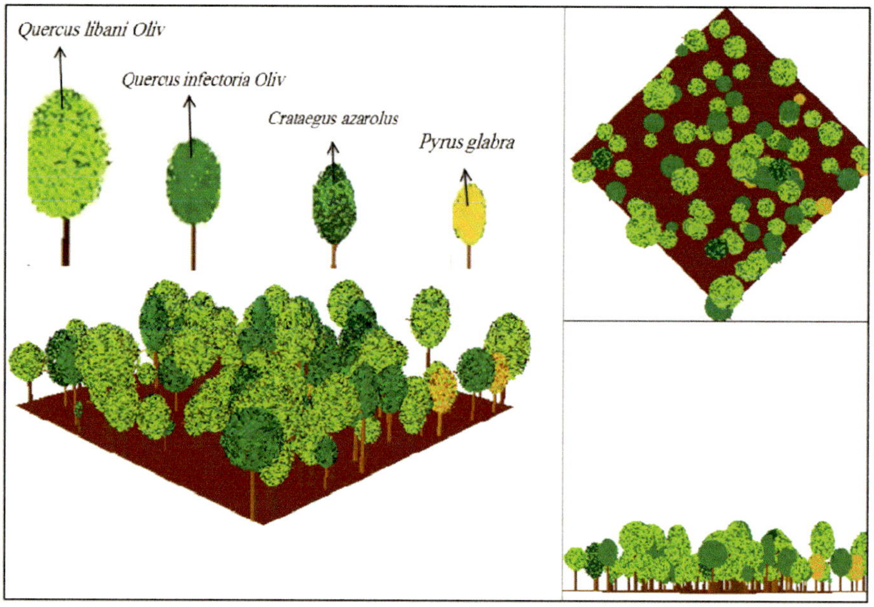

Fig. 4: The figure at the top right shows the horizontal structure, while the bottom right shows the vertical structure.

Regional Scale

At a regional scale, the number of trees, shrubs, herbs, their species, and their arrangements give the forest canopy structure (See Fig. 6).

At this large scale, forest canopy structure defines the biome and its conditions, and it tells us whether the forest is

Fig. 5: Vertical Forest Structure.

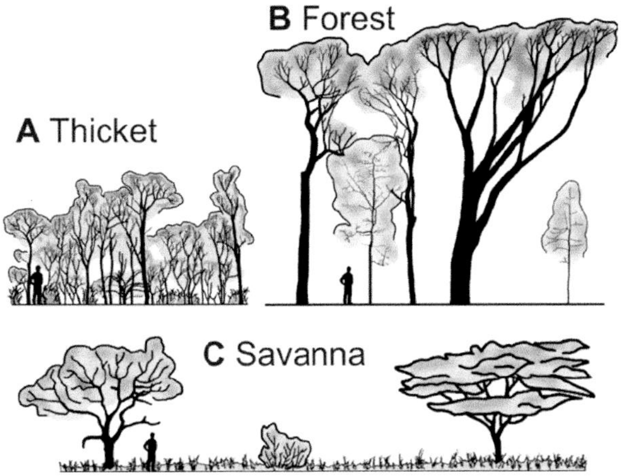

Fig. 6: Biome vegetation structure. A Thicket with a low closed canopy and a discontinuous layer of shade tolerant grasses; B Forest with a tall closed canopy, no grass layer, and a mid-storey of shade-tolerant trees; C Savanna with a continuous C4 grasses layer and a discontinuous tree cover.

- Continuous/patchy or pristine/disturbed.
- Coniferous, scrub, savanna, grassland, broadleaved, etc.

Remote sensing is the preferred method of collecting data about forest canopy structure at large scales.

Local Scale

At the local scale, forest stand attributes, such as tree girth, the shape of a crown, height, and tree architecture, are the defining characteristics of forest canopy structure. Here, the forest canopy structure can also refer to the total quantity of leaves in a forest. Ground based methods are used to collect data on local forest canopy structure.

Role of Forest Canopy Structure

Forest canopy structure determines many of the environmental factors and functions of a forest and is, therefore, important for all the species living in a forest.

- **Light interception:** Light distribution within the forest depends on the number of layers. The upper layers will get more light radiation, with little light coming through to the forest floor, if the canopy is dense. For this reason, there is little grass and herbs in well-developed and old forests. The amount of light reaching lower parts of a tree is also important as photosynthesis depends on it. So, there is more photosynthesis in the upper layers of a forest canopy.
- **Moderate temperature:** Temperatures within a forest vary with height. At the upper canopy layers, temperatures are extreme and fluctuate more, while in the forest interior, they are moderate and stable, making forest floors preferred homes of large animals like ungulates.

- **Wind and rain protection:** Forest canopies act as a windbreak and intercept rainfall making the forest interior more hospitable.
- **Produces niches:** The combination of environmental conditions at various heights creates a myriad of microclimates, which provide specialised niches for other plants, animals, birds, insects, etc. For example, epiphytes grow in branch nooks, and most insects live in the crowns of trees.
- **Structural diversity:** Besides the microclimates, the forest canopy with its branches and arboreal connections creates structural diversity which is vital for the survival of many animals. For example, many tree-dwellers escape ground predators by staying and moving in the canopy.
- **Biodiversity:** As a result of its influences, the forest canopy structure is necessary to maintain the biodiversity of a forest. Differing forest canopy structures of biomes increase the overall biodiversity of the earth.

Uses of Canopy Structure

Besides the importance of forest canopy structure for the forest ecosystem, it is useful for people in many ways and is therefore monitored. Some of the most important uses are as follows:

- **Ecosystem Productivity:** Forest canopy structure studies at the regional and local scales have often been used to determine carbon uptake, storage, and flux for carbon accounting and validation. Such productivity analysis is common in climate change mitigation studies.
- **Ecosystem Services:** It is the canopy structure of the forest that determines many of the ecosystem services we associate with forests. These could be:
    - Intercepting and helping in percolation of 60-90% of rainwater/ snow gently into the soil and producing a watershed region important for the hydrology of a region.
    - Providing people and animals with clean drinking water.
    - Producing litter (leaves, fallen branches, etc.) that maintain soil fertility and soil carbon banks.
    - Providing a canopy or ground cover that protects soil from the force of rain and a mat of roots that hold soil to conserve it.
    - Forest canopies hold a lot of the water that trees absorb and recirculate it into the atmosphere through transpiration, providing input for rain, so the water cycle is maintained. For example, one-third of the rainfall in the Amazons is due to the forest canopy.

- **Agricultural productivity:** Many cash crops have been traditionally cultivated as understorey crops and require a forest canopy to thrive. Examples are cardamom, coffee, or cacao.
- **Forestry:** Forest canopy structure is also regulated because many commercial forests provide timber and wood.
- **Non-timber products:** It is essential to maintain and monitor primary forests' canopy, as it provides non-timber products such as medicinal plants, nuts, fruits, resins, barks, etc. The tree crown provides most of these products.
- **Ecology studies**: Forest canopies are a reflection of available moisture and soil types, and differences in canopy structure can establish gradients in site quality of forests.

5

FLORISTIC COMPOSITION, SPECIES DIVERSITY AND SIMILARITY

INTRODUCTION

Diverse physiographic, altitudinal, climatic, and edaphic difference enables Ethiopia to have various types of vegetation ranging from alpine to desert plant communities which provide economical, sociocultural, and environmental benefits. Many studies confirmed that forests have an important role in maintaining the productivity of the environment; trees provide food for animals, serve as a standing cover to protect the land from wind and water erosion, stabilizing the water cycle, facilitate the process of evaporation, and keep the soil porous; they are also used for construction as well as for tools, furniture, fuel, medicine, grass, and herbage and for forage and provide edible fruits. They serve to absorb carbon dioxide to reduce global warming, give off oxygen, and renew the atmosphere. Plants also serve as a source of income by attracting tourists, serve as recreational facilities, prevent lakes and dams from silting, and clean, regulate, and distribute water resources.

Hence, forests sequester and store more carbon than any terrestrial ecosystem; that is, they store more than 80% of all terrestrial above ground carbon and more than 70% of all soil organic carbon. To do so stable climate is crucial but the climate of Ethiopia has been changing due to global and local effects of vegetation degradation. In Ethiopia different factors like deforestation, overharvesting, and permanent conversion to other forms of land use are leading to shrinkage of forest resources. In addition to anthropogenic activities, forest can be affected by different environmental factors such as altitude, slope, and aspect by affecting the patterns of tree species distribution.

The forests of Humbo Woreda have been cleared only within the last thirty to fifty years as the demand for energy, construction wood, food, fodder, and feed has increased as a result of the increase both in human and in livestock population. The main causes of deforestation and forest degradation at Humbo are tree cutting for wood fuel and construction materials production for sale especially to compensate food shortage created by erratic and scarce rainfall in the area over a long period of time. Oda forest was one of the forests of Humbo Woreda victimized by these anthropogenic activities as well as environmental factors. Awareness and knowledge on sustainable use and management were very much less even though the Humbo Carbon Project forests like Oda forest provide ecological services and socioeconomic values to local communities. Therefore, Humbo Community-Based Forest Management Project was established with the aim of sequestering CO_2 through regeneration of native forest utilizing the farmer managed natural regeneration technique.

The Humbo Community-based Natural Regeneration Project forests act as a carbon "sink" to mitigate climate change while at the same time building environmental, social, and economic resilience for future climate change impacts. Over 90 percent of the Humbo project area is being reforested from the stumps of previously cut down (but still living) trees. To date, 2728 hectares of degraded forest that was being continually exploited for wood, charcoal, and fodder extraction has been protected and is now being restored and sustainably managed. Of that, 340.04 hectares is covered by Oda forest.

The diversity, regeneration status, floristic composition, and vegetation structure are crucial elements to clearly visualize the anthropogenic activities as well as environmental factors affecting the vegetation of an area. Hence, for forest management accurate data on forest resources is considered to be an essential requirement, but as a result of being a newly established Carbon Project such a kind of information is not well documented in Oda forest. The lack of such basic information is one of the serious problems that hamper sufficient conservation and management of the forest. Therefore, this study intended to solve such a basic problem hindering observation of the next phenomenon.

The watershed forests are a vital role to protect watershed area in Myanmar because there are different kinds of challenges facing serious floods and landslides in some regions. It is clear that linkage between floristic composition, forest stand structure, soil physical and chemical properties results outcomes are useful to catchment regions' restoration measures in order to achieve sustainability of forests in long run.

Moreover, ecological data need to be obtained for the application of effective management practices in the watershed forests. Floristic composition and structure studies are essential for their values in understanding the extent of plant biodiversity in forest ecosystems as well as in watershed forests. This study was intended to provide for linkage between species composition, stand structure and soil properties of forest dynamics and floristic changes in the watershed areas. An assessment of composition and structure of selected forest types will follows, concluded with an assessment of the potential and possibilities for future forest management. Floristic composition, structure and natural regeneration of the forests in watersheds are critical to known in the present day.

Many currently successful forest management and restoration plans base on accurate and detailed characterization of the plant community, i.e. its species composition. This may include the number of species present, but often takes into account quantitative measures such as abundance, frequency, density, coverage, and basal area. Measures of diversity such as the Shannon and Simpson indices have also been useful for specifying and comparing forest communities. Soils are important to life on earth and the quality of the soil determines the nature of plant ecosystems and the capacity of land to support animal life and society. Soil properties are partially dependent on forest types which directly affect vegetation.

STRUCTURE OF VARIOUS FORESTS

Structural variation in forests revealed that density of trees ranged from 246 to 865 trees ha^{-1}. It was highest in Mixed Sal forest and lowest in Bamboo brakes. Density of trees in different forest types was in the order: Mixed Sal Forest > Mixed Forest > Mixed Teak > Teak Mixed > Bamboo Brakes. Average crop height and basal area of trees varied from 9.1 to 14.5 m and 12.1 to 29.9 m2 ha-1 , respectively. Mean tree height as well as basal area values were found to be highest in Mixed Sal forest, while these were lowest in bamboo brake and teak mixed forests, respectively.

Basal area in different forest types followed the order: Mixed Sal > Mixed Forest > Mixed Teak > Bamboo Brakes > Teak Mixed Forests. The patterns of structure, composition and diversity of vegetation recorded in the present study are closely resemble and comparable to other tropical forest ecosystems studied by earlier workers. Researchers in a study on moist tropical ecosystem in Amarkantak MP reported density of over storey and under storey vegetation accounted for 553.68 trees ha-1 (ranged from 424 to 752 trees ha^{-1}) and 5870 shrubs ha^{-1}.

Slightly a higher density of trees in the present study is attributed to better site quality and microclimatic conditions. It has long been recognized that soil fertility and environmental conditions are the major drivers that influence the structure and diversity of tropical forests. Basal area of tree layer in the present study are well within the range and consistent with the findings of earlier workers. Researchers reported that tree basal area ranged from 8.13 to 28.87 m^2 ha^{-1} in dry tropical forests of Barnowpara, Chhattisgarh. Researchers also reported 6.58 and 23.21 m^2 ha^{-1} basal area in Vindhyan range of U.P. India. Researchers in a recent study found that basal area of trees ranged from 8.33 to 29.93 m^2 ha^{-1} in moist deciduous forests of AABR, Madhya Pradesh.

Shorea robusta, Terminalia tomentosa and *Syzygium cumini* in Mixed forest, *Shorea robusta, Terminalia tomentosa* and Cleistanthus collinus in Mixed Sal Forest, *Tectona grandis, Diospyros ebenum* and *Courouptia guianensis* in Mixed Teak Forest, *Tectona grandis, Lagerstroemia speciosa* and *Diospyros melanoxylon* in Teak Mixed Forest, *Dendrocalamus strictus, Bahunia racemosa* and *Terminalia tomentosa* in Bamboo brakes were identified as predominant plant communities in tree layer. In shrub layer, *Smilax macrophylla, Gymnema sylvestris, Agave americana, Jatropha curcas, Nyctanthes arbor-tristis, Discorea hispida* etc associations were found as predominant plant communities in different vegetation types. *Shannon index* values in tree layer ranged from 0.61 to 3.07 among different vegetation types. It was highest in Mixed Forest, and lowest in Teak Mixed Forest. Simpson Index values ranged for 0.064 to 0.74, which was found to be highest in Bamboo Brake and lowest under Mixed Forest. The Evenness Index values ranged for 0.297 to 0.864. It was highest in Mixed Forest and lowest in Teak Mixed Forest. The Margalef' s Index values ranged from 5.30 to 16.9, which was highest in Mixed Forest and lowest in Teak Mixed Forest. Beta Diversity across vegetation types ranged from 2.96 to 3.60. Mixed Forest showed highest beta diversity (3.60) followed by Bamboo Brakes (3.33) and it was lowest in Mixed Sal Forest (2.96). In Shrub layer, Shannon Index values ranged from 0.54 to 1.54 across different vegetation types.

Diversity Values were higher in Mixed Teak Forest followed by Mixed Sal Forest and Teak Mixed Forest and lowest values were observed in Bamboo Brakes. Simpson Index values ranged from 0.267 to 0.7. The highest values recorded by Bamboo Brakes (0.7) and lowest values by Mixed Teak Forest (0.267). Evenness index values ranged from 0.49 to 0.861, the highest values was found in Mixed Teak Forest and lowest in Bamboo Brakes. The Margalef' s Index values ranged from 0.60 to 1.127. It was highest in Mixed Forest and lowest in Bamboo Brakes. Beta diversity values ranged from 2.14 to 3.75 among vegetation types. The diversity was highest in Mixed Forest and the lowest in Bamboo Brakes. In the herbaceous layer, Shannon Index values ranged from 0.624 to 2.01. It was higher in Teak Mixed Forest followed by Mixed Teak Forest and Mixed Sal Forest and lowest values were observed in Mixed Forest. Simpson index values ranged from 0.003 to 0.259.

The higher values were recorded in Mixed Sal Forest and lowest in Mixed Forest. Evenness Index values ranged from 0.271 to 0.89. Bamboo Brake recorded highest Evenness values, while it was lowest in Mixed Forest. The Margalef Index values in vegetation types ranged from 1.75 to 3.51. The highest value was recorded by Teak Mixed Forest and lowest by Bamboo Brakes. The Beta diversity values ranged from 1.5 to 3.03 in the vegetation types. It was highest in Mixed Forest and lowest in Bamboo Brakes.

Species diversity, the number of species in a community is ecologically important as it represents the spatial heterogeneity and niche differentiation for resource sharing among vegetation communities. The results revealed that species diversity varied across vegetation types of dry tropical ecosystem of Achanakmaar Amarkantak Biosphere Reserve (AABR). The diversity indices under this study are comparable to the indices reported in different tropical forests. Researchers recorded the Shannon-Wiener index values from 1.9 to 2.8, concentration of dominance from 0.18 to 0.75, species richness from 0.21 to 0.93, and beta diversity of 3.1 in Dry deciduous forests of Vindhyan ranges of Uttar Pradesh, India. Similarly, researchers reported the Shannon Index values ranged from 2.39 to 3.62, concentration of dominance from 0.12 to 1 in different tropical forest fragments of Barnowpara and Amarkantak. The present study also demonstrated inverse relationship between the diversity and dominance indices, which was corroborated by the several reports. The present study also showed highest (3.6) beta diversity in Mixed Forest and lowest (2.96) in Mixed Sal Forest clearly indicates the habitat heterogeneity level, as more number of species efficiently sharing resources and niche space in former compared to later forest type.

CLUSTER ANALYSIS

Cluster analysis can be used to identify groups of sites or samples that are similar in terms of their species composition. There are two main types: *hierarchical*, where samples are assigned to groups that are themselves arranged into groups, and *nonhierarchical*, where the samples are simply assigned to groups. Hierarchical methods are most commonly used. The results are usually illustrated by means of a *dendrogram* (Fig .1), in which sites or samples that are more similar in terms of species composition group more closely together. A wide variety of different algorithms can be used to produce dendrograms, which may produce different results. Cluster analysis can be done by many standard statistical analysis software packages, including SAS, SPSS, and GENSTAT, as well as specialist ecological software.

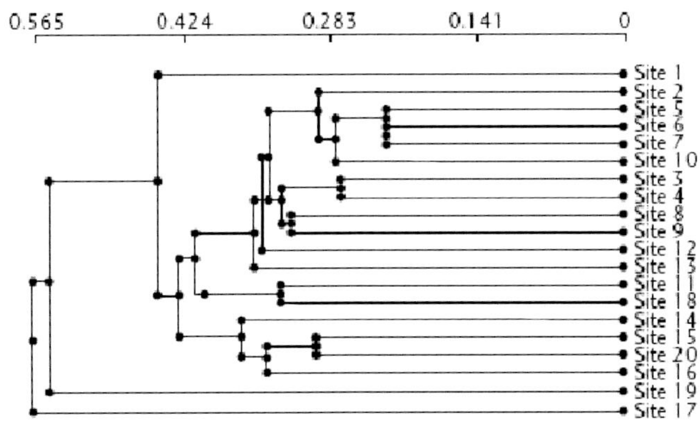

Fig. 1. A dendrogram showing the results of a cluster analysis. The dendrogram illustrates the degree of similarity between sites according to their species composition.

Two key decisions need to be made in cluster analysis: which measure of similarity (or distance) to employ, and which rules for cluster formation to adopt. The similarity measures described in section 3.9, such as the Jaccard and Sørensen indices, which are suitable for use with presence–absence data, can be used as a basis of cluster analysis. For continuous variables, Euclidean distance is typically used. With respect to the rules for cluster formation, the simplest method is to use single-link or nearest-neighbour clustering, in which clusters are joined based on the smallest distance between any pair of their component individuals. However, dendrograms produced by this method can sometimes be difficult to interpret. A number of alternative methods are described by researchers. UPGMA is a widely used alternative method that allows clusters to be joined on the basis of the lowest value of the average distance between clusters.

The main problem with cluster analysis, as with other classification techniques, is that the results can easily be misinterpreted. Definition of groups does not necessarily reflect genuine ecological differences. Cluster analysis will always divide the data into groups; even randomly generated data can produce a dendrogram. Another problem is that different clustering techniques can provide very different answers, and as no statistical significance is obtainable for this form of analysis, it can be difficult to decide which output is 'correct'. Careful consideration of the results is therefore required, to examine whether they make sense and have some genuine ecological meaning. Cluster analysis should be viewed as a tool for data exploration and hypothesis generation, and not as an end in itself.

TWINSPAN

All the species data, as well as the field plots, were used for the analysis. All classification methods were used on the raw data as recommended. No transformation of floristic data was used. The PC ORD ver. 3.05 was used for TWINSPAN classification and ordination analysis. R Package 4.06 was used for agglomerative clustering. For classification, floristic data were analyzed by TWINSPAN and agglomerative cluster analysis techniques.

For agglomerative clustering data were analyzed using different similarity indices and some of these use qualitative data, based on presence/absence, while others are quantitative and work on abundance data. Three similarity indices such as Jaccard's index (S7), Sørensen's index (S8), Steinhous index (S17) were used for agglomerative clustering (Legendre & Legendre, 1998). There are several methods of clustering algorithms (sorting strategies) available and in this study three different methods such as "UPGMA" ('Unweighted Pair-Groups Method using Arithmetic averages') Proportional-Link Linkage' by Sneath & Sokal, and Ward's method of minimum variance were used. Cophenetic correlations such as "Kendall's tau b cophenetic correlation" and "Pearson's r cophenetic correlation" were calculated for comparing the results of the agglomerative clustering methods.

The second method used for the classification of floristic data was TWINSPAN, a commonly used program in ecological studies. Researchers suggested that the presence of the species rather than its abundance should be dominant in numerical vegetation classification. Therefore, the analysis employs the presence/absence data and not the cover value of each species.

Classification by TWINSPAN was stopped at the 3rd level so that the size of stands would demonstrate ecological meaning through their floristic structure. This resulted in four distinct groups, which are presented in dendrogram (Fig. 2), together with the indicator species used by the software for every level of division. Although the dendrogram is not as detailed as the TWINSPAN table, it summarizes the results in a more convenient way.

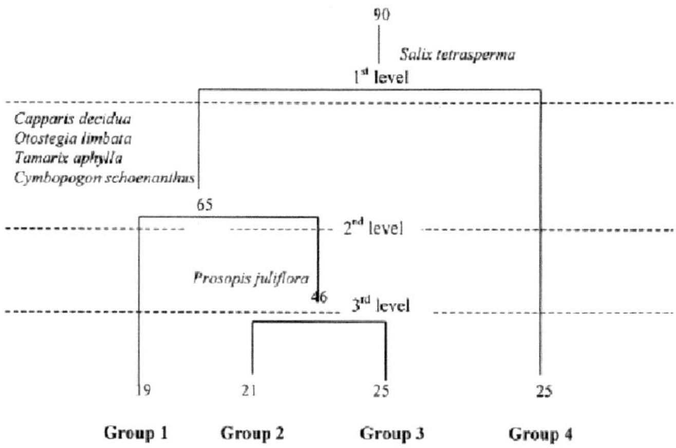

Fig. 2. TWINSPAN classification

Third method used was ordination analysis that employed abundance data using the 'Domin scale' without any transformation. Preliminary analysis using Principal Component Analysis (PCA) and Detrended Corenrpondance Aanalysis (DCA) suggested that ordination using DCA provided more robust and interpretable results than PCA and the length of first axis was greater than 3.0 and in terms of species turnover or standard deviation (s.d).

It is recommended by researchers that if species turnover is greater than 1.9 s.d then DCA is better choice of ordination. Therefore PCA was discarded from further analysis. Detrended Correspondence Analysis (DCA) were performed to describe compositional gradients in the vegetation. DCA was performed using a default value for detrending and rescaling. For detrending a default value "26" recommended by the ter Braak was used. Rare species were downweighted in DCA ordination.

The TWINSPAN analysis divided the field plots into four vegetation clusters, each cluster representing a specific community type with characteristic species (Fig. 2). Three clusters had a single characteristic species, while the other was characterized by the presence of four species which were restricted mostly to it. TWINSPAN classification initially separated all the plots into two broad groups: groups of 65 and 25 plots with Salix tetrasperma as a key indicator species. The second level of division separated the group of 65 plots into two groups of 19 and 46 plots. This division was based on the presence of Capparis decidua, Otostegia limbata, Tamarix aphylla, and Cymbopogon schoenanthus as the negative group. The third level of division separated the group of 46 on the left side of the dendrogram into two new groups of 21 and 25 plots where Prosopis juliflora was the positive indicator species

Four plant community types were recognized from the TWINSPAN and agglomerative clustering which were later used with the multispectral classification of SPOT XS data.

Capparis decidua-Eleusine compressa (referred to as Capparis-Eleusine community type) community type is heavily grazed and subject to uncontrolled cutting for a long period. Due to biotic pressures, all trees have been completely eliminated. This community type is represented by few open scattered shrubs, which include *Capparis decidua, Otostegia limbata, Tamarix aphylla, Periploca aphylla*, and *Ziziphus nummularia. Capparis decidua* seems to withstand the biotic pressure better than other species, so it dominates the shrubs. Grasses forms a more or less continuous association over appreciable areas. *Eleusine compressa, Chrysopogon montanus, Cymbopogon schoenanthus, Heteropogon contortus* and *Desmostachya bipinnata* were among the grass species. This community type was found mostly on gravelly, eroded hill slopes and tops and in the crevices of exposed rocks. Lantana camara and Broussonetia papyrifera were observed as invasive species in this community type.

Prosopis juliflora-Chrysopogon montanus (referred to as the Prosopis-Chrysopogon community type) is characterised by the presence of Prosopis juliflora, which covered a large area of Compartments 1 and 2 of the previously declared reserve forest. This community type is formed as a result of complete cutting of native forest vegetation for the construction of new housing schemes by the local development authority and is mostly confined to western aspects.

Prosopis juliflora is an invasive exotic but is naturalised in open and arid regions of Pakistan and constitutes a prominent feature of the vegetation of the plains. Prosopis juliflora does not depend entirely on the availability of surface moisture, but can extend its roots at least 20m into the soil to reach and tap the underground lenses of fresh water. Thus it resists the drought conditions and can survive provided the depth of soil is adequate in dry and arid tracts where vegetation of any other kind is scanty. *Desmostachya bipinnata* a grass of exposed and dry places, covering open ground is an indictor of moisture present in the deeper layer of the soils. In spite of higher rainfall compared to other drier areas nearby such as Attock and Chakwal, xeric conditions prevail due to the poor retention of moisture by the upper porous layers of the soil. *Desmostachya bipinnata* overcomes this by sending down its roots into the deeper layers of the soil. In this plant, the roots are 3 to 8 times the size of the aerial parts. *Prosopis juliflora, Lantana camara,* and *Xanthium strumarium* were three invasive species found in this community type.

Ziziphus nummularia-Malcolmia africana (Group 3 in Fig. 2 & referred to as the *Ziziphus-Malcolmia* community type) covers a large area of the reserve forest. Acacia modesta forms the main canopy. Due to heavy biotic pressure, such as grazing, unpalatable species such as Justicia adhatoda, Withania somnifera and Lantana camara are conspicuous with thorny *Gymnosporia royliana, Woodfordia fruticosa, Ziziphus nummularia* and *Carissa opaca*. The under-canopy of this community is not thick with open patches cover with grasses. Regeneration of *Acacia modesta* is not high. Among the tree species, scattered trees of *Dalbergia sisso*, and *Acacia nilotica* were observed. At some places, *Broussonetia papyrifera* in Compartments 3 and 4 and *Prosopis juliflora* in Compartments 7 and 8 have invaded the community type. *Olea ferruginea*, always found in close association with *Acacia modesta* and *Carissa opaca* had either been completely eliminated or was present as scattered, stunt, bushy trees; less than two metres in height. *Dodonea viscosa*, another important member of scrub forest, is completely destroyed. This community type was found mostly along west facing slopes on dissected rolling plains. *Chrysopogon montanus, Heteropogon contortus,* and *Desmostachya bipinnata* are among the main grasses recorded in this community. Broussonetia papyrifera, Lantana camara and *Prosopis juliflora* were three invasive species recorded in this community type.

Saccharum benghalensis-Salix tetrasperma (Group 4 in Fig. 2 & referred to as the Saccharum-Salix community type) was found growing along streams and river beds, adjoining slopes, edges, and flood plains that are annually/seasonally flooded. The vegetation is highly influenced by the flooding. *Saccharum spontaneum*, a plant along the streams, rivers and their banks, dominated this community. Signs of grazing were evident and large clumps of *Saccharum spontaneum*, cut at ground level also hinted at human interference. In some places away from the river/stream banks, *Cynodon dactylon* was the major component. It covers open places among *Saccharum spontaneum*. In other places that were more dry and sandy, *Desmostachya bipinnata* was found in abundance. Plots sampled closer to the main stream, *nallaha* and river banks were physiographically gravel bars comprised of gravel, pebbles of varying sizes and with a small proportion of sand.

Mostly these plots were the first to be submerged in flood periods and there had been lot of soil moisture in the soil and these plots were dominated by Saccharum spontaneum, which was found in small clumps accumulating, sand around it. These plots seemed to be in an advanced stage of succession as two woody species, viz., *Salix tetrasperma* and *Dalbergia sisso*, were reported. This habitat is suitable for the establishment of some other woody perennials too, such as *Ricinus communis, Acacia modesta* and *Acacia nilotica* etc., which were found growing around and might serve as seed sources. However it would be very difficult for these species to establish themselves due to the high competition from Saccharum spontaneum. These plots also support large number of annuals. Species like Saccharum benghalensis and Ricinus communis were also found among this community type. Close to the river and stream water, where high moisture was available, water loving species such as *Ranunculus arvensis, Rorippa nasturtiumaquaticum, Bacopa monnieri, Juncus* sp., *Anagallis* sp., and *Psammogeton* sp., *Fumaria indica, Convolvulus arvensis, Cyperus* sp., *Vicia sativa*, and *Verbena* sp., were found.

FLORISTIC COMPOSITION, SPECIES DIVERSITY AND SIMILARITY

Table 1 shows Diversity indices and evenness, DBH ~10 cm of two forest types. Table 2 shows value of coefficient of similarity by SORENSEN (Ks), WEIDELT (Kd) and BRAY and CRTIS (CN).

According to the results by calculating Shanon diversity index (H') and Simpson index (D), mixed deciduous forest is more heterogeneous than deciduous dipterocarp forest. The value of D decreases while diversity increases.

It can be observed that the species in mixed deciduous forest is more diverse than deciduous dipterocarp forest. Evenness refers to how the species abundance is distributed among the species. The values of Shanon evenness shows that the abundance of species are more even in mixed deciduous forest than deciduous dipterocarp forest.

Table 1. Diversity indices and evenness, DBH ~10 cm.

Diversity indices	MDF	DDF
Shannon index (H')	3.68	2.39
Simpson index (1-D)	0.97	0.83
Shanon evenness	91.16	74.28

Table 2. Coefficient of similarity indices value.

Coefficient of similarity indices	Value (%) (DBH ~10 cm)
Sorensen (Ks)	46.34
Weidelt (Kd)	42.49
Bray & Curtis (CN)	78.15

The results show slightly above a moderate floristically similarity between mixed deciduous forest and deciduous dipterocarp forest according to Sorensen's index of similarity. However similarity by researchers indicate that the floras of the two investigated forests are more similar with each other.

ABUNDANCE, DOMINANCE, FREQUENCY AND IVI OF BOTH INVESTIGATED FORESTS

Abundance, dominance, frequency and Importance Value Index of the lists of 9 species ranged based on their IVI for the investigated forests are shown in Table 3 and Table 4.

Table 3. Abundance, dominance, frequency and IVI of the species in MDF, DBH ~10 cm, 1 ha

Species name	Abundance	Dominance	Frequency	IVI
	(No. of trees)	(BA)	(%)	
Tectona grandis	30	4.40	100	27.01

Xylia xylocapa	23	3.02	100	20.33
Milettia pendula	17	2.12	100	15.57
Terminalia alata	11	2.52	100	15.17
Schleichera oleosa	10	1.14	100	10.25
Legerstromemia villosa	15	0.69	100	10.21
Mitragyna rotundifolia	12	0.67	100	9.26
Salmalia insignis	3	1.72	75	9.25
Delbergia cultrata	10	0.87	75	8.45
Others (48)	211	12.75	2000	174.50
Total (57)	342	29.89	2850	300.00

Table 4. Abundance, dominance, frequency and IVI of the species in DDF, DBH ~10 cm, 1 ha

Species name	Abundance	Dominance	Frequency %	IVI
Dipterocarpus tuberculatus	109	9.02	100	81.87
Shorea siamensis	27	3.70	100	32.09
Shorea oblongifolia	21	2.46	100	25.08
Terminalia alata	19	2.53	100	24.65
Xylia xylocapa	15	1.06	75	15.25
Melanorrhoea usitata	13	0.81	75	13.76
Schleichera oleosa	8	0.75	75	11.78
Delbergia cultrata	8	0.75	50	9.78
Buchanania lanzan	9	0.35	50	8.54
Others (16)	56	3.88	525	77.00
Total (25)	285	25.31	1250	300

Tectona grandis had the highest abundance, dominance, frequency and IVI in mixed deciduous forest. The dominance of Tectona grandis was 14.73% of the total basal area of all species. In deciduous dipterocarp forest, Dipterocarpus tuberculatus was the highest abundance, dominance, frequency, IVI and occupied 35.63% of the total basal area of all species. The second most abundance species was *Xylia xylocapa* 10.10% of total dominance of all species in mixed deciduous forest. The second most abundance species of deciduous dipterocarp forest was Shorea siamensis and 14.62% of total dominance of all species.

Termminalia alata and *Schleichera oleos*a species were occurred 8.44% and 3.82% in mixed deciduous forest and 9.99% and 2.97% in deciduous dipterocarp forest of total basal area. There are 35 species with relative abundance value below 2% in mixed deciduous forest and 12 species in deciduous dipterocarp forest respectively. Many species in the investigated forests have low abundance, low frequency and low dominance value. These species of both forests are rare casual and no major ecological or economic significance .

The most abundance species usually have high frequency values, i.e., they belong to the group of tree species with a regular horizontal distribution. It can be observed that only few species were most abundance and had high frequency values in the investigated forests but most of species are comparatively causal. In mixed deciduous forest *Tectona grandis, Xylia xylocapa, Milettia pendul*a were found to be the most abundance species with the high frequency values of 100% and Dipterocarpus tuberculatus, *Shorea siamensis, Shorea oblongifolia* in deciduous dipterocarp forest. This indicated favorable site condition for the development of them. It can also be assumed that these trees had regular horizontal distribution. It can be investigated that there is a close relationship between abundance and frequency values in both frorests.

In both investigated forests, it can be observed that rare species are present with a low abundance value. There are 11 species with only one individual in mixed deciduous forest and four species in deciduous dipterocarp forest. As abundance and frequency values are closely related to each other, rare species had low frequency values. The dominance of rare species is 3.49% of total basal area in mixed deciduous forest and 0.61% in deciduous dipterocarp forest.

DIAMETER FREQUENCY CLASS AND DISTRIBUTION OF INDIVIDUALS BY HEIGHT CLASS

According to this figure, the numbers of trees occurred in lower diameter classes than higher diameter classes. This indicated that the reserve of small diameter trees in moist tropical forest was in any case adequate to replace the losses in large diameter classes. Therefore natural sustained yield was therefore evidently completely assured.

Tectona grandis and *Dipterocarpus tuberculatu*s trees were the dominance species in terms of tree density and basal area in mixed deciduous forest and deciduous dipterocarp forest. The deciduous dipterocarp forest was less number of trees in the smaller diameter classes as compared to the mixed deciduous forest. It was depended on different forest types. The mixed deciduous forest occupied more growth stages and numbers of species than deciduous dipterocarp forest because of site conditions were favorable.

The distributions of individuals and species were similar in diameter classes. The larger numbers of species were concentrated in lower diameter classes up to middle diameter classes while only few numbers of species were found in higher diameter classes. It was found that 51% (29 species) in mixed deciduous forest and 72% (18 species) in deciduous dipterocarp forest were present in the class of dbh (10 - 20) cm.

NATURAL REGENERATION

The composition, distribution and density of seedlings and saplings reveal the future status of the forest. Seed production must be adequate to ensure more or less uninterrupted availability of viable seeds of several species. The other important factors for germination and establishment of seedlings are climatic and edaphic conditions. The total density of seedlings, saplings and trees were 1219, 531 and 342 ha^{-1} in mixed deciduous forest and 444 ,988 and 285 ha^{-1} in deciduous dipterocarp forest respectively.

FOREST SOILS

Physical Properties of Soils

Table 5 shows soil physical properties based on soil layers in mixed deciduous forest and deciduous dipterocarp forest. In both study sites a one-meter deep soil pit was dug to analyze and compare soil parameters. Soil samples were extracted from the following soil depths: 0 - 10 cm, 40 - 50 cm and 90 - 100 cm from each soil pit. The physical properties of soils determine their adaptability for trees and the level of biological activity that can be provided by the soil. Soil physical properties also mostly determine the soil's water and air supplying capacity to plants.

Mixed deciduous forest had a higher soil moisture content compared to deciduous dipterocarp forest. As the amount of clay-sized particles content of mixed deciduous forest were higher than deciduous dipterocarp forest, moisture content more retained in mixed deciduous forest. The deciduous dipterocarp forest had lower vegetation cover but higher bulk density than mixed deciduous forest. The average sand content of different layers (53%) was observed under the mixed deciduous forest and (70%) was recorded in the deciduous dipterocarp forest. The average silt fractions of the mixed deciduous forest and the deciduous dipterocarp forest were 25% and 12%, the clay fractions were 20% and 14%, respectively.

Chemical Properties of Soils

Table 6 shows soil chemical properties based on soil layers in mixed deciduous forest and deciduous dipterocarp forest. Soil chemical properties are the most important factors and determine the nutrient supplying power of the soil to the plants. The soil chemical reactions affect the processes that leading to soil development and soil fertility build up.

Minerals inherited from the soil parent materials overtime give up chemical elements that undergo various changes and transformations within the soil. Soil pH values of investigated forests were not significantly different. The maximum soil pH values were occurred in 90 - 100 cm depth and the minimum soil pH values were found in the depth of 40 - 50 cm of both investigated forests. The same values of soil pH were found in 40 - 50 cm of mixed deciduous forest and 0 - 10 cm of deciduous dipterocarp forest. However, pH values of every soil layers of mixed deciduous forest were higher than deciduous dipterocarp forest. The mixed deciduous forest had higher organic matter than deciduous dipterocarp forest in every depths of soil. The organic matter of mixed deciduous forest and deciduous dipterocarp forest were 3.53% and 2.07% in 0 - 10 cm. The organic matter of the two forests gradually decreased as the range of soil layers 0 - 10 cm, 40 - 50 cm and 9 - 100 cm. But the lowest organic matter 2.59% in 90 - 100 cm of mixed deciduous forest was higher than the highest organic matter 2.07% in 0 - 10 cm of deciduous dipterocarp forest. The mean total nitrogen concentrations of mixed deciduous forest and deciduous dipterocarp forest were 0.054% and 0.026% respectively. The mean values of phosphorus and potassium of mixed deciduous forest were higher than those of deciduous dipterocarp forest.

Table 5. Soil physical properties of mixed deciduous forest and deciduous dipterocarp forest

Forest types	Depth [cm]	Moisture	Bulk density	Sand [%]	Silt [%]	Clay [%]	Texture
Mixed deciduous forest	0 - 10	26.75	1.219	56.75	25.50	16.00	Sandy loam
	40 - 50	-	-	51.75	26.00	22.25	Sandy clay loam
	90 - 100	-	-	50.25	23.50	20.50	Sandy clay loam
Deciduous dipterocarp forest	0 - 10	19.37	1.32	69.00	12.25	13.50	Sandy loam
	40 - 50	19.37	1.32	69.50	11.75	15.25	Sandy loam
	90 - 100	19.37	1.32	70.25	11.50	13.50	Sandy loam

Table 6. Soil chemical properties of mixed deciduous forest and deciduous dipterocarp forest.

Forest type	Depth [cm]	pH	Total N %	Ava. P %	Exc. K %	OM %
Mixed deciduous forest	0 - 10	6.08	0.059	0.00030	0.0087	3.53
	40 - 50	6.05	0.054	0.00009	0.0047	2.81
	90 - 100	6.13	40.050	0.00015	0.0042	2.59
Deciduous dipterocarp forest	0 - 10	6.05	0.029	0.00018	0.0080	2.07
	40 - 50	6.03	0.025	0.00007	0.0047	1.74
	90 - 100	6.1	0.023	0.00006	0.0042	1.51

Discussion

Analysis of the forest inventory data of different vegetation types revealed that Tectona grandis had highest density, frequency and IVI in mixed deciduous forest and Dipterocarpus tuberculatus in deciduous dipterocarp forest because of these species were dominant species in respective forest types. In both vegetation types, the species with high IVIs showed good regeneration potential. This finding is in accordance with Whittaker's work, in which the dominance of one stratum may affect the diversity of another stratum. The diameter class distribution patterns for both forest types were in the form of reversed "J-shaped" curve, indicating the adequate representation of small trees in the lower diameter classes. The maximum numbers of species were encountered in lowest gird at breast height class (10 - 20 cm) while species numbers gradually decrease with increasing girth classes.

In vertical structure, most species and individuals were concentrated in lowest height class due to diameter and high of trees are closely related. The population structure characterized by the presence of adequate number of seedlings, saplings and young trees depicts satisfactory regeneration behavior, while inadequate number of seedlings and saplings of tree species in a forest indicates its poor regeneration. The regeneration potential of a species in a community can be assessed from the population dynamics of seedlings and saplings in the forest. In the two forests, the sufficient amount of seedlings, saplings and trees in the smaller diameter class indicate the ability of these two forests to recruit into adult phase. As the mean soil moisture contents were assumed to differ between mixed deciduous forest and deciduous dipterocarp forest, there was significant difference in soil moisture contents between both investigated forests. It is possible that the mixed deciduous forest retained denser canopy than deciduous dipterocarp forest and this finding is in accordance with the findings. Soil moisture content is higher under more woody vegetation than those in less woody vegetation due to the positive effect of shade from the canopies. Bulk density values were significantly different between the two forests. The lower soil bulk density occurs in mixed deciduous forest because much more organic matter (OM) which derived from different kinds of tree species' little decomposed humus exists at that type of forest. The soil OM and total N contents of mixed deciduous forest were more concentrated than deciduous dipterocarp forest.

Conclusion

Mixed deciduous forest is more diverse than deciduous dipterocarp forest. In mixed deciduous forest, total density of seedlings, saplings and trees were much more associated than deciduous dipterocarp forest. It is clear that large amount of OM associated with dense canopy cover of mixed deciduous forest ecosystem. Moisture content of MDF related to denser canopy of moist deciduous forest's crown cover. Thus this finding can be used as a good indicator in identifying the better way for effective management of watershed forests and in monitoring as well as in subsequent evaluation of the study area. In order to attain for fresh and clean water for the future prosperity and good life quality of people living around watershed forests, it is urgent need to prevent landslide, to protect forest degradation and soil erosion, and to protect this watershed areas for effective watershed ecosystem management without timber extraction in the future.

6

ASSESSING THE PRESENCE, CONSERVATION AND MANAGEMENT OF ENDANGERED PLANT SPECIES

INTRODUCTION

Of the estimated 350,000 vascular plant species, the c. 59,000 described trees species (Botanic Gardens Conservation International, BGCI; represent the bulk of biomass and are essential as ecosystem engineers housing and feeding millions of species. Furthermore, trees provide many of nature's contributions to people supporting the livelihood of virtually all humans, e.g., as sources of wood, food, shade, firewood, construction, and carbon sinks.

However, an increasing proportion of tree species are subject to anthropogenic threats. The global IUCN RL, arguably the most influential and comprehensive framework to estimate species risk with extinction, lists 67 tree species that are either extinct (EX) or extinct in the wild (EW). Many more species (n = 11,548) are currently listed as threatened with extinction meaning that they have been classified by experts to face extremely high to high risk of extinction in the wild, based primarily on criteria related to range size, population size, or population decline.

The RL provides detailed and verified information on species extinction risk and the potential threats; and is the basis for conservation policy [informing, for instance, e.g., Convention on Biological Diversity (CBD), Intergovernmental Science-Policy Platform on Biodiversity and Ecosystem Services (IPBES), and Convention on International Trade in Endangered Species (CITES)] and prioritization of millions of dollars of conservation funds worldwide. Yet, due to the standardized criteria, the expert-focused assessment process, and the required documentation, the IUCN RL assessments are data intensive and time-consuming, and therefore, the IUCN RL is taxonomically and geographically biased, and many species are classified as data deficient (DD) or have not been evaluated.

Hence, until early 2021, c. 30,000 tree species remained without an RL assessment, forcing conservation decisions based on incomplete and biased information. Furthermore, for efficient conservation measures, extinction risk assessments need to be repeated and updated regularly. Keeping extinction risk assessments up to date is a major challenge for the IUCN RL, because, in addition to the general data and time constraints of RL assessments, reassessments of already red-listed species are often less appealing and hence down-prioritized.

As an attempt to speed up the process of assessing species extinction risk for the RL, a variety of methods have been proposed to automatically approximate species extinction risk based on species occurrence data from online databases. All these methods have important caveats and cannot replace the rigorous RL assessments. Yet, they can provide an approximation of species extinction risk and offer the advantage of being scalable to potentially large number of species.

For trees, a separate effort to boost the proportion of species with known extinction risk assessment exists: the Global Tree Assessment (GTA). The GTA aims to assess the conservation status of all tree species following IUCN RL criteria to allow effective prioritization of conservation measures. As of late 2021, the GTA included, approximately 43,700 species, with about 20% of the known species yet to be assessed or classified. To achieve this remarkable result, the GTA used a combination of semi-automated assessments based on approximate species range size from available occurrence data (for approximately 10,000 species) and data from national assessments and new expert-based assessment. The process involved the coordination of a huge international effort that took 5 years of research involving 60 institutional partners and over 500 experts, which exemplifies the complexity of this approach. In contrast to the IUCN RL, the GTA identified 142 tree species as EW.

PREDICTING SPECIES CONSERVATION ASSESSMENT

We used the trained models to predict the extinction risk of the 21,691 unlabeled species in our dataset. The application of 100 Monte Carlo (MC) dropout replicates allowed us to measure uncertainty around predictions. We combined our predictions with the available RL assessments to summarize the estimated extinction risk within higher taxa and by region. Specifically, we computed the number and proportion of threatened species in each family to quantify the level of heterogeneity in conservation status among taxonomic groups. After assigning species to countries and biomes, we also computed the number and proportion of threatened species within these spatial entities.

Sensitivity Tests

We performed sensitivity tests to assess the extent of taxonomic and geographic bias among the species used for training our models, i.e., the species in the RL. Specifically, we looked at the fraction of evaluated species across plant genera, families, and orders assuming that a systematic bias would leave a signature in their distribution. For instance, if the assessments were carried out systematically by the taxonomic group, we would expect a bimodal distribution where the fraction of assessed species nears one in some groups and zero in others. Similarly, we quantified the fraction of assessed species across countries and biomes to estimate the level of heterogeneity in the available RL assessments.

We then calculated the cross-validation prediction accuracy for each country, to evaluate whether spatial biases in the distribution of RL-assessed species may impact the accuracy of our predictions. Specifically, we identified what tree species in our test sets (from the 5 cross-validation folds) were found in each country, based on the available geographic occurrences. We then approximated the prediction accuracy for each country as the fraction of species correctly classified out of all tree species occurring in the country.

We performed 100 predictions for all species using MC dropout probability as a measure of uncertainty around each prediction. This enabled us to identify the MC dropout probability above which the classified instances yield a predefined prediction accuracy. For instance, we could identify the MC dropout probability threshold such that instances classified with a higher probability yield a 95% test accuracy, while others will remain "unclassified". We performed this test to assess how many species could be classified with high confidence (accuracy > 90%) and whether the fraction of them assigned to the possibly threatened category changes compared with the full set of predicted species.

CONSERVATION METHODS OF ENDANGERED SPECIES

According to the International Union for the Conservation of Nature (IUCN), an endangered species is an organism or a population of organism that faces a high risk of becoming extinct due to loss of its habitat, high death rate or changes in environmental and prey-predator parameters. The death rate of such organism far outweighs its birth rate over a period of time resulting to a decline in its population. Some common examples of endangered animal species include African Elephant, Tiger, Lion and Cheetah. In Nigeria, of the 4600 plant species, 707 are said to be endangered, while some hundreds of both plants and animal species have since gone into extinction due to loss of habitat occasioned by environmental degradation and climate change. To save endangered species from extinction, adequate preservation and conservation of such species and their habitat is necessary. Two main methods of conserving biodiversity as reported by IUCN, 2009 are considered, this include: *ex-situ* and *in situ* conservation.

Ex-situ conservation

This is the process of protecting or preserving an endangered species of either plants or animal outside of its natural habitat either by removing whole or part of that population from the threatened habitat and placing it in a new environment which may be a wild area or within human control environment. In *ex-situ* conservation, species whose life is threatened in a particular habitat is removed and placed in a well-protected habitat and taken care of by man.

Researchers listed the oldest and best known methods of *ex-situ* conservation to include Zoos, Botanical Gardens, and Aquaria. Other modern methods of *ex-situ* conservation where reproductive parts of an endangered species are stored for future reproduction or propagation include Seed bank, Gene bank, Germplasm bank, and *In vitro* Storages.

Botanical Garden

These are areas set aside for propagation and preservation of plant species to ensure their continuous existence. An endangered plant species is harvested in the wild and grown in the garden under human care to prevent it from becoming extinct. Researchers reported that about 354,000 plant species are grown in 1500 botanical gardens around the world. The Royal Botanical Garden of England (Kew Garden) alone is said to house about 25,000 plant species out of which 2700 are said to be endangered.

In Nigeria, conservation of endangered plant species through botanical gardens is not receiving the desired attention. Rather some of the existing gardens are continually being destroyed to pave way for development of other projects. For instance, in 1896, the then famous Lagos Botanical Garden and Onitsha Botanical Garden were at various times in the 1950s cleared for the development of communication lines.

Aquaria

Aquaria are facilities used for captive breeding of fish and other aquatic animals. In the past, aquaria were only used for display (decoration) of environment. However, due to growing threat to freshwater species they are now being used for captive breeding programs. The World Conservation Union (IUCN) is said to be working on captive breeding programs for conservation of fishes in the Lake Victoria in Africa, the desert fishes of North America and Appalachian stream fishes where aquaria are being designed to depict the natural habitats for conservation of these fishes.

Seedbank

The term seed bank refers to a cryogenic laboratory facility in which seeds of certain plant species are preserved for a long time without losing their fertility. It can also refer to a special kind of arboretum where seeds are harvested and the crop is rotated.

In vitro storage

In vitro storage is a storage facility in form of glass tube or vessel containing nitrogen liquid which keeps the temperature in the vessel at about -150°C. Hence cuttings, stems or other reproductive parts of plant stored in these strict conditions remain viable for a long time without losing their fertility. Seeds or reproductive parts of endangered plant species stored in seed Bank or in in-vitro tubes can survive for a long time up to hundreds of years making the propagation of such plant species in future possible.

Gene Bank

These are cryogenic facilities used for the storage of living sperm, eggs or embryos for future reproduction. The eggs or sperms of endangered animal species are preserved through this method.

Inter-governmental panel on Climate change, reported that the zoological society of SanDiego has established a frozen zoo for the storage of eggs and sperms where more than 355 species of animals including mammals, reptiles and birds are said to have their reproductive cells stored.

Storage of animal reproductive cells in gene bank for future reproduction is applicable even in humans. In 2004 a woman from the United States of America was reported to have given birth to a baby boy through artificial insemination of sperm obtained from her late husband who died since 1973, the man's sperm was collected while he was alive, frozen and stored in a Cryogenic (Gene Bank) facility for 30 years before it was finally removed, de-frozen and used to fertilize his widow.

A plant or animal may become extinct today but if the seed, sperms or eggs of such species are stored in seed or Gene Bank such species can be brought back into existence through reproduction or propagation of the stored reproductive parts.

In situ conservation

This involves the conservation and preservation of species in their natural habitat in places where the species naturally occurs. Under this method the entire ecosystem is protected and maintained so that all the constituent species, both known and unknown are conserved. The main methods of in-situ conservations as stated by Researchers include:

- Strict nature reserve (SNR)
- Games Reserve
- National Park

Strict Nature Reserves

These are plant communities reserved in perpetuity in their natural state for economic, educational, scientific, cultural and aesthetic values. SNR are established in remote areas and given adequate protection.

Games Reserves

Game reserves are areas set aside for the propagation, protection, conservation and management of flora and fauna with fauna species been of main interest. In Nigeria there are several gazetted Games' reserves spread across the country. However research has shown that most of these reserves have been degraded while quite a number have totally been dereserved. Game reserves are statutorily owned and managed by state or local governments.

National Parks

These are areas exclusively set aside for the propagation, protection, conservation and management of flora and fauna as well as the protection of site, landscape or geological formation of a particular place for scientific or aesthetic values for the benefit of the general public. National parks are owned by the Federal Government.

In Nigeria, there used to be eight National parks spread across the different ecological zones of the country, but now the number has reduced to seven due to the takeover of the Yankari National Park by the Bauchi state government which has made it to lose the status of a national park. The act establishing parks stipulates that it must be owned and managed by the Federal government. According to the Nigeria National Park Service Brochure (NPS) 2010, the remaining seven national parks in Nigeria include:

i. Kainji Lake National Park-New Bussa Niger State
ii. Chad Basin National Park located in Borno and Yobe State.
iii. Cross River National Park-Cross River State
iv. Old Oyo National Park- Oyo State
v. Gashaka Gumti National park – cut across Adamawa and Taraba State
vi. Okomu National Park – Edo State
vii. Kamuku National Park –Kaduna State.

The above listed parks were operated as either games or forest reserves until 1991 when the General Ibrahim Badamasi Babangida led Military Government promulgated Decree 36 of 1991, upgrading the first five to the status of a Park. In 1999, Okomu and Kamuku were added to the list after Degree 36 of 1991 was replaced with Degree 46 of 1999 (Now Act 46 of 1999) on May 26th 1999.

Some of the objectives of establishing national parks in Nigeria include:

- Conservation of selective representative samples of wildlife communities in Nigeri
- Protection of endangered vertebrate species.
- Protection of crucial wetland and water catchments areas.
- Promotion and provision of education about wildlife and nature conservation etc.

CULTURAL CONSERVATION

Before the advent of modern conservation methods in Nigeria, some cultural beliefs had helped a lot in conservation of some wild plants and animal species. Not all animals were either killed or eaten by some tribes. For instance, it was forbidden for a Bini man to kill or eat Buffalo, Bushbuck, Leopard, Chimpanzee, Tortoise and black Snake. The tortoise was used for curing of some deadly or terminal illness and as such was believed to be capable of bringing back such disease to the family if eaten. The Buffalo on the other hand was believed to be a harbinger of ill luck. Therefore, anyone who kills or eats it is doomed to ill luck and penury. Similarly, it was a taboo for an Urhobo man or an Itsekiri man to kill or eat Crocodile, monitor Lizard or Bushbucks. The Yorubas despised the Vulture bird.

The Tiv people of North Central Nigeria do not kill the green snake locally called "kyarem". According to them it was the "kyarem" that helped their forefathers to cross a river on their way from Congo to Nigeria by lying across the river hence enabling the people to cross over as bridge. Based on this belief, the green snake was highly protected by the Tivs.

Cultural beliefs had also helped in conservation of some plants species. Some culture forbids the cutting down of certain tree species. For example, an Iroko tree, Calorophone excelsa , was never cut by the Binis. It was believed to protect the Bini Kingdom against witches and wizard. Also Mbakwe (1986), observed that trees like Parkia biglobosa,prosopis africana, Elaes guinensis, Irvingia gabonensis etc, were highly protected in many parts of the country particularly in northern Nigeria due to cultural beliefs and traditions.

LAWS AND REGULATIONS TO PROTECT ENDANGERED PLANTS

The Convention on International Trade in Endangered Species of Wild Fauna and Flora

This Convention (more commonly known by its acronym) is an agreement between governments worldwide to ensure that international trade in wild plants and animals does not threaten their survival. CITES began in 1975 with 80 participating nations. Now, 169 nations participate and the agreement protects more than 30,000 species of plants and animals.

International trade in wildlife and plants is worth billions of dollars annually and includes hundreds of millions of plant and animal specimens. The trade is diverse, ranging from live plants and animals to a vast array of products derived from them, including foods, exotic leather goods, wooden musical instruments, timber, tourist curios, and medicines. The extensive international trade of some plants and animals, together with factors like habitat loss, threatens to deplete some to the point of near extinction.

Species protected under CITES are placed on one of three lists (appendices) depending on the degree that international trade threatens their existence:

1. Appendix I - includes species that are threatened with extinction and CITES generally prohibits commercial international trade in these species. However, trade may be allowed under exceptional circumstances, e.g. for scientific research. In these cases, trade may be authorized if permits are obtained from both the exporting and importing countries.
2. Appendix II - includes species that are not threatened presently with extinction but may become so unless trade is closely controlled. International trade in

Appendix-II species requires a permit from the exporting country; no import permit is necessary for these species. Permits are only granted if the exporting country is satisfied that certain conditions are met; above all, that trade will not be detrimental to the species' survival in the wild.

3. Appendix III - includes species that a country already regulates and the country needs the cooperation of others to prevent unsustainable or illegal exploitation. International trade in Appendix-III species requires a permit from the exporting country.

The U.S. Fish and Wildlife Service is responsible for implementing CITES in the United States. Information about CITES and permits is available at the .

Examples of plants protected from international trade under CITES include orchids, cacti, pitcher plants, some cycads and palms, ginseng, goldenseal, and some tropical timber trees. Some of these plants grow on national forests and grasslands where they are managed to ensure their perpetuation as viable populations and, where possible, for sustainable commercial use.

Altered fire regimes

Fire is a primary agent of recurrent disturbance in the seasonally dry landscapes of the SWAFR, consuming biomass and interacting with climate and geology to influence plant population dynamics and community assembly. As is the case for other fire-prone Mediterranean-type climate regions, plants in the SWAFR's fire-prone ecosystems possess traits that provide resilience to periodic burning. Variation in traits relating to survival, reproduction, dispersal, recruitment, competitive ability and longevity contribute to a diversity of responses and makes some species vulnerable to decline under particular fire regimes.

Consequently, changes in the fire regime relative to historical variability, brought about by patterns of human settlement and land-use, may result in species decline. Those species that are killed by fire and rely on seed germination after fire for population persistence are most at risk from fire interval extremes. These obligate-seeding species face two threats in relation to fire interval: immaturity-risk, where intervals are shorter than the time needed to establish a seed bank; and senescence-risk, where intervals are longer than the life span of plants and their seed bank.

Inappropriate fire regimes are identified as a major threat of growing importance to threatened plant species in the SWAFR. The nature of this threat is likely to vary across the landscape because of the influence that population density and patterns of land-use have on ignition and factors which affect flammability such as landscape fuel continuity.

For example, in peri-urban landscapes with moderate population densities, increases in ignitions and the frequency of fire at the wildland–urban interface, may expose threatened plant species to immaturity-risk. In contrast, in heavily cleared agricultural landscapes (see above section) with relatively low population densities, many fragments of native vegetation have a considerably lower probability of burning than analogous continuously vegetated landscapes because of low fuel continuity, reduced likelihood of ignitions and fire suppression, potentially exposing threatened plant species to increased senescence-risk.

Over half of the 242 species were obligate seeders, and consequently, have population dynamics particularly sensitive to fire interval and frequency. Their study also found differences in fire history across nine bioregions in the SWAFR and identified areas where current fire regimes may be detrimental to threatened flora. The greatest risk appears to be in the highly fragmented agricultural landscapes of the wheatbelt, where the absence of fire may be limiting opportunities for regeneration of obligate seeder species. In contrast, in some areas of continuous vegetation, fires may be too frequent for species with long juvenile periods, increasing the risk of local extinctions.

New approaches to fire management of threatened species based on their fire responses and other critical attributes such a juvenile periods and plant longevity are being developed to use fire specifically to conserve and regenerate threatened flora and their habitats in the SWAFR.

Landscape context and the extent of weed invasion are also important considerations when using fire to regenerate threatened species and their habitats. For example, in highly altered remnants of native vegetation or at the edges of larger remnants where there is nutrient enrichment and weeds are abundant, the introduction of fire will not be appropriate unless there is post-fire control of the weeds that flourish, smother and outcompete seedlings of native species.

GRAZING BY INTRODUCED AND NATIVE ANIMALS

Introduced animals (such as sheep, goats, or rabbits) and native animals (such as kangaroos, quokkas or possums) impact on threatened plant populations by damage through trampling and/or reducing plant biomass and plant reproductive ability through grazing. In addition, weed invasion can be facilitated through animal vectors and increased soil nutrients from animal droppings. As habitat is fragmented through land clearing there is a reduction in available habitat for native animals which can result in increased pressure on remaining areas of vegetation from grazing by native animals. Management of threatened flora impacted by grazing is primarily through fencing of sites to exclude herbivores.

ALTERED HYDRO-ECOLOGY

A consequence of the wide scale land clearing of the SWAFR is alteration of regional hydro-ecology. Vegetation removal at such a significant scale has resulted in saline ground water tables rising and increased salinity of the ground water (salts stored in the soils are dissolved into the ground water as it rises). Water logging and increased soil salinity has impacted on the native vegetation across whole water catchments threatening some otherwise stable species. Large scale revegetation plantings (e.g. Gondwana Link) and engineering works to redirect saline water (e.g. Toolibin Lake) are required across whole water catchments, to stabilise water tables and reverse hydrological trends, in order to prevent the extinction of many plant species.

DROUGHT/CLIMATE CHANGE

Superimposed on the threats outlined above is anthropogenic climate change. Mean annual temperatures in the SWAFR have increased during the 20th century, and since the 1970s there has been a significant decline in autumn and early winter rainfall. This decline is largely a consequence of changes to synoptic patterns and changes in sea level atmospheric pressure, which are likely to be due at least in part to increases in greenhouse gas emissions from human activities. The consensus among global climate models is that temperature will continue to increase and rainfall will continue to decline in the SWAFR. These changes are projected to have a substantial influence on species geographic ranges and impacts may be exacerbated by fragmentation and other threatening processes. Phylogeographic analysis of both common and rare species indicates that species have persisted *In situ* through cycles of aridity associated with Quaternary climate change in the region suggesting some level of resilience in the flora. Research on refugia and genetic adaptation are yielding important insights into vulnerability and providing a basis for management.

It is high time that the state and the central governments come forward for recovery of the rare, endangered and threatened (RET) tree species.

Before taking scientific measures to ensure their conservation and cultivation, it is very much necessary to identify them, assess their natural distribution and study their population status.

Pristine forests have been fragmented because of intense developmental activities over the past few decades. More than 100 tree species of high economic importance have become threatened and critically endangered in the Western Ghats.

Their small population size is considered to be the major threat. It cannot sustain them due to inbreeding and loss of genetic variability. It has become urgent to carry out scientific conservation programmes for recovering these species.

Unless urgent measures are taken to restore critically endangered species, they will become extinct. Species recovery is the process through which the decline of a threatened species is arrested or reversed and threats removed so that the survival of the species in the wild can be ensured.

Many countries have initiated plans to address the resurrection of the RET species. Species recovery programmes have been carried out successfully in the United States, Canada, United Kingdom and Australia.

In the United States, there is special legislation such as the Endangered Species Act, 1973, (ESA) for carrying out species recovery programmes. The Act, that was implemented in 1973, has provisions for listing the species as 'endangered', developing recovery plans for each species and designating critical habitats.

So far, 47 species have been stabilised through different recovery processes and have been excluded from the recovery programmes. A gradual increase in the population size, habitat restoration and captive breeding or population stabilisation have been achieved through recovery programmes.

TREE SPECIES IN THE WESTERN GHATS

India's 1.34 billion people exert a heavy pressure through encroachment, raising commercial plantations and other developmental activities. Due to anthropogenic activities like excessive harvesting and habitat destruction, many of the economically important tree species are under serious threat.

Syzygium travancoricum, an economically important tree species, is reported to exist with a population size of only 15-20 individuals. Similarly, *Dipterocarpus bourdillonii*, another endangered species has only 14 individuals occurring in three patches in the Kodagu district of Karnataka.

Since tree species require decades for regeneration of the optimum population, if there is lack of regeneration or habitat, their present population cannot be considered healthy.

Increased inbreeding because of limited pollen and seed dispersal flow caused by fragmentation of populations can impact regeneration of the species. In case of *Dysoxylum malabaricum*, an endangered tree species in the Western Ghats, inbreeding between related individuals has caused reduced regeneration. Hence, if urgent action is not taken to restore the population of these species, they may be irrecoverably lost. In India, recovery programmes for a few plant species have been taken up.

For instance, *Paphiopedilum druryi*, a slipper orchid, has multiplied through tissue culture and has been reintroduced in the Agasthiamalai hill ranges of the southern Western Ghats.

Out of the 387 Indian plants listed under the International Union for Conservation of Nature's Red List, 77 have been enlisted as 'critically endangered', six are 'extinct' and two are 'extinct' in the wild. The IUCN is an international organization working in the field of nature conservation and sustainable use of natural resources.

The 77 critically endangered species can be prioritised for recovery programme and the balance can be taken up subsequently. A systematic species recovery programme is thus the need of the hour to restore the populations of these species.

While taking earnest measures for conservation and recovery of RET species, collection of data about their population sizes, identification of the specific threats and developing mitigation strategies are to be attempted systematically.

The Vallanadu Black Buck Sanctuary and the Grizzled Giant Squirrel Wildlife Sanctuary in Tamil Nadu; Aghanashini Lion-tailed Macaque Conservation Reserve in Karnataka and Gibbon Sanctuary in Assam have been established for conserving specific taxa.

But, till date, no area has been specifically protected for any single endangered plant species, except for species groups like the Varsey Rhododendron Sanctuary in Sikkim and the Sessa Orchid Sanctuary in Arunachal Pradesh.

Combined and collective efforts are required on the part of the forest department as well as the forest stakeholders.

Promulgation of specific acts and framing of rules and regulations are urgently needed to protect the threatened species by reaching an agreement between the department and the stakeholders.

Documentation of the RET species, their threatened status, surveying all known populations and mapping their locations, identifying the extrinsic and intrinsic factors that drive the species to threatened status and assessing the genetic variability of the species are to be carried out scientifically on war footing.

Based on the above strategies, long-term monitoring programmes are to be developed for assessing the population changes periodically.

7

CHARACTERIZING FOREST DISTURBANCE REGIMES

INTRODUCTION

Reciprocal interactions between disturbances and forested landscapes have been a cornerstone of landscape ecological research and modeling for decades. A current frontier in this line of research is the spatially explicit investigation of disturbance interactions across spatial scales. Within forested ecosystems, natural disturbances, anthropogenic disturbances, and climatic and anthropogenic drivers interact across a range of spatial scales to shape forested landscapes in term of patterns, processes, and functions. Modeling disturbance interactions can be particularly challenging when and where the cumulative effects of such interactions, including non-linear and threshold behavior, result in catastrophic mega disturbance. Hence, a combination of empirical and modeling studies is needed to understand forest ecosystem dynamics that emerge from the interactions of multiple disturbances as well as biophysical and demographic drivers within forested landscapes.

Natural disturbances (e.g., wildfire and insect outbreaks) have traditionally been investigated separately within different sub-disciplines of ecology (fire ecology and entomology, respectively) and further separated from the effects of human disturbances, focusing therefore on the properties of "disturbance regimes" (patch size distributions, severity, frequency or rotation length) and with an emphasis on stochasticity. Despite the apparent stochastic nature of natural disturbances, disturbance regimes generally emerge from feedbacks between internal system processes and external drivers across scales in time and space.

Comparatively, anthropogenic disturbances (e.g., harvesting) have been traditionally viewed as deterministic (i.e., under human control). Yet, human systems are subject to analogous uncertainty and surprise caused by economics, social pressures, and political change that directly impact our ability to implement harvest and other land use plans at the temporal scale of forest rotations. Interactions among natural disturbances, human disturbances, and vegetation responses to those disturbances further influence system predictability. Proactive and adaptive management practices that embrace system uncertainty are therefore needed to respond to emerging "surprises" in disturbance regimes.

Several conceptual advances have been proposed to help disentangle the emergent properties of disturbance interactions. These conceptual advances underscore the interplay between pattern and process in natural disturbance dynamics, deterministic and stochastic elements of anthropogenic disturbance, and uncertainty due to climate change that need to be accounted for in modeling frameworks of processes and disturbances across scales. Here, we review the conceptual advances and empirical approaches that help disentangle the apparent complexity of disturbance interactions. A class of models known as forest landscape models (FLMs) dominated this field of study.

We therefore overview the general design of FLMs, showing how recent developments have shifted from statistically-based disturbance regimes (i.e., static) to process-based methods where disturbance regimes and ecosystem responses are emergent behaviors (i.e., dynamic), and further expanded the ability to choose the degree of system feedbacks depending on the question at hand. Studies were then categorized according to the specific disturbance interaction questions investigated (as clarified by recent conceptual advances) and the relative balance of static to dynamic model components across disturbance types. We argue for minimum standards in documentation – particularly as model complexity increases – for increased transparency and confidence in model results. We conclude highlighting the current limitations, frontiers, and directions in the understanding and modeling of disturbance interactions at landscape levels.

CLIMATE INFLUENCES ON FOREST DISTURBANCES

A review of how each disturbance is influenced by climate, affects forests, and might be exacerbated by climate change provides a background for examining ways to cope with the impacts of climate change. The effects of each disturbance are partly tempered by prior adaptations. For example, species present in a forest reflect past disturbances. Droughty sites typically support species that survive well under dry conditions with uncertain rainfall. Sites that have frequent fires contain gymnosperm species with serotinous cones. Thus, if climate change alters the distribution, extent, frequency, or intensity of any of these disturbances, large impacts (such as loss of species regeneration) could be expected. The effects on species or communities already at the margin of their range may be particularly severe.

FIRE

The frequency, size, intensity, seasonality, and type of fires depend on weather and climate in addition to forest structure and composition. Fire initiation and spread depend on the amount and frequency of precipitation, the presence of ignition agents, and conditions (e.g., lightning, fuel availability and distribution, topography, temperature, relative humidity, and wind velocity).

Fire effects on forests include acceleration of nutrient cycling, mortality of individual trees, shifts in successional direction, induced seed germination, loss of soil seed bank, increased landscape heterogeneity, changes in surface-soil organic layers and underground plant root and reproductive tissues, and volitalization of soil nutrients. Erosion can occur where soil disturbance accompanies fire (e.g., during fire fighting or timber salvage operations). Fire affects forest value for wildlife habitat, timber, recreation, and, through smoke, human health.

The rapid response of fire regimes to changes in climate can potentially overshadow the direct effects of climate change on species distribution and migration. Modeling results predict great variation in future fire–weather patterns for the northern portion of North America. The seasonal severity rating (SSR) of fire hazard increases over much of North America under both the HADCM2SUL and the CGCM1 scenarios. The wetter Hadley scenario produces some small decreases in SSR for the Northern Great Plains, and increases are generally less than 10% over most of the rest of the continent. Some fire history studies suggest that the frequency of fire can decrease despite warmer temperatures because of increased precipitation.

The warmer and drier CGCM1 produces a 30% increase in SSR for the southeastern United States and Alaska, with about 10% increases elsewhere. These scenarios suggest an increase in fire intensity and a 25%–50% increase in the area burned in the United States. In addition, recent results from the MC1 model, which is described, show an increase in area and biomass burned under both scenarios. This model includes an interaction with CO_2 concentrations, which, through increased CO_2 fertilization and increased water-use efficiency, produces more biomass and thus more fuel, contributing to more and larger fires under a highly variable climate that has dry years interspersed with wet periods.

DROUGHT

Droughts occur in nearly all forest ecosystems. Drought effects are influenced by soil texture and depth; exposure; species present; life stage; and the frequency, duration, and severity of drought. Droughts occur irregularly in forests of the humid regions east of the Mississippi River and in the superhumid Pacific Northwest. Droughts occur annually at the end of the growing season in forests at the midcontinental prairie–forest border, where annual precipitation ranges from 600–1000 mm, or within humid regions that have shallow or rocky soils. Seasonal summer droughts are experienced by western interior dry forests that depend on winter precipitation, such as forests in the semiarid plains and intermountain regions of the western United States. In some regions, droughts last several years.

The primary immediate response of forests to drought is to reduce net primary production (NPP) and water use, which are both driven by reduced soil moisture and stomatal conductance. Under severe conditions, plants die. Small plants, such as seedlings and saplings, are usually the first to die and can succumb under moderate conditions. Deep rooting and stored carbohydrates and nutrients make large trees susceptible only to severe droughts. Secondary effects also occur. When reductions in NPP are extreme or sustained over multiple growing seasons, increased susceptibility to insects or disease is possible, especially in dense stands. Drought can also reduce decomposition processes, leading to a buildup of organic matter on the forest floor that may increase fire frequency or intensity or reduce nutrient cycling.

The consequences of drought depend on annual and seasonal climate changes and on whether the current drought adaptations are sufficient to confer resilience to new conditions. Forests tend to grow to a level of maximum leaf area that nearly fully uses soil water during the growing season. A small increase in growing-season temperature could increase evaporative demand, triggering moisture stress. New results from two models, MAPSS and MC1, suggest that this mechanism may cause future increases in drought stress in the Southeast, southern Rockies, and parts of the Northwest. The MC1 model indicates that the Prairie Peninsula and Great Lakes region, parts of the Northwest, and the Gulf Coast could experience drought stress within two decades, even though these regions may become wetter in later decades.

INSECT AND PATHOGEN OUTBREAKS

Climate influences the survival and spread of insects and pathogens directly, as well as the susceptibility of their forest ecosystems. Changes in temperature and precipitation affect herbivore and pathogen survival, reproduction, dispersal, and distribution. Indirect consequences of disturbance from herbivores and pathogens include elimination of nesting trees for birds and negative effects on mycorrhizal fungi. Other indirect effects include the impacts of climate on competitors and natural enemies that regulate the abundance of potential pests and pathogens.

Changes in the intensity and frequency of herbivore and pathogen damage in forests can have a range of effects. Most tree species support a community of other organisms, so the loss of any tree species can significantly reduce overall biodiversity. Such a loss occurred when chestnut blight almost completely eliminated chestnut trees; the die-off of Fraser fir (*Abies fraseri*) caused by balsam woolly adelgid (*Adelges piceae*) also raises concerns. Herbivore and pathogen damage to trees can increase understory plant diversity; the overall abundance and diversity of birds; and the diversity of predators, parasitoids, and detritivores.

Because climate change can both directly and indirectly affect herbivores and pathogens through various processes, the ultimate effects on patterns of disturbance include increased disturbance in some areas and decreased disturbance in others. For example, an increase in the interannual variation in minimum winter temperatures is expected to favor more northerly outbreaks of southern pine beetles but could reduce more southerly outbreaks. Similarly, decreased precipitation and increased evapotranspiration should boost tree secondary chemical metabolism (and, therefore, resistance to pests) in forests that currently suffer modest growing-season water deficits.

If global warming shifts species abundances, there may be associated shifts in herbivory. Compared to the cooler Paleo cene, the Eocene had a greater diversity of herbivores and higher attack rates on the most abundant tree species. Increased warming would most likely increase the diversity of insects at higher latitudes. Because insects typically migrate much faster than trees, many temperate tree species are likely to encounter nonnative insect herbivores that previously were restricted to subtropical forests.

INTRODUCED SPECIES

Introduced species can affect forests through herbivory, predation, habitat change, competition, alteration of gene pools via hybridization with natives, and disease (as either pathogens or vectors). Introduced species can alter the diversity, nutrient cycles, forest succession, and fire frequency and intensity of some ecosystems. The effects of introduced species should be considered concurrently with changes in native species distribution and abundance that occur as a consequence of climate change. The impact of introduced species on ecosystems is influenced by such climatic factors as temperature, drought, and cloud cover. Invasion biology is not yet adept at forecasting impacts of invasions. The complex interactions among introduced species, native communities, managed and intensely harvested forests, and climate change compound this forecasting problem.

The ultimate ranges of introduced species are largely determined by climate and human activities. Climate change will modify the distributions of many introduced species. Developmental rates will be modified by temperature change. For example, laboratory studies of balsam woolly adelgid growing under various temperature conditions provided the basis for simulations that suggest that temperature-induced changes in the population dynamics of the insect significantly affect Fraser fir survival.

The great majority of introduced species do not survive. Many fail because the climate is unsuitable at their points of arrival. Thus, a changed climate will lead to a different mix of surviving and failing species. In general, one might expect a larger fraction of survivors when the climate is warmer; introduced species comprise a far larger fraction of the biota in the warmer areas of the United States.

Increased CO_2 can directly influence introduced plants through enhanced photosynthesis, but at different rates for different species. Resistance of trees to introduced herbivores is sensitive to both climate and CO_2 concentrations. Climate change, in concert with CO_2 concentration and nitrogen deposition, affects leaf nitrogen, which in turn influences herbivory.

HURRICANES

Hurricanes disturb forests of the eastern and southern coastlines of the United States, as well as those of the Caribbean islands and the Atlantic coast of Central America. Ocean temperatures and regional climate events influence the tracks, size, frequency, and intensity of hurricanes. An average of two hurricanes make land every 3 years in the United States. Global warming may accelerate the hydrologic cycle by evaporating more water, transporting that water vapor to higher latitudes, and producing more intense and possibly more frequent storms. However, other variations may override possible increases in hurricane frequency.

Changes in the global hydrologic cycle and temperature will influence hurricane formation, but we cannot yet predict the direction and magnitude of change. Sea-surface temperatures are expected to rise, with hotter temperatures expanding to higher latitudes. Most studies point to an increase in hurricane frequency. However, even if frequency does not increase, it is likely that intensity and possibly duration of individual storms will increase because of the warming of the air and ocean, sources of energy for a hurricane.

The effects of hurricanes on vegetation include sudden and massive tree mortality, complex patterns of tree mortality (including delayed mortality), and altered patterns of forest regeneration. These changes can lead to shifts in successional direction, higher rates of species turnover, and opportunities for species change in forests, which can in turn increase landscape heterogeneity, produce faster biomass and nutrient turnover, and result in lower aboveground biomass in mature vegetation. Hurricanes can also result in buried vegetation and carbon sinks.

WINDSTORMS

Small-scale wind events are products of mesoscale climatic circumstances and thus may be affected by climate changes, although the type and amount of alteration in windstorm characteristics cannot be predicted because these smaller-scale events are below the resolution of today's GCMs. Yet, tornadoes, downbursts, and derechos (a series of storm cells along a squall line) are probably the most important agents of abiotic disturbance to eastern deciduous forests. These disturbances can create very large patches of damage: A windstorm on 4 July 1999 in the Boundary Waters Canoe Area of Minnesota flattened roughly 250,000 acres of forest.

Windstorms can cause heavy mortality, produce canopy disruption, reduce tree density and size structure, and change local environmental conditions. Consequently, the disturbance may trigger advance regeneration, seed germination, and accelerated seedling growth. These effects can change successional patterns, gap dynamics, and other ecosystem-level processes. The relationship between wind strength and severity of disturbance is not constant across different forests and species; although shallow-rooted species and thinned stands may be especially vulnerable to wind events, multiple factors influence tree response to high winds.

Researcher suggests that increased intensity of all atmospheric convective processes will accelerate the frequency and intensity of tornadoes and hailstorms. Consistent with this view, researchers found that the proportion of precipitation occurring in extreme thunderstorm events increased in the United States from 1910 to 1990, and researchers further suggest that the climate of the United States has become more extreme (in terms of temperature and precipitation anomalies) in recent decades. Thus, it appears that the thunderstorm conditions that contribute to tornado formation have increased and are likely to continue increasing under projected climate changes. Furthermore, researcher found a positive correlation between monthly tornado frequency and mean monthly temperature in western Canada, and inferred that this relationship suggests increased tornado frequency under a warmer climate scenario. Despite these inferences about tornado frequencies and the direct data on thunderstorm trends, understanding of tornado genesis is still inadequate to allow a direct forecast of how climate change will affect the frequency or severity of windstorms in the next century.

Chronic effects of wind on trees: acclimative responses

The effects of wind on trees range from chronic to acute, incorporating a spectrum of sub-lethal to lethal effects. Airflow across the surface of leaves facilitates transpiration and the exchange of carbon dioxide and oxygen between the leaf and the atmosphere. Short-duration displacements of branches and leaders by wind leads to thigmomorphogenetic responses such as formation of flexure wood, increased secondary thickening of stems and structural roots, and reduced shoot extension. These growth responses produce shorter, thicker and, therefore less slender, trunks and branches which better resist deflection, and improve root anchorage.

Trees are self-designing structures that experience increasing loads as they grow larger. They also compete for light and soil resources with neighbouring plants, and must allocate photosynthates efficiently. This optimization leads to the conformation of trees with some general design principles. Bending of a tree under its own weight, or because of lateral loading by wind, produces stress in the outer fibres and the cambium. The portion of photosynthate allocated to structural increment would be most efficiently used if the cambium produced new wood in a manner that equalizes the distribution of stress along the outer surface of the stem, and this is known as the 'constant' or 'uniform stress hypothesis'. While there is observational and experimental evidence in support of this hypothesis, there is ongoing debate on the universality of the theory. Researchers note that while it is logical to assume that natural selection has lead to biomass partitioning patterns that achieve a balance between bending and counter-resisting moments, leaf, stem and root biomass partitioning patterns must also accommodate the simultaneous performance of other functions.

In combination, the growth responses in roots, stems and branches to deflection and bending enable tree crowns to maintain what researcher refers to as a genetically determined equilibrium shape – where wind exposure is not excessive. Under higher, directional, sustained winds, the spectrum of responses extends to the formation of reaction wood, asymmetric growth of stems and root systems and flagged (windswept) crowns. In examining the literature concerning windswept crowns, researcher distinguishes between redirection of plant form under constant unidirectional wind (a biophysical response) and mechano-perception coupled with thigmotropism (a growth response). He concludes that redirection of branches is the more probable mechanism for the windswept crown form.

The terms 'acclimative growth' and 'adaptive growth' are used by windthrow researchers in referring to the process by which acclimation to wind and other mechanical loading develops in trees, and the result of this process. Acclimation to wind loads includes thigmotropism, deformation and sub-lethal mechanical impacts (abrasion, failure) and their collective effects on the form and biomechanical properties of roots, stems and branches. Since deformation and loss of tissues via impact do not constitute growth, it would be more accurate to use the broader term 'acclimative responses' to describe the full spectrum of effects.

Acute effects of wind on trees and stands

If wind loads exceed the resistance of stem or root/soil systems, trees break (stem snap, windbreak) or uproot. These modes of failure are often lethal, but trees may survive, particularly where they are capable of resprouting. In either case, this disturbance affects stand and soil conditions. Uprooting and stem breakage are incorporated in the term 'windthrow' or more colloquially 'blowdown' and 'windfall'. In the case of hurricanes, there are gradients of wind speed across the hurricane track, with resulting gradients of damage to trees and stands. These gradients provide insights into the hierarchy of tree and stand-level responses to increasing acute wind speeds.

Stem or root system failure may be preceded during extreme wind events by partial defoliation or branch failure. The loss of foliage and branch shedding may be an adaptive strategy in storm-exposed landscapes, since it reduces drag and the potential for stem breakage and uprooting. Partial or complete failure of branches also occurs as crowns of neighbouring trees collide during more routine wind events. Periodic pruning of branches or branch tips via collision can lead to gaps between adjacent tree crowns, termed crown shyness, and has implications for tree growth and understory development. At higher wind speeds, as trees within stands begin to uproot or experience stem breakage, they may knock off branches, break or uproot adjacent trees as they fall through the canopy. The extreme winds and flying debris during hurricanes and tornadoes can strip the bark from stems.

Stand-level damage ranges from the creation of small canopy gaps via failure of individual or small groups of trees, through intermediate levels of failure of overstory trees, to complete and extensive failure of the overstory canopy. Damage patterns within stands can be variable or uniform. The propagation of damage through a stand during a high wind event, and the uniformity of this damage, depends in part on the uniformity of trees within the stand, along with uniformity in terrain and soil conditions that affect wind exposure and root anchorage.[6,38]

Windthrow mechanics, stand-level instability

Windthrow results when the wind-induced drag force on the tree crown, multiplied over the lever-arm of the stem, results in a turning moment that exceeds the bending resistance of the stem, or the root anchorage. Tree design appears to balance core rigidity with peripheral flexibility. The bending and reconfiguration of branches, and consequently tree crowns, during high winds streamlines foliage and reduces the frontal area.

Over the range of wind speeds at which this reconfiguration occurs, drag on tree crowns is directly proportional to wind speed rather than to the square of wind speed, as is the case for solid bodies. The resistance of the stem to failure is proportional to the cube of the stem radius, so small increases in radial growth add substantially to stem strength.

Researchers found that the critical moments for trees that fail via stem breakage were similar to the critical moments for trees that failed via uprooting for trees of similar size and form. Root anchorage is a function of root architecture and soil properties, and is empirically very tightly correlated with stem mass. The effects of acclimative growth on tree stability in Sitka spruce plantations in Britain were examined by researchers. who found that anchorage increased as wind exposure increased.

The stability of trees is affected by precipitation and temperature. Soil moisture affects root anchorage, and uprooting often occurs when high winds coincide with saturated soils. Researchers found that all of their winched trees failed via stem breakage when soils were frozen, and with higher critical moments than for similar sized trees winched during non-frozen conditions.

Above-canopy wind speeds attenuate rapidly within dense stand canopies. While open-grown trees are exposed to and acclimate to routine wind events, stand-grown trees are partially sheltered by neighbouring trees. Stand grown trees are also in competition with these neighbours for growing space, and photosynthate is allocated preferentially to shoot growth and fine root production, and in diminishing priority to storage, radial increment and protective chemicals. Shade-grown trees allocate fewer resources to radial increment than trees grown in full sunlight, even when mechanical loads are artificially increased. In stands grown at high densities, stems become increasingly slender, crown centre of gravities shift higher and higher and individual trees become increasingly unstable. Crop planning tools such as stand growth models or stand density management diagrams enable projection of height and diameter growth, slenderness, and in some cases crown length. These can be coupled with windthrow models to enable prediction of stand stability for different planting density or thinning regimes.

At some critical stand height, there is the potential for uniform stands to become mechanically unstable. This likely reflects both the increased leverage for tall, high crowned trees, but also the potential for trees to domino fall during damage events rather than simply falling into and being supported by their neighbours, as happens in short, dense stands. The mechanics of this are demonstrated in mechanistic windthrow models, albeit without representation of tree collisions. The phenomenon of critical stand height has been observed in planted Sitka spruce (*Picea sitchensis* (Bong.) Carr.) in the UK, and can be observed in natural stands of black spruce (*Picea mariana* (Mill.) BSP) in eastern Canada, mountain beech (*Nothofagus solandri* var *cliffortioides* (Hook.) f.) in New Zealand, and likely underlies the recurrent stand-replacing windthrow observed in even-aged stands of southern beech (*Nothofagus pumilio* Poeppig and Endl.) in Tierra del Fuego and western hemlock (*Tsuga heterophylla* (Raf.) Sarg.) and associated species in coastal Alaska.

ICE STORMS

Ice storms are caused by rain falling through subfreezing air masses close to the ground; those air masses supercool the raindrops, which freeze on impact. Ice accumulation can vary dramatically with topography, elevation, exposure, and areal extent of the region over which conditions favor glaze formation. Ice storms occur throughout the United States except along the southwestern borders and parts of the plains, but the frequency and severity of ice storm events increase toward the northeastern US borders. However, the historic record of ice-storm events over large areas has not been consistent or precise, with rigorous measurements of ice accumulation.

Ice storms affect trees, forests, and forested landscapes in different ways. Ice damage to trees can range from severing a few twigs, to bending stems, to moderate crown loss, to outright breakage of trunks. Depending on stand composition, amount and extent of ice accumulation, and stand history, damage to stands can range from light and patchy to total breakage of all mature stems. Effects on forest stands include shifts in overstory composition in favor of more resistant tree species, loss of stand growth until leaf area is restored, and damage to stem form. Damaged stems are then more susceptible to the impacts of insects and disease. Recently thinned stands can be highly vulnerable because crowns have spread into the new space but branch strength has not developed. Several tree species can survive within areas frequented by ice storms. Though weather conditions producing ice storms are well understood, it is unclear how changes in climate will affect their frequency, intensity, regional location, or areal extent. However, atmospheric warming will most likely shift the locations of prevailing ice storms northward.

LANDSLIDES

Both slow and rapid movements of soil, rock, and associated vegetation are triggered directly by climate factors and indirectly by climate-influenced processes (e.g., stream-bank erosion) and by nonclimate factors such as earthquakes and volcanism. Triggering climatic events include snowmelt and intense rainfall, including that associated with hurricanes. Landslide frequency and extent are influenced by precipitation amount and intensity; snow accumulation, melt rate, and distribution; and roads and other land uses. The potential for a site to slide is influenced by slope steepness, properties of soil and rock, and hydrologic factors. Vegetation influences the likelihood of sliding through the soil-stabilizing effects of root systems and the effects of vegetation structure and composition on hydrology. Landslides remove soil and vegetation from steep slopes and damage forests on gentler slopes where landslide deposits come to rest. Landslides in forest landscapes can also damage aquatic resources and threaten public safety. Yet it is important to recognize that landslides are natural components of terrestrial and aquatic ecosystems.

Climate-change effects on landslides reflect changes in the delivery of water to soils through altered precipitation and snow hydrology. Because climate change is expected to vary geographically and with elevation, landslide responses will vary with similar complexity. Landslides are expected to be less frequent in areas where GCM scenarios predict reduced overall precipitation or reduced snowmelt because of warming trends, limiting snow accumulation. In the Pacific Northwest, much of the small, rapid landsliding occurs during rain-on-snow events in a broad elevation band where snow accumulates and melts several times in an average year. A simple warming without change in overall annual precipitation would be expected to result in reduced sliding by limiting the amount of snow (and its associated snowmelt) available to augment the rainfall reaching the soil. The most socially and ecologically significant landslides are triggered by intense precipitation. Thus climate change that increases storminess, and hence soil saturation, will increase landslide occurrence.

INTERACTIONS AMONG DISTURBANCES

Many disturbances are cascading. Drought often weakens tree vigor, leading to insect infestations, disease, or fire. Insect infestations and disease promote future fires by increasing fuel loads, and fires promote future infestations by compromising tree defenses. Increased fire intensity or extent would enhance the potential for landslides. Also, changes in land use, forest management, and atmospheric chemistry can interact with these natural disturbances. For example, harvest and road establishment in landslide-prone areas coupled with increased wetness could result in more landslides. In the southern Appalachians, ozone exposure coupled with infestations of exotic insects and climate change may increase Fraser fir mortality and red spruce stress. In some cases, however, the combination of disturbances may ameliorate impacts. Under droughty conditions, stomata tend to close, reducing the effects of high ozone exposure.

Nevertheless, when ecosystems experience more than one disturbance, the compounded effects can lead to new domains or surprises. A new domain is entered when the system has not recovered from the first disturbance before a second perturbation occurs, leading the system to a new long-term condition. For instance, the combination of climatically driven wildfires, fragmentation caused by agricultural settlement, and logging in the boreal forest has resulted in significant and unprecedented changes in forest composition. Invasive nonnative species are sometimes able to modify existing disturbance regimes or introduce entirely new disturbances. Under climate change, these compounded interactions may be unprecedented and unpredictable. They are likely to appear slowly and be difficult to detect because trees live for so long.

STRATEGIES FOR DEALING WITH FOREST DISTURBANCES

Coping strategies for forests are influenced by the value of the forest, the naturalness of the disturbance, and the range of acceptable management options. Often the least ecologically disruptive response after a disturbance is no action at all, but managers or society usually call for some type of cleanup or restoration, even when such action may retard recovery. The value and management goals for the forest dictate how many resources can be allocated to its management. These values can change, as is illustrated by the revision of burn policy to recognize fire as a natural part of forest development that should not always be controlled.

The ability to manage for these eight disturbances varies greatly. However, current understanding of the disturbance nearly always provides some guidance for management under a future changed climate.

Coping strategies for one disturbance type are often appropriate management responses to other disturbance types. For example, the removal of dead or dying trees and downed woody debris can reduce the risk of fire as well as alter insect and disease dynamics. Density management can reduce drought stress as well as alter insect population dynamics, but it could make forests more susceptible to wind. Thus, management effects are not always positive.

Strategies for coping with disturbances in forests may also vary regionally. No matter where they are carried out, however, these practices often take 50–100 years to convert a landscape, and they are difficult to implement on inaccessible sites or in reserves.

We organize the coping strategies into several categories: managing before the disturbance, managing the disturbance itself, managing the recovery, and monitoring for adaptive management. These options are presented independent of climate-change effects but with the understanding that climate may alter the disturbance regime.

Managing before a disturbance

Before a disturbance occurs, forests can be managed to reduce vulnerability or to enhance recovery. In both cases, management actions can alter the structure or the composition of the forest. In situations where the goal is to reduce the chance of future disturbances, adjustments to forest structure can be useful. For example, species or individual trees susceptible to ice or wind storms can be removed, as is common in cities. In addition, tree spacing and density can be altered to reduce susceptibility to drought.

However, dead woody debris has numerous benefits, and its extensive removal can affect the biota and nutrient cycling. Managers can also change species composition to reduce the vulnerability of forests to disturbances. Tree species that are less vulnerable to fire, droughts, wind, insects, or pathogens can be planted or maintained.

For example, the colonization of phloem-feeding insects, such as bark beetles, is partially controlled by the ability of the tree to produce oleoresin, which is under genetic control. So, planting selected tree species and genotypes with relatively high oleoresin could limit insect outbreaks.

Landscape structural changes can also reduce the chances that future disturbances will damage the forest. The pattern of clear-cutting influences the potential for windstorms to blow down trees, because destructive winds are more prevalent along the edge of a cut. And the placement of roads can influence the likelihood of future landslides and the spread of wildfire.

Management can be designed to reduce the opportunity for disturbance to occur. Examples are regulations that limit the introductions of nonnative species, the imposition of burning restrictions, and the use of controlled burns to reduce fuel loads. Trees can be planted that are less susceptible to disturbance. Species that promote disturbances can be removed.

Density of trees can be managed to reduce the potential for future insect outbreaks or storm damage. Finally, roads can be designed to reduce the potential for landslides.

Other management actions can enhance forest recovery. Forest structure can be modified to speed up the successional process in the event of a disturbance. Alternatively, species composition can be adjusted to promote recovery. For example, in areas likely to experience a disturbance, trees with salvage value can be planted.

Managing the disturbance

Some disturbances, such as fire, insects, disease, and drought, can be managed during the disturbance through preventive measures or manipulations that affect the intensity or frequency of the disturbance. Alternatively, the disturbance can be managed to reduce its impact. A common way to control outbreaks of the southern bark beetle is to be on the alert for sites experiencing some beetle damage, then to cut those trees quickly to reduce the size of the area affected. Fire control is another example of a management action to reduce the impact of a disturbance.

Managing the recovery

Recovery efforts can focus either on managing the state of the system immediately after the disturbance (e.g., salvage logging) or managing the ongoing process of recovery (e.g., planting and reseeding). Recovery efforts need careful consideration of the long-term impacts because such actions can damage soils and residual trees. Stands can recover naturally without any removal of the dead or damaged trees.

Recovery actions can be designed to speed recovery. In the aftermath of a disturbance, recovery can be enhanced by adding structural elements that create shade or other safe sites necessary for reestablishing vegetation or that serve as perches for birds (and thus places where seeds would be dispersed). Alternatively, late successional species can be planted to speed up succession. Finally, additions of water or nutrients can reduce environmental stress and facilitate restoration. Recovery can also be managed to reduce vulnerability to future disturbances.

Monitoring for adaptive management

A monitoring program should be used to determine how disturbances affect forests and to continually update our understanding of how climate change is potentially influencing the disturbance regimes. Monitoring can be designed to measure the state of the forest with and without disturbance under different management activities or to identify potential risks of forest disturbances. Such information is used to inform management of the potential outcomes of management actions.

Although many coping strategies associated with these disturbances could be incorporated into current forest-management practices regardless of climate change, the potential changes in climate may create a novel disturbance. For example, climate change may allow the migration of nonindigenous species into a forest, and current understanding of interactions and coping strategies may not apply to the resulting competitive interactions between nonindigenous and native species. Adaptive management approaches management as a continual learning process. The continued monitoring of ecosystem structure and function could be part of the coping strategy to address the likely surprises. The impacts of insects and pathogens are already monitored through the Forest Health Program, and weather and fuel moisture are monitored to assess the risk of fire during the fire season. However, few surveys quantify the extent and severity of damage from wind and ice storms or landslides.

Information from monitoring programs could be used to update risk assessments in management plans and prescriptions in an adaptive-management sense. A risk-ranking system could identify aspects of the forest most susceptible to disturbance under a changing climate. In conjunction with spatially explicit modeling of the site under various scenarios of disturbance impacts, a risk map could be created to identify sites most in jeopardy.

RESEARCH NEEDS

A key feature of this analysis is the realization of our lack of knowledge in many critical areas. The numbered aspects of Figure 4 depict major interactions about which more information is needed. We determined the key research needs for each disturbance and then organized the questions that must be resolved into the six topics discussed below. Such research will lead to better management decisions.

UNDERSTANDING CLIMATOLOGICAL CONDITIONS THAT INITIATE DISTURBANCE

Accurate projections of climate effects of disturbances require improved climate and weather forecasts. The projections should include not only average climate conditions but also their range and variance. Short-term weather forecasts will be needed to predict drought occurrences for existing forests. For long-term climate change projections, improved resolution in climate models is needed so that regional patterns can be projected.

We have limited understanding of what climatological conditions lead to some disturbances. Improved understanding of local meteorological events that spawn tornadoes is needed, as well as improved projections of conditions that foster thunderstorms. Our ability to predict the occurrence of fires and hurricanes has benefited from research that allows managers to focus their attention on sites most likely to be disturbed.

However, some disturbances result from interactions between ecological and climatological conditions that are often poorly understood. For example, better monitoring is needed to improve the characterization of ice accumulation in relation to storm characteristics and associated weather, especially the delineation of areas by amount of ice accumulation.

Once the relationship between climate and disturbances has been quantified, more-accurate predictions of disturbances can be developed to minimize their impact.

UNDERSTANDING THE EFFECTS OF DISTURBANCES ON MICRO-CLIMATE

Because land-cover patterns can affect atmospheric circulation and cloud formation, changes in forest structure in the aftermath of fire, wind or ice storms, hurricanes, landslides, drought, and pest outbreaks may alter weather or climate conditions. This interaction needs to be studied and better understood.

Quantifying impacts of disturbances on forests

There is a paucity of basic information on the frequency, intensity, and spatial extent of some disturbances and their impacts on forests. This problem is especially severe for landslides, ice storms, and small wind events. For example, reconstructive studies should be done to determine the long-term influence of successive ice storms on forests. Such analysis also allows exploration of interactions between disturbances and delayed responses.

Research should identify herbivores and pathogens that are likely to be key agents of forest disturbance in the next 50 years. Integrated continental surveys are needed to determine the sensitivity of different types of pests and diseases to environmental change and the potential for increased outbreaks of insect herbivores and pathogens at the margins of their existing ranges.

INTERACTIONS BETWEEN FOREST DISTURBANCES AND MANAGEMENT

Our ability to manage forests now as well as under climate change rests on our understanding of how forests respond to multiple disturbance events. A better understanding of interactions among fire, hurricanes, and biological disturbances (such as insects, pathogens, and introduced species) would improve our long-range predictions about forest succession and ecosystem dynamics and would lead to better prediction of conditions under which one event would affect the response to a subsequent one. This understanding, however, is complicated by the diverse goals of forest management (e.g., fiber products, wildlife habitat, biodiversity, and recreation).

Some management practices have been developed to cope with the physical disturbances of droughts, hurricanes, and wind events. However, additional research could expand options for management. Research is needed on the mitigation of hurricane impacts (i.e., how to hurricane-proof landscapes and how to design protected areas, for example, determining what their area, shape, and distribution should be).

Forest ecologists and land managers are exploring the prospects for tailoring forest management regimes to the range of ecosystem conditions and wildfire disturbance regimes observed in the past, and in some cases to those anticipated under future climate conditions. For drought, new field experiments could test forest sensitivity to specific climate-change projections in combination with changes in the concentration of atmospheric trace gases. How the genetic diversity of host plants will determine the future epidemiology of forest pathogens needs further exploration. Critical evaluations of known patterns of species change and yield following past climate changes are needed, along with models of succession that incorporate disturbance processes.

Conclusions

Over geologic time, changes in disturbance regimes are a natural part of all ecosystems. Even so, as a consequence of climate change, forests may soon face rapid alterations in the timing, intensity, frequency, and extent of disturbances. The number and complexity of climate variables related to forest disturbance make integrated research an awesome challenge. Even if changes cannot always be predicted, it is important to consider ways in which impacts to forest systems can be mitigated under likely changes in disturbance regimes. The task for the next decade is to understand better how climate affects disturbances and how forests respond to them. Improved monitoring programs and analytic tools are needed to develop this understanding. Ultimately, this knowledge should lead to better ways to predict and cope with disturbance-induced changes in forests.

8

SEED BANK IN FOREST REGENERATION

A seed bank is a place where seeds are stored to preserve genetic diversity for the future. They are usually flood, bomb and radiation-proof vaults holding jars of seeds from different plant species.

The seeds are typically kept at low humidity and in cold conditions - around -20°C. This helps to preserve the seeds, ensuring they can still grow when they are needed later.

Seed banks have played the biggest role in the conservation of endangered wild plant species in the world. The potential role of seed in the regeneration of plant species is threatened by a rapid decline in the longevity and viability of the seeds in their natural habitats. This study addresses the key role of seed banks in the conservation of floristic diversity. The increasing popularity of the seed bank as a tool in the conservation of wild plant species can be attributed to different significance, including; provision of immediate access to plant samples, allowing researchers and conservation biologists to evaluate them for properties such as new sources of medicines, nutrition, and genes. Also, wild plant seeds conserved in seed banks are immune to habitat destruction, diseases, and predators. The stored seeds can be used to reinstate species into existing, suitable habitats where they were once present, or to augment the diversity of small, genetically depleted populations.

Conservation efforts, as sources for population reintroduction and restoration, have been cited as an important justification for seed banks. Thus, it is recommended that seed banks with state-of-the-art storage facilities should be established at the national and local levels; and all of such banks should be networked so that materials, knowledge, and expertise on particular wild plant species is available on a global scale.

Most, if not all, of our hardwood trees have reproductive processes, in addition to seeds, which allow them continue to exist on a site in the woods. One such strategy is sprouting. While the ability can decrease as the tree ages, dormant buds at the root collar (where the stem connects to the roots) can sprout in response to damage or disease. We see a lot of stump sprouts after a tree has been harvested, and many of the trees in the current forest have sprouts as their start - red maple is rather successful at this strategy.

Other trees can send sprouts up from their root networks - beech brush is an example of root sprouts. Another strategy used by some species is the ability to "layer." As a branch curves down to the ground, it may become covered by leaves and duff, eventually enabling that branch to grow roots and create a separate clone that will survive if something happens to the parent tree - some spruces have this ability and a few non-native shrubs as well. And then we've got a few plants that negatively impact our forests which can grow new plants from root fragments or rhizome fragments (rhizomes are subterranean stems that can send out both new roots and new stems - trees don't have rhizomes, but problematic plants like Japanese knotweed do).

An important consideration for all of these "vegetative" reproduction strategies is that the new tree is a clone of the parent. If the parent is lost due to insect or disease, the sprouts will have the same susceptibility.

Seeds are a tree's potential offspring from sexual reproduction strategies. If you'll remember from your high school biology classes, genes are given to the seed from each parent. These genes will express differently as they combine, leading to different physical characteristics in an individual. While some tree species can self-pollinate, wind and insects are the prime distributors of tree pollen to the female flowers that will result in seed production.

Just as trees have different strategies to pollinate their flowers - wind and pollinators - so too do they have different strategies to disseminate the seed. Wind is an important seed disperser for light seeds with wings (maples, ash, yellow poplar, pines, black birch); gravity also does its part to move seeds. Overland water flow is responsible for movement of lighter seeds. Small mammals move larger seeds as they cache food for winter survival; however, many squirrels and chipmunks tend to take bites out of seeds, limiting their ability to start to grow. Blue jays are winners in distributing oak seeds - they have been recorded moving acorns several hundred to thousands of yards from parent trees. As humans, we also play a role in seed distribution - think about those seeds that attach to your clothing, or land on your car to blow off at random times.

Different seeds have different triggers that prompt a seed to germinate. Light and moisture are key for most species. Some require a period of exposure to cold, over-wintering, before they start to grow. Others germinate within days of falling to the ground. Some prefer bare mineral soil; others need to be covered in the duff of the forest floor to grow. There are also those that can hang out buried in the forest floor for years or decades before they respond to perfect conditions to grow - these seeds are part of the seed bank.

The seed bank has an air of mystery around it. It's an easy assumption to make that there are seeds on site that will respond to a harvest or natural event. But not all the trees that we want to continue to exist on a site have that ability.

Yellow poplar has one of the longer-lived seeds in the seed bank at eight years. White ash and black cherry can bank for up to three years. The real winners in the seed bank, with seeds remaining viable for 50 to 100 years, are pin cherry (also called fire cherry) and black raspberry. Knowing the history of a site will help you understand what may come back after disturbance. Unfortunately, there are a few bad actors with the ability to bank their seed. *Paulownia tomentosa* (princesstree) seed can survive for two to three years in the forest soil. Japanese stiltgrass seed can remain viable for three years. If you're struggling with these non-native invasive species, you may have to keep fighting for multiple years to exhaust the seed bank - an important consideration for your invasives treatment planning.

Endangered tree species that have direct and indirect economic uses to mankind are conserved in seed banks (Figs 1 and 2). The increasing popularity of the seed bank as a tool in the conservation of wild tree species can be attributed to a number of factors: seed banks provide immediate access to plant samples, allowing researchers and conservation biologists to evaluate them for properties (such as new sources of medicines, nutrition, and genes). The availability of plant germ-plasm in seed banks facilitates research that could provide helpful information for conserving the remaining natural populations of the species. Perhaps, most important, plants conserved in seed banks are immune to habitat destruction, diseases, and predators. They can be used to reinstate species into existing, suitable habitats where they were once present, or to augment the diversity of small, genetically depleted populations.

Fig. 1. Seeds Conserved in Labeled/sealed Bottle

Strict Nature Reserve (SNR) is prominent among the methods for in situ conservation of biodiversity in Nigeria in particular and the world in general. Other common methods in Nigeria include Biosphere Reserves, Game Reserves, Regeneration Plots, Permanent Sample Plots, and Sacred Groves. SNRs are created to protect representative samples of natural ecosystems for the preservation of biodiversity and ecological processes, scientific study, environmental monitoring, education and the maintenance of genetic resources in a dynamic and evolutionary state.

Fig. 2. Seeds Stored in Iron & Plastic Containers in Modern Seed Bank

SEED BANKS IN THE WORLD

There are some 1500 seed banks around the world, including the Svalbard Global Seed Vault, located on the remote Norwegian island of Spitsbergen in the Svalbard archipelago, some 1300 kilometres north of the Arctic Circle. It is considered the world's largest germplasm centre. Also known as the 'vault at the end of the world', this dystopian-universe building was opened in 2008 in the middle of the ice and functions as a huge warehouse where more than one million seed samples from all over the world are stored.

CHARACTERISTICS OF THE SOIL SEED BANK

Soil seed bank density and seed dormancy

The success of a seed bank depends on the seed density ready to germinate, when replacement of a plant is necessary and when the environmental conditions for establishment are favourable. The longevity of seeds represents a major mechanism of survival of certain weed species, and this leads to a continuous source of emergency. The seed longevity represents a major mechanism of survival of certain weed species, and this leads to a continuous source of emergency [8]. The seed longevity in the soil varies among species characteristics of the seeds, burial depth, and climatic conditions. A study with weed species buried and placed to germinate in different times of the year After 40 years, the species, *Amaranthus retroflexus, Ambrosia elator, Lepidium virginucum, Plantago major, Portulaca oleracea* and *Rumex crispus* were the originated seedlings.

The seed dormancy is another characteristic that affects the seed bank reservoir. The seed populations of several vegetable species behave in different ways with respect to germination; the weeds produce polymorphic seeds, with a certain proportion that is dormant while the others are not.

Several internal and external factors prevent seeds from germinating. Among the internal factors are: the presence of a seed coat, which is a barrier to the penetration of water and oxygen; presence of a biochemical inhibitor in the seed; and immature embryo. Among the external factors, the most common are soil water content and temperature.

The term innate dormancy (primary) and induced dormancy (secondary) characterise the development of the dormancy in the mother plant and after the dissemination in space, respectively. The term enforced dormancy, has been used for the inability of the seeds to germinate due to an enviromental restriction, like water deficit, low temperature and poor aeration.

However, some seed physiologists do not consider the induced dormancy as an actual dormancy but responds only to favourable conditions for germination. This situation is more conveniently referred to as a case of quiescent seeds. The dormancy represents a mechanism of species preservation in the seed bank, distributing the germination through the year. It can guarantee the species survival in the form of seeds, under adverse conditions, even when the population of plants is completely eliminated.

Dispersal of seedbank

Most species are capable of long-distance seed dispersal, despite morphological dispersal syndromes that would indicate morphological adaptations primarily for short-distance dispersal. Assessments of mean migration rates found no significant differences between wind and animal dispersed plants. Long-distance migration can also be strongly influenced by habitat suitability suggesting that rapid migration may become more frequent and visible with rapid changes in habitat suitability under scenarios of rapid climate change. The discrepancy between estimated and observed migration rates during re-colonization of northern temperate forests following the retreat of glaciers can be accounted for by the underestimation oflong-distance dispersal rates and events. Nevertheless, concerns persist that potential migration and adaptation rates of many tree species may not be able to keep pace with projected global warming. However, these models refer to fundamental niches and generally ignore the ecological interactions that also govern species distributions.

THE NEED FOR FOREST RESTORATION

Deforestation can be defined as the conversion of forest ecosystem (natural and secondary regrowths) to a preferred permanent non-forested land use such as agriculture and urban development. Deforestation is of major concern for the developing countries of the tropics because it causes rapid reduction of the tro pical forests causing loss of biodiversity and enhancing greenhouse effect. A plantation of trees established primarily for timber production can also be referred to as a forest according to Food and Agricultural Organisation and therefore does not classify natural forest conversion to the plantation as deforestation (but still records it as a loss of natural forests).

Deforestation has served a major threat to the forest ecosystems from time immemorial, causing a major loss of biodiversity as well as global climate change. These loss can be attributed to illegal logging of trees for fuelwood and charcoal, wildfire, unregulated burning for harvesting logs, urbanization and natural disasters.

Forest degradation occurs when the ecosystem functions of the forest are degraded but where the area remains forested rather than being cleared. Logging opens mature forest area for colonization. Also, new roads are constructed in the forest during logging. This has the potential of increasing spontaneous colonization of logged forest by agricultural colonist. The ability of such tropical forest ecosystems to recover is limited as high and excessive logging has negative effects on the availability of quality seed germplasm for natural regeneration. Professor Norman Myers, one of the foremost authorities on rates of deforestation in tropical forests described rate of world's deforestation thus: "the annual destruction rates seems set to accelerate further and could as well double in another decade". Hence, an urgent need for forest restoration.

THE IMPORTANCE OF SEED BANKS IN BIODIVERSITY CONSERVATION

Conservation of Plant Species Seeds

There is an urgent need for the conservation of wild plant genetic resources, but it is largely impractical to conserve the very large number of plant species and their wild relatives in their natural habitats. A viable alternative is to conserve whole seed in seed banks. The seeds are germinated to raise plants from them for use in plant improvement. The important role seed banks play in the conservation of wild plant genetic resources is now globally recognized. Various techniques have been developed for preserving seeds of wild plant species; retaining their viability for longer periods. As temperature and humidity are very critical factors, cleaned seeds are stored at around -20°C, often using silica gel in the seed containers to reduce humidity.

Conservation of Endangered Wild Plant Species Seeds Seed banks serves to conserve species that otherwise might be lost. It also stores seeds of regional populations so that regionaldiversity is conserved and restoration oflocal populations can be achieved in future. It serves to conserve species that otherwise might be lost. It also stores seeds of regional populations so that regional diversity is conserved and restoration of local populations can be achieved in future.

Conservation of Genetic Stocks

Gene banks are type of biorepository which preserve genetic material. For plants, this could be by freezing cuttings from the plant, or stocking the seeds (e.g. in a seed bank). For animals, this is the freezing of sperm and eggs in zoological freezers until further need. Plant genetic material in a 'gene bank' is preserved at -196° Celsius in Liquid Nitrogen as mature seed (dry). Seed banks can be used for a range of purposes, including the following: 'Species bank' – to conserve species (generally back-up to in situ measures); 'Population bank' – to conserve genetic diversity within species (this sometimes includes population restoration); 'Genome bank' or 'gene-pool bank' – to conserve sets of related species that may be able to share genomes in a single gene-pool; 'Breeding bank' – to conserve selected families of interbreeding genotypes.

RECOVERY OF BIODIVERSITY IN TROPICAL FORESTS

As defined by the Convention on Biological Diversity, "biological diversity" means the variability among living organisms from all sources including terrestrial, marine, and other aquatic ecosystems and the ecological complexes of which they are part; this includes diversity within species, among species, and of ecosystems. Biodiversity may include the variety of life at multiple scales of ecological organization, including genes, species, ecosystems, landscapes, and biomes.

Biodiversity can be referred to as the variety that exists among all levels of life forms. One of the major roles of regeneration from soil seed bank is recovery of biodiversity in tropical forests as observed in Porto Trombetas, Para state, Brazil after a major disturbance as a result of anthropogenic impact. The role of biodiversity in the resilience of forest-carbon dynamics is evidenced by the specialized species that have evolved and characterize the distinctive forest ecosystems found in the major climatic and forest domains tropical, temperate, and boreal. Over time, evolution results in new plant traits, through the filter of natural selection, and aided by ecological processes such as dispersal, result in forests comprising species that function optimally under the climatic conditions and disturbance regimes prevalent in each domain. The recovery-ability of vegetation after disturbance is believed to lie mainly in the buried seed population. Resilience in ecosystems is related to the biological diversity in the system and the capacity that it confers to maintain ecosystem processes.

SEED VIABILITY, GERMINABILITY, AND DORMANCY

Seeds are the sexual regeneration stage of plants. They are a means of plant dispersal in space and in time, allowing plants to exploit suitable habitat when and where it is available. Seed dispersal in space and in time can also reduce sibling and parent–offspring competition. We include here all seed-like structures, for example, achenes from apomictic plants.

Seed germination can occur whenever the germination requirements are met, such as suitable environmental conditions (i.e. temperature and moisture: some species require rather specific light or temperature regimes). However, a seed or dispersal unit lying on the soil may not germinate for one of four reasons:

i. The seed is inviable because the seed never formed an embryo and is in effect an empty case;
ii. The seed is inviable because it died at some point;
iii. The seed is viable but conditions are not suitable for germination; or
iv. The seed is viable but dormant, that is, the seed has some innate mechanism that inhibits germination under conditions that are otherwise suitable.

Dormancy is geographically and phylogenetically widespread. Dormancy can be imposed by external conditions or be an innate property of the seeds themselves, for example via an undeveloped embryo or an impermeable seed coat (hard-seededness). It can be broken by, for example, light (the red : far-red ratio is particularly important), scarification of the seed coat, temperature cycles, or extrinsic chemical signals such as smoke. Some species produce a mixture of dormant and non-dormant seeds, the ratios of which may be affected by climate. Other forms of global change such as nitrogen deposition can also affect the proportion of dormant to non-dormant seeds.

Understanding seed viability, dormancy, and germination is important for understanding plant community responses to disturbance, population dynamics, and competitive interactions. If the dormancy mechanism is known, then the proportion of seeds germinating before and after a dormancy-breaking mechanism has been applied can be compared. In the context of climate-manipulation experiments, changes in the proportion of seeds which are viable can tell us about reproductive fitness and changes in the proportion which are dormant tell us about adaptation to changing environmental conditions.

Treatment conditions mimicking a range of climate projections may be replicated either in the laboratory or in the field. The effect of increased nitrogen can interact with climate effects: for example, researchers show that increased nitrogen widens the thermal range at which *Buglossoides arvensis* will germinate. For precise climate manipulations, laboratory experiments may be preferred over the imprecise or noisy conditions of field sites.

What and how to measure?

Seed germinability

Seed germinability is usually measured using emergence tests. Seeds are sown onto growth medium (e.g. 1% agar or moist filter paper) or damp compost, kept in appropriate conditions for germination, and the proportion of seeds which produce a radicle is recorded.

The number of seeds used for germination experiments depends on availability and varies between studies, but numbers around 100 per treatment and species are advisable. Experiments of this nature require a good understanding of the behaviour of the seeds of your study species and a good deal of forward planning. Researchers provide a list of guidelines for laboratory studies of germination.

Collect seeds at maturity

How to collect seeds depends on the architecture of the seed-bearing part of the target species and descriptions of the plants can be found in floras. The timing of seed collection is important and should be done at the time when the seeds have ripened, but ideally before they disperse. The outer part of the dispersal unit will often dry out or change colour when the seeds start to ripen, particularly for animal-dispersed fleshy fruits.

For annuals, biennials, and perennials which make seasonal stems, the whole flowering structure or even plant is likely to be dying back. For seeds dispersed from capsules, the capsules split open or the seeds start to rattle. For seeds that are wind dispersed and have a pappus (e.g. Asteraceae), the inflorescence can be covered with netting to prevent the seeds dispersing before collection because dispersal can be very sudden and happen immediately after maturity is reached (Fig. 3).

Fig. 3. A net bag placed over a flower of Leontodon autumnalis to collect seeds.

Add the bag only after the flower has been pollinated and begins to senesce. In wet areas the seeds can start to become infected by fungus if they are covered so the bag technique may not be appropriate. In species where dormancy is broken by a dry or cold season, or by fire, it is possible to collect seeds after the dormancy-breaking event has passed provided that they are large enough to find (or remain on the parent plant, e.g. *Pinus sylvestris* or other serotinous cones, but note that some mortality may have occurred during that time.

Check seeds for the presence of an embryo

This is tricky to do non-invasively but seed cases which are completely empty shells, or are relatively easily squashed between the fingers in comparison to healthy seeds, should be discarded.

Test and use seeds immediately after harvesting

This is not always possible, for example in the middle of a field campaign, in which case the seeds should to be stored under optimal conditions. It is important to know the suitable storage conditions for different species as some seeds do not tolerate drying: these are called recalcitrant seeds (this can be checked in Baskin & Baskin, 2014 or on the Kew Seed Information Database). Keep non-recalcitrant seeds cool (ideally below 5 °C but at least below 25 °C), air dry, and in the dark. Recalcitrant seeds should be kept cool and dark and slightly damp (but not soaking wet). Longer term storage can be attained for non-recalcitrant seeds by drying to 5–7% moisture and freezing.

Check that seeds can imbibe water

If your seeds sit in water without any hint of softening or swelling, they may either be dead or be hard-seeded (physical dormancy). If they are physically dormant, the coating may need to be broken gently, for example with a needle or some sandpaper.

Use intact natural dispersal units

This is because the pericarp (parts of the dispersal unit formed by the parent plant) first protects the seed, for example from drying out, and second may be rather securely attached such that detachment damages the seed. Removing the pericarp can also change the germination rate – which is acceptable for viability testing but not for testing germination of the seeds under "natural" conditions. It is of course difficult to access some seeds, for example, the fruit of an almond, without removing the tough endocarp but in this case collection and storage should be in the natural dispersal units, only removing them when the seed is ready to be tested.

It is vital that the conditions the seeds are placed under are suitable for their germination and that they are given adequate time. Suitable germination conditions can be implied from conditions at a site where new individuals of the species are frequent or determined experimentally. These may include fluctuations in photoperiod or temperature or both.

Regarding duration, researchers caution against running experiments for longer than four weeks, but this recommendation relates to finding the dormant proportions and where dormancy could be broken by germination conditions. Researchers monitored their Arctic seeds for 13 weeks and incubated alpine seeds for 48 weeks, albeit with a temperature-induced period of germination suppression during the experiment.

Establishment of maximum germinability (for use in population models) takes considerably longer than recording relative germinability (for example, for comparing between manipulation treatments). For maximum germinability experiments, decide beforehand a length of time during which no new germinations will trigger the end of the experiment. Ungerminated seeds can be tested for viability.

If the mechanism for breaking dormancy is known, then that should be applied before starting the germination test. For example, for temperate, high-elevation and high-latitude areas, vernalisation is the most common dormancy-breaking mechanism applied – seeds should be kept slightly damp and refrigerated for four months (this can be shorter for climates with warm winters; alternatively seeds can be gathered after winter).

Seeds with a hard coat can be scarified by gently rubbing the seeds with medium sandpaper on a flat surface until chips of testa (seed coat) can be seen. Some seeds need to be kept warm and dry. A fresh seed control can be used to determine dormancy fractions (proportion of the seeds that are dormant and non-dormant). If the mechanism for breaking dormancy is not known, some caution should be applied to subsequent germination data.

SEED VIABILITY

Seed viability is a more specific test of whether the seed embryo is still alive. There are three main groups of methods: chemical, physical, and non-invasive. The most common **chemical method** is to expose the embryo to tetrazolium solution. The method works best on large grass and tree seeds, but is challenging with very small seeds due to difficulties in handling and visual assessment (although not impossible). The seed coat is cut open very carefully with a scalpel, without damaging the embryo (practice on spare material). Then -2,3,5triphenyl tetrazolium chloride (usually %1–0.1 w/v) is dropped onto the embryo. If the embryo goes pink or red, there is living tissue in the embryo which has reduced the tetrazolium chloride (colourless) to formazan (pinkish red) via dehydrogenase enzymes. The International Seed Testing Association protocol (ISTA) is the ***gold standard*** for viability tests, but that protocol requires 2500 seeds per test.

Physical methods involve poking the seed to see if the embryo is fleshy rather than hollow inside, although note that some plants produce hard and hollow seeds. It is therefore critical to cut the seed open and examine the embryo under a stereo microscope and so poking and cutting could be considered the ***bronze standard*** viability test, although note that the method is destructive and that the structure of the seeds of the study species should be studied and understood beforehand. **Non-invasive methods** such as the use of X-rays (Foucat et al., 1993; Dell'Aquilla, 2007) are either under development or very expensive and outside the scope of this protocol.

Special cases, emerging issues, and challenges

In all cases where material is handled in the laboratory, care must be taken to minimise artefacts of the laboratory conditions, such as by accounting for soil temperature differences relative to air temperature, or properly replicating the diurnal cycle of light and temperature. Likely, warming is the easiest global change variable to manipulate, which is appropriate given the propensity toward temperature as the dominant control of dormancy and germination. Germination tests from different treatment populations in identical laboratory conditions may fail to account for altered dormancy effects from treatments, but can still capture population-level changes in other seed traits that influence viability.

NATIVE TREE REPRODUCTIVE STRATEGIES

- Beech has limited seed dispersal. Blue jays are an important distributor. Most seeds germinate the first spring after seedfall, after over-wintering. Beech is a prolific root sprouter and can also stump sprout.
- White ash wind-blown seeds are viable for three to four years, and require exposure to cold before germination. Younger trees can resprout from the root collar, but the ability declines with age.

- Red maple seeds are wind-dispersed. Seeds germinate soon after they fall, with 95% of viable seed germinating within the first ten days. Red maple can stump sprout.
- Black cherry seeds must overwinter. Animals contribute to the seed dispersal. Black cherry can delay germination for up to three years, lying dormant in the forest soil until conditions are prime.
- Northern red oak acorns must overwinter before they germinate. The acorns fall in autumn and germinate in spring. Blue jays and mammals spread the seed. Oaks in the red oak group can stump sprout prolifically.
- White oak acorns germinate soon after they fall. Many white oak acorns are damaged by insects or seed eaters. The blue jay is a primary disperser. Oaks in the white oak group can stump sprout, but not quite as prolifically as those in the red oak group.
- Yellow poplar seed is wind-dispersed. Seed can remain viable in the forest floor for up to eight years.
- Sugar maple seed is wind-dispersed. In nature, few persist as viable seed for more than one year. In the northern part of its range, sugar maple can sprout prolifically; it sprouts less vigorously in the southern part of its range.
- Chestnut oak acorns, as part of the white oak group, germinate soon after they fall. Mammals help spread the seed. Chestnut oak stump sprouts more frequently as compared to other oaks.
- Hemlock seeds are gravity- or wind-dispersed. The seeds germinate the first spring after falling. Seed is the hemlock's only method for regeneration.
- Black birch seed is wind-dispersed. Seeds germinate the spring after dispersal. It is a less prolific stump sprouter.
- Black gum seed is dispersed by gravity and birds. The seed germinates the spring after seed fall; it can also stump sprout.
- White pine seed is dispersed by wind and animals. Seed germinates the first year after seed fall.
- Hickory seeds are dispersed by mammals. The nut must overwinter before germination. Hickories can stump sprout but that ability declines with age.

NON-NATIVE TREE AND PLANT REPRODUCTIVE STRATEGIES

- Tree of heaven has a tremendous ability to root and stem sprout. In fact sprouting is its most common reproductive method. Any threat or injury will cause tree of heaven to sprout prolifically. Its wind-dispersed seed must overwinter before germinating, but it does not persist in the seed bank.

- Paulownia seed is wind-dispersed and can persist in the seed bank two to three years.
- Asiatic bittersweet has tremendous sprouting ability - sprouting from the roots, root fragments, and the root collar. Any form of damage encourages sprouting. Its seeds are spread by birds, but are short-lived in the seed bank.
- Japanese knotweed also has tremendous ability to sprout, sprouting from the stem, rhizome, or rhizome fragments.
- Japanese stiltgrass can seed bank up to three years. Gravity, water movement, and human or animal carriers spread its seed.

9

ECOLOGICAL MODELS FOR CONSERVATION, RESTORATION, AND MANAGEMENT OF FORESTS

INTRODUCTION

The term *ecological models* is used here to refer to all forest dynamics models originating from an ecological perspective or simulating ecological processes or characteristics of forests. Globally, there are myriad situations in which people aim to conserve, restore, or manage forest ecosystems at small spatial scales of 50 ha or less. To inform management, forest dynamics models provide an increasingly diverse and valuable portfolio of tools for projecting forest change under different management and environmental conditions.

Here, following Hope, two main types of ecological model are differentiated:

- Process models focus on modelling physiological processes such as photosynthesis and respiration. Process models generally do not depict changes in species composition or stand structure, and therefore have relatively little relevance to forest conservation or ecological dynamics. However, some models of this type are of value for exploring the provision of ecological services.
- The other main group of models may be described as succession models, as they focus on simulating the ecological dynamics of forest communities; it is these models that are considered below. This type of model may be further subdivided into gap and transition models.

DEFINITION

An ecosystem model, also called an ecological model, is usually a mathematical representation of an ecological system. It is a simplified form of a highly complex ecosystem in the real world. The ecological system can range in scale from an individual population to an ecological community, or even an entire biome.

MODEL BUILDING

Based on data gathered from the field, ecological relationships—such as the relation of sunlight and water availability to photosynthetic rate, or frost resistance to cold temperature and the length of growth period—are derived, and these are combined to form ecosystem models.

Due to the complexity of ecosystems, it is impossible to capture every possible element of a system in a model. Thus, modelers need to take decisions about which of its features to include and which to disregard. These decisions are usually guided by the aims that the modeler is attempting to achieve. It is desirable to make an ecological model to be general (the model can be applied to a wide range of real-life systems), realistic (the model's conclusions have a close match to a real-life system) and precise (the model's predictions for a specific set of circumstances have little or no uncertainty). Trading-off one desideratum in order to achieve greater performance in the other two is often involved in model-building processes.

MODEL APPLICATIONS

Ecological models enable researchers to simulate large-scale experiments that would be too costly or unethical to perform on a real ecosystem. They also enable the simulation of ecological processes over very long periods of time (i.e., simulating a process that takes centuries in reality, can be done in a matter of minutes in a computer model). More importantly, ecological models can be used to predict the state of ecosystems under future climates.

Ecosystem models have applications in a wide variety of disciplines, such as natural resource management, ecotoxicology, and environmental health, agriculture, and wildlife conservation. Applications of ecological models in forest resources management to mitigate the negative impact of climate change and help forest trees to adapt to a rapidly changing environment have become particularly important.

At present, our understanding of the relationships ecosystems and the changing environment is relatively weak. The vulnerability of ecological systems, and the services provided by them, to the impact arising from climate change remain largely unknown. Ecological models have been widely used to assess the impact of climate change in terms of productivity, carbon storage, and vulnerability to disturbances, and to provide information for developing forest adaptive strategies, such as assisted migration at the species level and assisted gene-flow at the population level.

TYPES OF MODELS

There are two major types of ecological models:

1. Analytic models and
2. Simulation models.

Analytic models

Analytic models are typically relatively simple (often linear) systems, that can be accurately described by a set of mathematical equations whose behavior is well-known. Analytic models are also called correlative models or empirical models.

For example,

A model of tree height growth of populations originated from different climates that assumes the transfer distance of populations affects the tree height growth in a quadratic relationship as:

$$Y = b_0 + b_1 * MAT\_td + b_2 * MAT\_td^2$$

Where Y is the tree height; MAT_td is the transfer distance in term of mean annual temperature; MAT_td^2 is the quadratic form of the transfer distance; and b_0, b_1, and b_2 are parameters to be determined during the model fitting process.

A model can be acceptable if the amount of variance explained by the model is high (or the relationship is strong) based on the statistical significance ($P < 0.05$), such as the one illustrated in Fig. 1.

Although the solution to the above simple question is fairly transparent, analytical models describing more complex systems can often become fairly complicated. However, for those comfortable with mathematics, an analytical solution does provide a concise preview of a model's behavior that is not as readily available with a simulation model. We will focus on this type of models in this course, and more details on analytic or empirical models will come in later lectures.

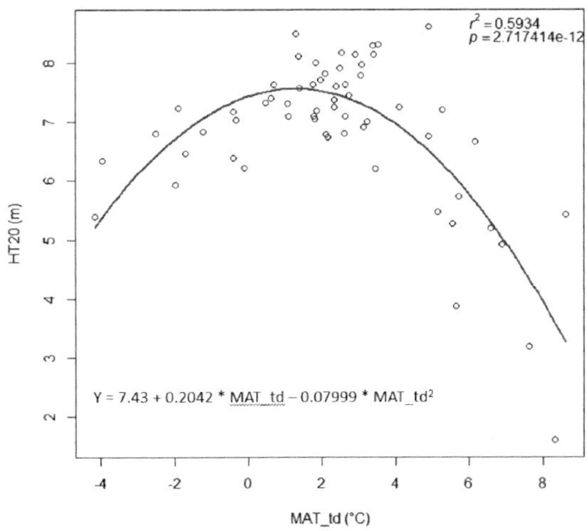

Fig. 1. Illustration of an analytic model based on the relationship between climatic transfer distance (MAT_td) and the height growth (HT20) of tree populations.

As the name stated, a process-based model can simulate the process or dynamics over time. For example, many process-based models can simulate the tree growth from seedlings to harvest, or simulate the stand performance over rotations.

Empirical and process-based models in climate change applications

Both of Empirical and process-based models are applied in modeling the impact of climate change on species distributions, productivity, and vulnerability to stress.

The empirical approach relies on correlative relationships in line with mechanistic understanding, but without fully describing system behaviors and interactions. For example, the model represents the relationship between the transfer distance and tree height of populations. We can see that a strong relationship is represented by the model, but the model does not show the physiological process driving this relationship.

The process-based models, on the other hand, can be more comprehensive and incorporate mechanisms explicitly. For example, many process-based models include the process of how radiation, water, and temperature affect the rate of photosynthesis, and how the biomass generated from the photosynthesis allocates to different parts of the plants, as shown in Fig. 2. However, the accuracy of the model predictions is often lower than an empirical model. This is because there are many factors affecting the performance of trees, populations, or an entire ecosystem, and many of them cannot be incorporated into the process-based model.

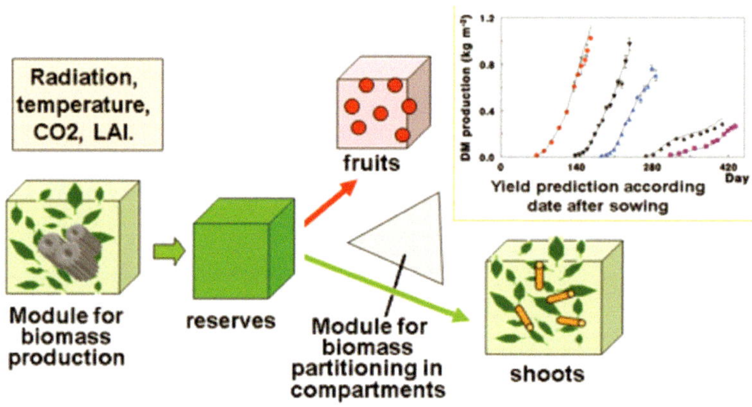

Fig. 2. Illustration of a typical process-based model application. The biomass production model computes the biomass produced by the leaf compartment from the LAI and environmental conditions. The partitioning model then splits the biomass between the fruits and the other organ compartments. The model then loops for a new development cycle.

Empirical models based on the occurrence data of a species, particularly with machine-learning algorithms, have demonstrated clear superiority over process-based models in predictions of species distributions. These models are often called species distribution models (SDMs). The occurrence of the species resulted from the process of long-term local adaptation to environmental conditions, interaction with other species, and many other factors. These processes are difficult to model by a process-based model. In contrast, the empirical models do not need to consider these processes, but just relate the occurrence locations with its environmental variables.

MODEL BUILDING

As empirical models are relatively easy to build and will be described in detail in the latter lectures; thus, this section is mostly focused on process-based models.

Model design

Ecological systems are composed of an enormous number of biotic and abiotic factors that interact with each other in ways that are often unpredictable, or so complex as to be impossible to incorporate into a computable model.

Because of this complexity, ecosystem models typically simplify the systems to a limited number of components that are well understood and relevant to the problem that the model is intended to solve. Thus, the process of model design needs to begin with a specification of the problem to be solved or the objectives of the model.

The process of simplification typically reduces an ecosystem to a small number of state variables and mathematical functions that describe the nature of the relationships between them. The number of ecosystem components that are incorporated into the model is limited by aggregating similar processes and entities into functional groups that are treated as a unit.

Another important factor in the ecosystem model structure is the representation of the space used. Historically, models have often ignored the confounding issue of space. However, for many ecological problems, spatial dynamics are an important part of the problem, with different spatial environments leading to very different outcomes.

Spatially explicit models (also called "spatially distributed" or "landscape" models) attempt to incorporate a heterogeneous spatial environment into the model. A spatial model is one that has one or more state variables that are a function of space or can be related to other spatial variables.

Model validation

After construction, models are validated to ensure that the results are acceptably accurate or realistic.

One method is to test the model with multiple sets of data that are independent of the actual system being studied. This is important since certain inputs can cause a faulty model to output correct results.

Another method of validation is to compare the model's output with data collected from field observations.

Model predictions almost always involve errors. Researchers frequently specify beforehand how much of a disparity they are willing to accept between parameters output by a model and those computed from field data.

PROCESS-BASED MODELS

In process-based models, the environment variables and vegetation are linked up upon the comprehensive physiological processes and functions between them. They often have a high requirement for data input. The process-based approach focuses on simulating detailed physical or biological processes that explicitly describe system behavior. They can be comprehensive and mechanism explicit. All process-based models include some empirical information. Uncertainty in process-based model outputs could be higher than for the empirical approach due to greater model parameters and data inputs to represent the many processes in the system. The application of physiological rules makes process-based models possible to mimic the vegetation growth and display the future species distribution, as well as the forest dynamics.

We will introduce several process-based forest ecological models here.

3-PG

The 3-PG (Physiological Principles Predicting Growth) forest growth model has been most widely used. The 3-PG model is a simplified physiological processes-based stand growth model developed by Landsberg and Waring in 1997.

A 3-PG model calculates the radiant energy absorbed by forest canopies and converts it into biomass production. The efficiency of radiation conversion is modified by the effects of nutrition, soil drought (the model includes continuous calculation of water balance), atmospheric vapor pressure deficits, and stand age.

3-PG is a deliberate attempt to bridge the gap between conventional empirical growth and yield models, and complex process-based carbon-balance models. It can be applied to plantations, or to even-aged, relatively homogeneous forests. It is a generic stand model, in the sense that its structure is not site or species-specific. However, parameterization is necessary for improving model effectiveness when it is applied for different tree species. The 3-PG model is usually more powerful and effective after the parameter values have been adjusted for aimed tree species.

3-PG consists of five simple sub-models:

- The assimilation of carbohydrates,
- The distribution of biomass between foliage,
- Roots and stems, the determination of stem number,
- Soil water balance, and
- Conversion of biomass into variables of interest to forest managers.

The 3-PG model has been used to stimulate forest dynamic and productivity and estimate future migration direction and distribution under changing climate. A 3-PG model can also be applied together with other tools for a more comprehensive and convincing performance.

FORECAST

FORECAST is a management-oriented, stand-level, forest-growth, and ecosystem-dynamics model. The model was designed to accommodate a wide variety of silvicultural and harvesting systems and natural disturbance events (e.g., fire, wind, insect epidemics) in order to compare and contrast their effect on forest productivity, stand dynamics, and a series of biophysical indicators of non-timber values. The model was developed by Kimmins and his colleagues at the University of British Columbia.

FORECAST can project stand growth and ecosystem dynamics. The projection is based upon a representation of the rates of key ecological processes regulating the availability of, and competition for, light and nutrient resources (a representation of moisture effects on soil processes, plant physiology and growth, and the consequences of moisture competition is being added).

The rates of these processes are calculated from a combination of historical bioassay data (such as biomass accumulation in plant components and changes in stand density over time) and measures of certain ecosystem variables (including decomposition rates, photosynthetic saturation curves, and plant tissue nutrient concentrations) by relating 'biologically active' biomass components (foliage and small roots) to calculated values of nutrient uptake, the capture of light energy, and net primary production.

Using this 'internal calibration' or hybrid approach, the model generates a suite of growth properties for each tree and understory plant species that is to be represented in a subsequent simulation. These growth properties are used to model growth as a function of resource availability and competition. They include (but are not limited to):

1. Photosynthetic efficiency per unit foliage biomass and its nitrogen content based on relationships between foliage nitrogen, simulated self-shading, and net primary productivity after accounting for litterfall and mortality;
2. Nutrient uptake requirements based on rates of biomass accumulation and literature- or field-based measures of nutrient concentrations in different biomass components on sites of different nutritional quality (i.E. Fertility);
3. Light-related measures of the tree and branch mortality derived from stand density and live canopy height input data in combination with simulated vertical light profiles. Light levels at which mortality of branches and individual trees occur are estimated for each species.

The FORECAST model has been widely applied as a long-term management evaluation tool in a variety of forest ecosystems. The FORECAST model simulates the effect of light and nutrient availability on forest productivity but it has no explicit representation of moisture or temperature on ecosystem processes. Thus, the model is not directly driven by climate data. FORECAST Climate has been developed to incorporate these two factors. The foundation for FORECAST Climate was established through the creation of a dynamic linkage with a forest hydrology model including direct feedback to the core processes driving forest ecosystem production.

Tree & Stand Simulator (TASS)

The Tree and Stand Simulator (TASS) is a biologically based, spatially explicit, individual tree model. By simulating the potential effects of silvicultural decisions and prescriptions, TASS provides yield projections for timber supply reviews and helps resource managers identify investment opportunities. The model is developed and maintained by the Ministry of Forests, Lands, Natural Resource Ministry of Forests, Lands, Natural Resource Operations & Rural Development.

TASS currently exists in 3 main forms:

1. **TASS III** is the all-new public-release Windows™ version, which begins to extend TASS into more complex stand structures with multiple-species and -age cohorts. The initial release is limited to lodgepole pine (Pinus contorta var. latifolia) and white spruce (Picea glauca).
2. **TASS II** (commonly referred to as TASS) is the well-established, in-house version described below. Although, the concepts largely apply to TASS III, as well.
3. TIPSY (Table Interpolation for Stand Yields provides direct operational access to yield tables generated by TASS II. For scenarios outside the limited range of TIPSY, a custom TASS run service is available by contacting the Growth and Yield Application Specialist.

Photos in Fig. 3 show trees remaining after variable retention harvesting by Weyerhaeuser Company Limited (the real world), and the TASS simulations. They are amazingly similar.

Fig. 3. Illustration of TASS simulations of forest stands.

The TASS is a height-driven, biologically oriented, spatially explicit individual tree model. It was designed to produce potential growth and yield tables for even-aged managed stands. It simulates the growth of individual trees in a three-dimensional space with competitions. Tree height, bole diameter, branch length, and crown form and foliage are updated annually. These measures can be affected by inter-tree competition, thinning, pruning, animal damage, site characteristics, defoliation, mortality, and fertilization, as well as genetic variation (or genetic treatments) in tree growth.

TASS can predict silvicultural treatment response by modelling individual tree crown dynamics and their relationship to bole growth and wood quality. This makes TASS particularly well suited for predicting response to treatments such as

- Espacement,
- Fertilization,
- Pruning,
- Pre-commercial and commercial thinning.

However, effects of climate change have not been incorporated into TASS models. Scientists are working on functions and parameters to link population response to climate into the modeling algorithms.

LANDIS-II

The LANDIS-II forest landscape model simulates future forests (both trees and shrubs) at decadal to multi-century time scales and spatial scales spanning hundreds to millions of hectares. The model simulates change as a function of growth and succession and, optionally, as they are influenced by a range of disturbances (e.g., fire, wind, insects), forest management, land-use change. Climate and climate change affect processes throughout the model. LANDIS-II is highly customizable with dozens of libraries ('extensions') to choose from.

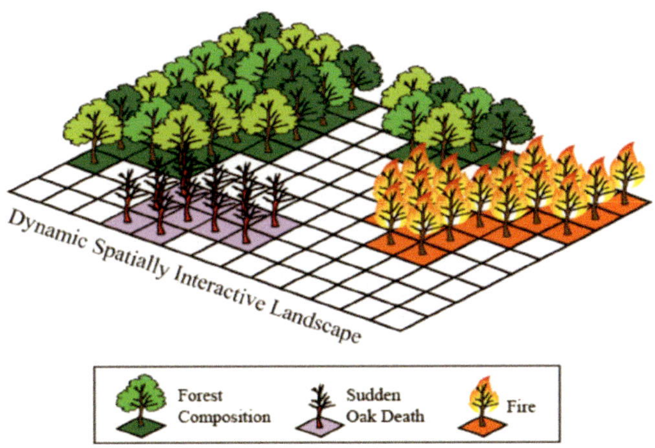

Fig. 4. Illustration of Landis-II model simulations of tree growth spatially explicit.

LANDIS-II is optimized for the simulation of spatial processes and the interactions between spatial processes and patterns. Landscapes within LANDIS-II are represented as a grid of interacting cells with a user-defined spatial resolution (cell size) and extent. Practicable cell sizes can range from a few meters up to a kilometer. All LANDIS models assume that individual cells have homogeneous light environments. Cells are aggregated into ecoregions with homogeneous climate and soils and a user-defined extent, thereby creating a hierarchy of spatial interactions.

The LANDIS-II community is very active with 100s of users and developers worldwide. Everyone is welcome to join the LANDIS-II community and to contribute. All model components are free and open-source. There are active bulletin boards for Users and Developers. And there are meetings and training held every year in various locations across the US and Canada.

USING EMPIRICAL MODELS IN A CHANGING ENVIRONMENT

Empirical models (EM) for forest management are generally calibrated on forest inventory data or data from long-term forest experiments and are consequently considered as being unable to incorporate the effects of changing environmental conditions on tree and stand growth. However, in contrast to the classical «static» approach of EMs, recent empirical modeling approaches can actually accommodate the dynamics of environmental conditions, including climate change, and are thus capable of reflecting the effects of changing conditions on natural as well as management-driven forest dynamics, at least within the historical range of variability.

Dynamic state-space approach

More flexible EMs can be achieved using the dynamic state-space approach. In this concept, site productivity effects are incorporated implicitly through a combination of common standspecific parameters at any stage of stand development and the possible interactions between site, tree growth and management actions. By doing so, they account implicitly for (1) temporal variations in site and stand productivity, and (2) the combined effect on stand dynamics and growth as a function of site potential, the genetically determined potential for volume growth, and possible management effects.

The state-space approach assumes that variables describing the current state of a given system at any time include the information needed to predict the future behavior of the system. It is assumed that the ndimensional state vector at some point in time, x(t), can be predicted by a transition operator F of the state vector, x(t0), and a vector of input variables, U at some other point in time. Current additional outputs from the model, y(t), are deduced from the current state by a function g:

$$y(t) = g\,[x(t)]$$

$$x(t) = F\,[x(t_0), U, t - t_0]$$

The state-space approach thus predicts any future states of the system from the initial state, $x(t_0)$, through iteration. For example, an initial observation of the two-dimensional state vector of height and basal area may be used to predict height and basal area after one period. The new estimates of the two state variables are then re-entered into the model to predict the state after one more period, and so forth. Abrupt changes in, for example, basal area due to thinnings are handled by simulating the shifts in the state vectors (U) and are seen as shifts between different growth paths. Mortality, growth and stand development may thus be modeled through iterations based on simple site and stand variables combined with numeric information on management actions.

In contrast to classical (i.e., static) EMs and most PBMs, models based on the dynamic state-space approach rely on a minimum of assumptions regarding allometric relations and management effects. For example, no assumptions on mortality due to selfthinning or on the relationship between height growth and basal area growth are needed. The use of plot- or stand-specific calibration for operators F and g is fundamental, and it ensures that the model adapts to changing site, stand and management conditions at any time, as they are manifest in simple mensurational variables. Simultaneous estimation procedures are used to account for the joint effects of the variables employed to describe stand dynamics. Updates are possible whenever new data become available, which renders the approach adaptive.

PRODUCTIVITY-ENVIRONMENT RELATIONSHIPS FOR GROWTH AND YIELD MODELS

Site index - or top height of a stand at a given base age - is a traditional and popular proxy for site fertility in even-aged forestry. It is a key input to most growth and yield simulators that are applied in forest management. Environmental change, however, challenges the assumption of constant site quality, which underlies the use of the site index concept, and thus the use of these traditional models. Because EMs remain an accurate tool for yield prediction, there has been a renewed interest for uncovering the environmental determinants of site index, based on regression models of site index against soil, climate and other environmental indicators.

Such productivity-environment relationships have to be designed at scales much larger than individual stands, to cover a wide range of environmental conditions, ranging from regional to national. They are therefore especially relevant for forest management. They further encompass environmental gradients that often cover a considerable fraction of the species' range, including northern and southern margins, and this is a key advantage for the sound anticipation of species productivity levels in a future climate. At national scales, forest inventories (NFI) were found to be a major support tool for providing comprehensive growth and environmental data. The specification of climate-productivity relationships should not include implicit climate dependencies. Hence, the use of geography (e.g., latitude/ longitude) and topography proxies (e.g., altitude, slope) alone or in combination with climate indicators should be avoided, despite their usually high predictive power in regression models.

Because empirical growth models are principally well suited for the investigation of a wide range of management alternatives, their coupling with productivity-environment models based on large-scale ecological gradients provides a cost-effective and accurate alternative approach for the prediction of timber production in the context of environmental change.

HYBRID FOREST MODELLING

The term "hybrid modelling" refers to approaches that are grounded in both empirical and process-based concepts of forest dynamics, thus trying to capitalize on the advantages of each approach. Specifically, the underlying idea is to benefit from the predictive ability and parsimony in the calibration data needs of empirical approaches as well as the explicit environment-dependence of process-based formulations.

To date, hybrid modeling strategies addressing the issue of environmental change have been explored in two ways: (i) coupling of existing EMs and PBMs that were developed for the same forest context. This coupling relies on the development of environment-productivity signal-transfer functions from simulations of the process-based model, that are then incorporated into the empirical model; (ii) the development of hybrid models sensu stricto, based on concepts from both empirical and process-based modeling that are embodied within the same piece of computer code. Such models should hence be able to cope with environmental changes while keeping the predictive and parsimonious properties of EMs.

Signal-transfer modelling

In a hierarchy of models, signal-transfer modeling designates the transfer of input/output relationships (signal functions) established from smaller-scale process models into larger spatial- and temporalscale models of ecological and economic phenomena. Signal-transfer modeling approaches have been developed to incorporate the effects of environmental changes in empirical growth and yield models, based on their assessment in more detailed process models. As signal-transfer functions are calibrated once for all, the approach avoids the computational complexity of applying process models to large regional forest contexts, and allows forest management alternatives to be addressed in the efficient way of traditional growth and yield models.

In the calibration of signal-transfer functions, an indicator of forest productivity common to each class of model is selected (e.g., tree volume increment, site index). The response surface of the indicator to environmental factors is calibrated from the process-based model outputs based on a high-dimensional simulation design. Interactions can also be taken into account, for example to extend the effect of climate scenarios to enlarged local site conditions of regional or national case studies. The approach thus results in a multidimensional space that can be queried e.g. in a database to yield the appropriate value of the response variable.

This type of approach has been implemented for the main tree species of Finland to estimate future resource use in industrial and energy wood under different management and environment scenarios of temperature and CO_2. It was based on FinnFor (a PBM) and Motti (a growth and yield model), and annual stem volume increment was the transfer variable. An original aspect of the approach was the effective incorporation of environmental scenarios in a simulation system optimizing management scenarios based on economic indicators.

Another example is provided by the coupling of PTAEDA2 (EM) with MAESTRO (PBM) for *Pinus taeda* across 13 States in the USA, to address changes in five environmental factors (precipitation and temperature, atmospheric CO_2 and ozone concentrations, nitrogen deposition). The objective was to render site index, which is the driver variable of the dynamics in PTAEDA2, adaptive to changing environmental conditions.

Inherent to the approach, short-term responses of physiological processes to environmental drivers are extrapolated to larger time scales, although some extrapolations such as the long-term stimulation of NPP by CO_2 may be questionable. Also, because growth and yield models are often calibrated in regional or national contexts, the response of processes to environmental factors may not apply outside these areas.

10

REPRODUCTIVE ECOLOGY AND GENETIC VARIATION

INTRODUCTION

The Earth's forests and associated biodiversity supply vast quantities of ecosystem services and products, including clean water, human and wildlife habitat, erosion control, carbon storage, medicines, and wood, chemical and energy products. Increases in the global human population (projected to grow from 7.1 to 8.3 billion by 2030) are expected to dramatically increase demand for forest products, such as timber by 129 million m^3. Resource extraction has traditionally come at the cost of forest loss and degradation, with a loss of *c.* 13 000 ha yr^{-1}, which, in turn, drives biodiversity loss and declines in ecosystem services. Tree plantations are recognized as a partial solution, as they usually have higher rates of wood volume accumulation than natural forests. Approximately 35% of the world's wood supply is generated by plantations, yet they comprise only 5% of the global forest area.

Tree breeding, increasingly accompanied by genomic methods, is ubiquitous in plantation management and is a major contributor to high rates of plantation productivity. In recent years genetic engineering (GE) – defined here as the direct and asexual alteration of DNA in living organisms – has been proposed as a new tool for tree breeding. Because GE can be used for changing native genes or transferring in novel functions, it has the potential to produce a variety of modified or novel traits. A large body of field research has shown that GE methods appear capable of delivering improvements to plantation productivity, wood processing, pest resistance or chemical products.

In some cases, the improvements can be substantial, such as elevating productivity in the field by *c.* 20% in poplar. However, the movement of seeds and pollen from GE plantations to feral and wild populations (hereafter termed 'gene flow') has been highly contentious in its regulatory treatment, public acceptance and possible impacts on markets. For example, pollen 'contamination' of non-GE products, an important economic problem with GE crops, may be even more difficult with GE plantation trees as a result of their typically longer dispersal distances. Pollen from GE eucalypts (that might contaminate honey products) was cited as a concern related to the growing commercial use of GE eucalypts in Brazil – where GE reproductive modification is illegal – and was a contributing factor to vandalism of company glasshouses.

In order to mitigate the social and biological problems of gene flow, GE methods that would reduce or prevent gene flow have been developed, but their potential impacts on biodiversity and society have received limited attention. Removal of pollen and/or seeds could obviously impact pollinators, as well as nectar- and seed-consuming organisms, and thus deserves careful consideration for each potential application so that this potential fix to the problem of gene flow is used judiciously and does not needlessly reduce the benefits of GE traits, or exacerbate rather than attenuate global impacts of plantations on biological diversity.

The primary goals of this paper are to evaluate the potential ecological impacts that modifications to forest tree reproduction might have on biodiversity, contextualize these impacts by comparing them to effects of other forms of forest management on reproductive activity, and examine how reproductive modification technology is expected to affect public, interest group and government responses.

We focus primarily on the potential biodiversity and social impacts of GE reproductive modification; the impacts of other types of GE plantation modifications and the potential consequences of gene flow from GE plantations (when allowed or encouraged) have been discussed elsewhere, including in a recent compendium volume on GE forest biotechnologies.

GENERAL REPRODUCTIVE STRATEGIES

Sexual Reproduction

In developing from seed to maturity, plants pass through a series of stages. This paper discusses six: seed production, seed dispersal, overwintering and long-term storage of seeds, seed germination, early seedling survival, and plant establishment and later growth.

Seed Production

Seeds are produced following pollination and fertilization of the female flower. The success of these two processes (measured by the number of viable seeds produced) is dependent on the age and vigor of the parent plant, as well as environmental factors such as weather. Age is important because a plant must be sexually mature before it can produce seeds. However, the age of sexual maturity differs dramatically among plant species.

Oaks, for example, become sexually mature around age 40, while alders mature around age 10. The vigor of a parent plant is directly dependent upon the environment in which it grows. For example, a harsh environment might produce a tree or shrub with poor vigor. This often results in a poor seed crop because so much of the plant's resources are tied up in processes related to survival. However, some plants respond to a harsh environment in the opposite manner, producing a heavy seed crop in order to generate viable offspring. Environmental factors such as weather are also important to the reproductive process. For example, cool, damp weather at the time pollen is shed plays a tremendous role in pollination because it limits dispersal by wind and restricts the activity of pollinating insects.

Seeds of broadleaved trees and shrubs are borne within fruits that come in a variety of shapes and sizes. In the Pacific Northwest, the following types are especially common.

Fleshy Fruits (those in which the seeds are enclosed inside a soft, fleshy ovary)

- **Berry:** Multiple seeds loosely arranged inside a fleshy fruit. The fruits of salal, huckleberries, elders, snowberries, and Oregon-grapes are all berries.
- **Drupe/drupelet:** Hard, usually single, seeds surrounded by a fleshy covering. The fruits of cherries, plums, Indian-plum, devilsclub, Californialaurel, and cascara buckthorn are all drupes. Drupelets are merely small drupes; the fruits of blackberries are actually aggregates of drupelets.
- **Pome:** Multiple seeds borne inside papery capsules, surrounded by a fleshy fruit. The fruits of apples, pears, hawthorns, mountainashes, and serviceberries are pomes.
- **Aggregate**: Multiple seeds, with each seed borne inside its own fleshy covering. The fruits of raspberries, blackberries, and magnolias are aggregate fruits.

Dry Fruits (those in which the seeds are enclosed inside a dry, often papery, ovary):

Dry fruits are further categorized by whether their outer covering splits open at maturity to release its seeds (dehiscent), or does not split open (indehiscent).

Dehiscent (splitting open along predetermined lines at maturity):

- **Legume or pod:** A multiple-seeded fruit that develops from a single carpel. When mature, the carpel splits open along two predetermined sutures, often explosively expelling its seeds. The fruits of Scotch broom, gorse, locusts, and honeylocusts are legumes.
- **Follicle:** A multiple-seeded fruit that develops from a single carpel. When mature, the carpel splits open along one side. The fruits of ocean spray, ninebark, spirea, and milkweed are follicles.
- **Capsule:** A multiple-seeded fruit that develops from two or more fused carpels. When mature, the carpels open in various ways. The fruits of rhododendrons, mockorange, willows, cottonwoods, boxwood, and Ceanothus species are capsules.

Indehiscent (not splitting along predetermined lines at maturity):

- **Nut/nutlet:** A single-seeded fruit with a hard outer shell. The fruits of California hazel, tanoak, golden chinkapin, and all oaks are nuts. Acorns are nuts capped by scaly or bristly caps (involucres). Nutlets are merely small nuts that may be winged, as in the alders and birches, or unwinged.
- **Samara:** Dry, single-seeded fruit with a prominent wing. The fruits of ashes are single samaras; the fruits of maples are typically double samaras.
- **Achene:** Dry, one-seeded fruit that is not winged but may have a feathery or hairy tail. The fruits of bitterbrush, sagebrushes, rabbitbrush, and mountain-mahoganies are achenes.

Seed Dispersal

Once mature, seeds are dispersed in a variety of ways. Although it is common to focus on what happens from the time seeds leave their parent until they reach the ground (primary dispersal), redistribution after they land (secondary dispersal) often determines the resources available to them for germination and early growth. The most common dispersal mechanisms for seeds are gravity, wind, water, and animals, often working in conjunction with one another.

- **Gravity:** Gravity plays a role in the dispersal of all seeds, but it is especially important for those that are heavy and unwinged, like acorns and other large nuts. Although gravity's role is often thought of in conjunction with primary dispersal, it is also an important secondary dispersal mechanism, especially on steep slopes.
- **Wind:** Wind is especially important for the primary dispersal of winged and plumed fruits such as samaras and achenes, carrying them far from their parents. It is also an important secondary dispersal mechanism, redistributing seeds and some dry fruits after they hit the ground.
- **Water:** Water is typically a secondary dispersal mechanism, carrying seeds away from their primary landing spot to new locations. The cottony tufts of willows, cottonwoods, and aspens are all redistributed by water, as are the samaras of ashes and maples and the winged nutlets of some alders.
- **Animals:** Animals are important primary and secondary dispersers for both fleshy and dry fruits. Birds and many small mammals have the ability to collect seeds either directly from the parent plant or after they hit the ground. Once gathered, the seeds may be swallowed immediately or stored for later consumption. When swallowed, some seeds are destroyed by the digestive process, but others pass through the animal's system intact and may be deposited far from where they were gathered. If stored, many seeds are never rediscovered by their hiders, leaving them free to germinate when conditions are right. Large fruits like acorns, hazel nuts, and walnuts are commonly stored, but so are small seeds like those of manzanita and Ceanothus.

Seed Germination

In order to germinate, seeds must undergo specific physiological changes during their storage period. To overcome dormancy, all seeds need water. In addition, some may need one or more of the following stimuli:

- **Light:** For example, red alder seeds need exposure to infrared light before they will germinate.
- **Heat:** Some species of Ceanothus need exposure to intense heat to break their seed coats. As a result, Ceanothus seeds often germinate abundantly following wildfire or prescribed burns.
- **Cold period:** Seeds of California hazel and bigleaf maple require a cold storage period lasting 2 to 6 months before germination will occur.
- **Scarification:** Either mechanical (scraping or scratching of the seed coat) or chemical (chemical breakdown of the seed coat). In a laboratory, soaking dogwood seeds in sulfuric acid accelerates germination; mechanically removing or cracking the hard wall around the seed aids germination of California-laurel. Outside the lab, animals play a significant role in scarification by passing seeds through their digestive tracts. Trampling by hooves also plays an important role in the germination of some species.

VEGETATIVE REPRODUCTION

In addition to reproducing by seeds, most broadleaved plants have the potential to propagate vegetatively. Sometimes new members arise directly from the base of the "parent" plant, as in basal sprouting, and sometimes they arise at quite a distance, as with root suckers and rhizomes. In each case, resources from the parent (called an ortet) are used to support the "offspring" (called a ramet). Over time, ramets are often separated from the ortet, becoming independent plants that are genetically identical to their parent. Collectively, an ortet and its ramets are called a clone. Some scientists call this process vegetative reproduction; others prefer to call it vegetative propagation, reserving the term "reproduction" for events involving sex. Regardless of what the vegetative process is called, it allows plants to rapidly occupy new territory, capturing available water, light, and nutrients. In plants that reproduce vegetatively, sexual reproduction is still important for moving species long distances and for maintaining genetic diversity within the plant population.

As plants capable of vegetative reproduction grow, they often produce a surplus of buds (sometimes called a bud bank) that have the potential to develop into new shoots or roots. These buds may be stored above or below ground—on stems, at the root collar, along underground stems (rhizomes), or along roots. In addition, buds may develop along existing stems, branches, or roots following some form of disturbance; these are called adventitious buds and can also give rise to new plants. Together, bud banks and adventitious buds result in the formation of new shoots and new plants. Five common methods of vegetative reproduction are rhizomes, layering, basal sprouting, root suckering, and shoot separation. Some species have the capacity to reproduce using several of these methods, while others are limited to one.

Mechanisms for Vegetative Reproduction

- **Rhizomes:** Rhizomes are stems that grow horizontally either partially or totally below the ground's surface. Like all stems, they produce buds that have the potential to give rise to new stems and leaves. In the case of rhizomes, most of these buds remain dormant, but some develop into new above-ground plants. The new aerial stems and the roots that develop to support them (collectively called ramets) have the potential to become independent ortets. Ramets and ortets are genetically identical to one another. Rhizomes, and the stems and roots that arise from them, are capable of forming a network of plant biomass so dense that the establishment of other species may be inhibited. Salmonberry, thimbleberry, salal, and Oregon-grape are several Pacific Northwest species that reproduce by rhizomes.
- **Layering:** Layering occurs when adventitious roots form along stems or branches that come into contact with the soil for long periods of time. After roots have become established, latent buds that have been stored along the branch or adventitious buds that form following root development may develop into new aerial stems. Over time, these new stems may become separated from the parent plant and grow independently. Layering enables a single plant to occupy and dominate a large area over time. Vine maple often spreads via layering.
- **Basal sprouting:** Basal sprouting occurs when a plant sends up new shoots (called sprouts) from buds located either at the base of the stem or on below-ground burls. Sprouting most commonly occurs when the original stem is killed, injured, or disturbed. Sprout clump size depends on the size and age of the parent plant. A large parent is capable of producing a tremendous clump of sprouts, while a small parent may produce only a small clump. Some species that commonly basal sprout are capable of producing sprouts even if the parent plant appears to be healthy and undisturbed. This sprouting ability allows a plant to maintain itself on a site under a wide range of conditions. Tanoak, bigleaf maple, California black oak, and Pacific madrone are several Pacific Northwest species that sprout profusely from their bases. The location of a species within its range may also influence its ability to sprout. Some species that sprout near the centers of their ranges will not sprout near their extremes; the pattern for other species is just the opposite.
- **Root suckering:** Root suckering occurs when the roots of a parent plant are capable of producing buds that sprout into new shoots. Almost any type of

disturbance to the root system, from heavy machinery to animal hooves, can trigger a suckering response; suckering also occurs in the absence of apparent disturbance. Quaking aspen is the Pacific Northwest species best known for root suckering, although it also occurs in cottonwoods and locusts

- **Shoot separation from the parent plant:** Some shoots have the ability to sprout new roots and shoots when severed from the parent plant. Humans take advantage of this attribute by taking "cuttings" and stimulating them to produce new roots and shoots; however, small branches can also become separated from the parent plant through damaging agents such as wind, flood, or animals. These separated shoots, when carried away from the parent plant by wind or water and deposited in a new area, can then become established if environmental conditions are favorable. Black cottonwood is one Pacific Northwest species that utilizes this strategy. In fact, its ability to abscise short, leafy shoots may help it colonize sand and gravel beds along streams.

Factors Affecting Vegetative Reproduction

Nearly all broadleaved trees and shrubs have the ability to reproduce vegetatively. In sexual reproduction, age and vigor of the parent plant are primary influences on seed production. In vegetative reproduction, primary influences are the species involved, the vigor of the plants, and the type of disturbance that occurs.

Most plants that reproduce vegetatively first produce a "bud bank"— buds stored along their rhizomes, roots, stems, or branches. Bud banks located on rhizomes and roots continuously expand as the rhizomes and roots extend laterally; buds stored on stems, burls, or root collars increase in number as the plant grows. In addition, the plant must have enough vigor (energy) to respond to disturbance. And finally, the plant must be stimulated to respond, either through disturbance to the plant or to its immediate environment. If these conditions are met, the plant may respond by producing several to several dozen new stems per plant, which may translate into tens or hundreds of thousands of new plants per acre, depending on plant density.

Even in the absence of disturbance, some vegetative reproduction is likely to take place (e.g., shoots from rhizomes will continue to be produced; layering will occur). In addition, as older above-ground stems senesce and die, buds in the bud bank may be released and sprouting may occur. As the forest canopy closes, sprouting plants may die back and revert to smaller forms with smaller bud banks, waiting until an opening occurs to sprout again.

IMPACTS OF REPRODUCTIVE MODIFICATION

We begin by outlining methods for GE of reproduction in forest trees. Then, to help interpret the significance of GE modifications to reproduction, we discuss the extent that forest management itself impacts tree reproduction. Next, we discuss the extent that various organisms utilize reproductive resources in our target genera. We conclude by considering the social factors that could influence acceptance or rejection of GE reproductive modification.

GE reproductive modification

There are a variety of potential and demonstrated options for GE-based reproduction modification in trees. In addition to the goal of containment, reproductive modification also has been sought as a means to improve vegetative growth by increased allocation of photosynthetic resources, reduce the production of allergenic pollen and decrease nuisance reproductive tissues in urban trees. Most published studies of GE have focused on male-sterility because pollen is the main source of long-distance spread and seed propagation would allow for continued sexual breeding. One method for obtaining male-sterility, genetic ablation (i.e. cell elimination), was first developed and commercialized in agriculture, and is effective and reliable in field trials of *Populus*, *Eucalyptus* and *Pinus*; Fig. 1).

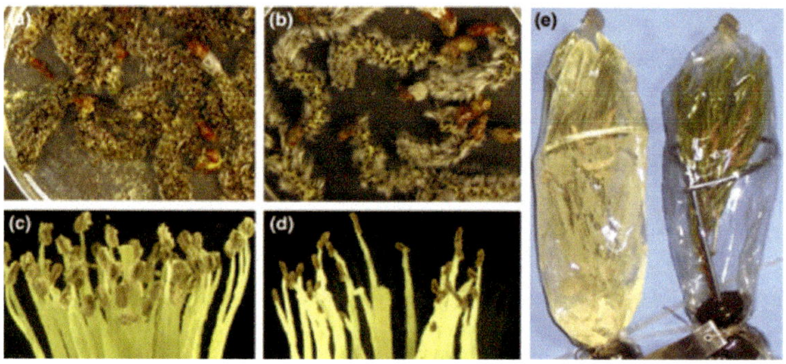

Fig. 1. Barnase expression leads to male sterility in field-grown trees. Male catkins of Populus tremula × tremuloides clone INRA 353-53 showing pollen release in Petri dishes from (a) control but not from (b) barnase trees. Eucalyptus occidentalis flowers of (c) control and (d) barnase (note shriveled anthers that are not releasing pollen). (e) Pinus rigida × taeda male cone clusters of GUS control (left) and barnase (right) in pollen isolation bags. Control flowers shed large quantities of pollen grains, whereas pollen grains were absent from barnase flowers.

However, because many intensively grown forest trees, most notably poplars and eucalypts, have seeds that can travel large distances by wind, water or animals, research also has focused on achieving complete bisexual (male and female) sterility. Both poplar and eucalypt plantations are widely established via vegetative cuttings, such that loss of seed production will still allow for tree propagation and establishment using current methods. Published results for achieving female or bisexual sterility have emphasized suppression of key flowering genes by RNA interference (Fig. 2), a method that can be highly stable in field-grown trees.

Direct editing of floral genes by use of CRISPR-Cas nucleases or similar approaches should be even more stable as entire genes can be deleted, but to our knowledge, gene-edited trees have yet to be tested in the field. Other methods that have been proposed for trees, but not yet demonstrated in the field, include the use of developmentally inducible recombinases to excise (i.e. to selectively remove from the genome) GE-modified genes during the formation of gametes.

This approach would allow normal flowering yet little or no release of transgenes. Genes that suppress the age-related onset of reproduction also are potential targets for breeding or GE, as a wide array of floral suppressor genes has been identified in basic plant research.

A common concern is that containment will be incomplete or unstable, or that asexual forms of dispersal will still allow for gene flow. In poplar, which has been widely studied, field research and simulations show that even incomplete or unstable male and female sterility would still provide high levels of genetic containment in various scenarios, including when linked to a strong fitness-promoting gene such as one giving resistance to an important herbivore of poplar.

Vegetative spread was also found to be a weak means of dispersal compared to pollen and seed. It is possible to modify organisms to prevent or slow vegetative spread, such as by preventing rooting except in the case of gene induction.

However, to our knowledge, such methods have never been shown to be effective and vegetative spread can often be readily controlled by forest management rather than genetic methods (e.g. by scouting for and removing root and branch sprouts within and near to plantations). It also appears feasible to engineer post-flowering forms of containment by targeting genes required for pollen tube growth or fertilization, or genes that take part in embryogenesis.

Many such genes have been characterized in model plants, which might enable the production of floral and fruit organs with essentially normal nutritive value for biodiversity, but without the possibility of propagule growth and development.

Figure 2 RNAi of floral genes is effective at altering tree floral morphology and fertility in field trials. (a) Wild-type (WT) female poplar catkins with well-developed carpels,

(b) female sterile RNAi-LFY poplar catkin lacking externally visible carpels, (c) female sterile RNAi-AG poplar catkin with replicated carpels (unpublished data, Strauss Lab), (d) WT female sweetgum flower with numerous white stigmas, (e) female sterile RNAi-AG sweetgum with stigmas converted to leaf-like projections, (f) WT apple flowers with five petals, 20 stamens and five styles, (g) RNAi-AG apple flowers with anthers and styles converted to petals and reduced male and female fertility

However, to our knowledge, such methods – although widely discussed for crops and for which there are a number of patents for reversible forms – have not been developed or field-tested in crops or trees. Given the wide scope for continued genetic innovation as knowledge of genes, development and methods for genetic modification grow, we can expect to see many refinements and new forms of genetic containment technologies in the future.

POLLINATION ECOLOGY

Pollination biology is a broad discipline embracing evolutionary ecology and taxonomy, as well as animal behaviour. Here, only a small selection of methods is presented, focusing on those field-based techniques relevant to *in situ* forest conservation. The significance of pollination ecology for conservation lies in understanding the processes influencing gene flow and reproductive success, and therefore the evolutionary viability of populations.

Practical conservation challenges include low production of viable seed by populations of threatened species, perhaps caused by loss of native pollinators, and the problem of reestablishing populations of species in areas where specialist pollinators are absent. Where threatened species are dependent on specific pollinators for reproduction, conservation action must obviously address the pollinators as well as the plant species of concern.

Tagging or marking flowers

In many pollination studies, flowers need to be marked or tagged. It is important not to use a method that might alter the attractiveness of the flower to the pollinator (for example, the use of red tags may attract hummingbirds). Inconspicuous tags can be made by tying sewing or embroidery thread on to individual flowers; it is also possible to write on the flowers themselves with felt-tip markers, indelible ink, or paint. Jewellery tags (made out of small rectangles of stiff paper), or gummed labels wrapped around the stem or pedicel, can also be used to label flowers. Alternative methods of marking flowers include thin copper wire, lengths of plastic drinking straws slit lengthwise, and coloured waterproof tape.

Pollen viability

One potential cause of reproductive failure is that the pollen is not viable. Direct tests of pollen viability involve depositing pollen on receptive stigmas and observing whether seeds are produced. Although it gives an unequivocal answer, the method is time-consuming. Indirect methods focus on correlating the ability to fertilize an ovule with some physiological or physical characteristic that can be determined relatively rapidly.

The most widely used and reliable indirect method for assessing pollen viability is the *fluorochromatic reaction (FCR) test*. The reagent is prepared by placing 10 ml of freshly made 15% sucrose solution in a transparent vial. A solution of 20 mg fluorescein diacetate in 10 ml acetone is prepared, and added drop by drop (1–3 drops in total) to the sucrose solution until it turns a light milky or greyish colour.

Dehydrated pollen grains should be stored for 10–30 min under high relative humidity before the test, to enable membrane recovery. The pollen sample is dispersed in a drop of the fluorescein diacetate solution; the microscope slide is placed in a Petri dish lined with wet filter paper for 10 min and then the drop is covered with a coverslip. The drop can then be examined under a fluorescent microscope, through a violet exciter filter.

Pollen grains with bright golden-yellow fluorescence can be scored as viable; undeveloped or empty grains will not fluoresce. To record percentage viability, first count the grains under white light then assess how many grains remain visible after switching to fluorescent light. Observations should be completed within 10 min after placing the coverslip over the sample.

Alternatively, pollen grains can be stained with a vital dye such as methylene blue (1%), neutral red (1%), or aniline blue (1%). A sample of the pollen grains is placed in a droplet of the dye and covered with a coverslip; the dye is replaced with water or glycerol after 5 min.

The percentage of dyed pollen grains can be counted. This method should be used with caution or only as a preliminary assessment, because values obtained may depart significantly from the actual value of pollen viability.

Pollen germination ability for many species can be examined in aqueous solutions of sucrose. Germination can be compared in solutions of different sucrose concentration (0–60%) as percentage by mass (g sucrose/100 g solution) to determine the optimum concentration for pollen germination, and the maximal germination rate as an indicator of pollen viability. Pollen grains are left in the sucrose solution for 24 h at room temperature, then examined under a microscope with a small drop of methylene blue. Germination of some species will be low unless they are kept under high humidity (95%) for a period beforehand (for example 30 min).

Pollen viability can also be assessed by examining the germination of pollen grains on the stigma, which also provides an indication of the stigma's receptivity. Pollen tubes can be detected in the style by using the following method. A solution of FPA is prepared (formalin 40%, concentrated propionic acid, 50% ethanol 5 : 5 : 90 by volume). The excised stigma and style is washed in FPA for 24 h, then stored in 70% ethanol. The style is then washed in tap water, softened for approximately 5 h in sodium hydroxide, then rinsed again in tap water for 1–3 h. Stain with 0.1% aniline blue in potassium acetate for 4 h, then squash the stained style under a coverslip and observe under a fluorescent microscope equipped with a filter set (maximum transmission 365 nm). Pollen tube walls and callose plugs should display bright yellow to yellow-green fluorescence. Alternatively, germinating pollen and pollen tubes can be stained with a hot solution of dye, composed of 150 mg of safranin O and 20 mg aniline blue in 25 ml hot (60°C) glacial acetic acid, which should be filtered before use. The styles are first hydrolysed in 45% acetic acid at 60°C for 10–60 min (until the style becomes soft enough to be squashed), then split longitudinally before staining. Pollen grains are stained blue, but the ends of pollen tubes stain red.

Pollen Dispersal

Understanding patterns of pollen dispersal is important for studies of gene flow. However, direct measurements of pollen movement are difficult to obtain. One option is to measure foraging distances travelled by pollinators, although it should be noted that not all pollen is transferred to the next flower visited, and carryover to a sequence of flowers can be extensive. Studies of the direct measurement of pollinator movements have recently been transformed by the development of radio tracking methods suitable for use with small birds and insects, providing some remarkable new insights into pollination ecology. Techniques include the use of *miniaturized VHF radio-transmitters*, *harmonic radar transponders*, and *passive integrated transponders* (PIT tags). Further technological developments are likely in this area with continuing progress in the miniaturization of transmitters.

Mating system

Plant mating systems have a major influence on patterns of genetic variation within populations, by affecting the extent of gene flow. There are five main types of mating system, although these are best considered as points along a continuum rather than discrete categories:

- Predominantly selfing
- Predominantly outcrossing
- Mixed mating
- Partial apomixis
- Partial selfing of gametophytes (in ferns).

The mating system can vary between populations of the same species, and even within a population there may be differences between individuals in the amount of selfing or outcrossing that occurs. Within individual trees, variation in the amount of selfing has been recorded in different parts of the crown. These sources of variation should be borne in mind when sampling populations or individuals for analysis.

Excluding flower visitors is a useful method of determining whether the plant is self-pollinating or not. This is usually achieved by enclosing the flower in bags made from fine-mesh cloth, gauze, cheesecloth, or nylon mesh (such as mosquito netting). Synthetic fabrics such as nylon are often preferred because they shed water more easily. Paper and cellophane have also been used to construct pollination bags, but are not very durable. The bags can be tied on to the plant with string or wire ties. Birds or bats can be excluded from flowers by using chicken-wire cages.

It is important to be aware that the use of bags or cages can alter the microclimate around the flower, which may alter patterns of flowering phenology. If there is a possibility that thrips are acting as pollinators, and there is a need to remove them, they can potentially be treated by regular applications of an insecticide such as malathion. Toothpaste has also been successfully used to exclude insects but not hummingbirds from tropical passionflowers.

To test for wind pollination (anemophily), airborne pollen can be excluded by using bags with a very fine mesh, made out of nylon stockings or cotton fabric, or even plastic bottles (Fig. 3).

Fig. 3 Pollination bag being used to exclude pollinators from flowers of the tropical tree Cordia alliodora

To determine whether pollen is being carried by the wind, microscope slides coated with silicone grease or glycerine jelly can be used as pollen traps. These may be placed both inside and outside bags placed around flowers, to test whether the bag is effective at excluding wind-borne pollen. If there is a possibility of both wind-borne and animal-borne pollen entering a flower, the relative importance of each pollination mechanism can be observed by comparing fertilization and seed-set with a variety of different bags designed to exclude different pollen sources. Bags with a mesh of 0.25×0.25 mm up to 1×1 mm are regarded as generally suitable for enabling airborne pollen to penetrate, but dense enough to exclude pollinators (Dafni 1992) - with the possible exception of thrips.

Exclusion of pollinators before anthesis, by using bagging or caging techniques, can be used in combination with hand pollination methods (see below) to identify whether flowers are able to self-pollinate and self-fertilize. Open-pollinated emasculation treatments (involving removing the anthers before anthesis) should be included in such investigations to differentiate between facultatively outcrossing and facultatively selfing plants.

Results from bagging and hand-pollination experiments provide information on the potential methods by which seed set can occur in a particular species, which can be analysed in a variety of ways. Researchers suggested the following index to measure self-incompatibility (ISI):

$$ISI = \frac{\text{Fruit set from self-pollination}}{\text{Fruit set from cross-pollination}}$$

Values of ISI can be interpreted as follows:

- >1=self-compatible
- >0.2<1=partly self-incompatible
- <0.2=mostly self-incompatible
- 0=completely self-incompatible.

IMPACTS OF FOREST MANAGEMENT ON REPRODUCTIVE ACTIVITY

Intensive forest management practices often strongly influence reproductive activity. Perhaps the most extreme examples are short rotation bio-energy plantations that are managed on a one- to few-year coppice system. Trees such as poplars are densely planted and literally cut-off near ground level during annual or semi-annual harvests, allowed to re-sprout from the roots, and then cut and re-sprouted for multiple harvest cycles. Given this early harvest, trees are kept in a largely juvenile phase of fast growth with little or no flowering. However, even in longer rotation forest plantation systems, trees normally do not become reproductively active for several years, and the high planting density common to most systems (to suppress weeds, promote height growth and stimulate branch pruning), further delay the onset of reproduction and reduce the intensity of flowering and fruit production.

Thinning, or low natural stocking, has been associated with increased reproductive activity in a number of studies. For example, researchers reported that Aleppo pine (*Pinus halepensis* Mill.) showed two- to six-fold increases in the number of new cones produced per tree in thinned stands compared to unthinned stands. Thinning also led to a four-fold increase in the percentage of newly reproducing trees. Benefits of thinning can last for 5 yr or more, as in *Pinus taeda*. This common trend (i.e. negative correlation between reproduction and stand density) was supported in 15 of 18 studies on conifer species. Exceptions to this trend were observed in both wild stands and seed orchards, perhaps because densities already were quite low, or as a result of other environmental factors known to affect flowering intensity (e.g. fertilization and irrigation in seed orchards).

Intensive genetic selection for rapid growth can further reduce the onset or intensity of reproduction, especially where propagation systems do not bias propagules toward those that flower early and most intensively (e.g. young seed orchards). In maritime pine (*Pinus pinaster*), researchers reported that a single cycle of selection for timber yield decreased the proportion of reproductive trees (51% in selected group vs 66% in control group), delayed the age and size at first reproduction, lowered the reproductive-to-vegetative allocation ratio, and affected the proportion of protandrous vs hermaphroditic individuals.

STAND-LEVEL BIODIVERSITY IMPACTS

Modifying tree reproduction and changes in resources

In order to interpret the potential impacts of flower modification on biodiversity, we need to be able to interpret how the impairment of plant reproduction can influence resources, which may then influence communities and ecosystem processes in a diversity of ways. Impairing flower reproduction may alter the availability of pollen, nectar, fruits and seeds. Understanding both how tree reproductive resources are used by animals, and the nature and timing of changes in reproductive activity, are important for interpreting potential effects and may vary widely among taxa (Fig. 4). For example, alterations of male vs female flowering, changes in the time of year that flowering occurs and changes in the year of onset during a tree rotation will have differential impacts on dependent organisms. Genetic methods can bring about different floral modifications, including a complete absence of flowers, pollen-less flowers, and (at least in theory) ecologically normal flowers, fruits and seeds whose pollen and seed are unable to give rise to viable progeny, or lack transgenes.

Fig. 4. Reproductive structures of Pinus, Populus, Pseudotsuga and Eucalyptus. Pistilate cones and catkins of (a) Pinus taeda, (b) Pseudotsuga menziesii and (c) Populus alba.

Staminate cones of (d) P. taeda, (e) P. menziesii and staminate catkins of (f) P. deltoides. (g) Eucalyptus globulus produces fruit and perfect flowers.

The relative impacts of floral modification by GE will depend on the plantation system in place and on natural environmental variation. Even without GE floral modification, resources from flowering can be extremely rare, either due to short-rotation harvests occurring largely before reproductive onset (e.g. bioenergy plantations) or when flowering occurs late in a rotation or at a low level due to high stand density, as discussed above.

In these situations, resources from reproduction are limited or absent in existing plantations, such that flower modification may have negligible consequences for altered resource distribution. In many plantations, resources from reproduction such as pollen, nectar and short-lived seeds may only be available during brief times of year, such that those resources are only available to certain members of the community and for a restricted period of time. Across plantation tree species, the type of reproductive output, its quantity (e.g. seedfall or biomass of pollen) and its nutritive value, will vary.

Natural environmental variation can also exert strong effects on reproductive resources, both due to variation in weather (e.g. moisture stress) and climatic cycles. Many tree species fail to reproduce for many years then flower heavily in others (masting), with flowering associated with complex climate and weather variation. Many conifers, including our focal genera *Pinus* and *Pseudotsuga*, have well-known masting cycles. In addition to this temporal variability, monoculture plantations also represent a spatial aggregation of floral resources. In agricultural contexts, monocultures of mass-flowering crops can impact pollinators both positively and negatively, implying that floral modification in mass-flowering tree plantations may also have impacts on pollinator populations, pollination services and honey production.

RESOURCE CONSUMERS AND BIODIVERSITY CHANGES

Although searches including terms for GE reproductive modification yielded > 400 results, we found no empirical studies of impacts of GE reproductive modification on biodiversity, thus all of the inferences below are based on studies of reproduction–biodiversity relationships in conventional plantations or wild stands.

Classifying animals into functional groups, rather than considering them taxonomically, facilitates understanding of the potential effects of resources changes on biodiversity. We analyze functional groups in terms of the morphological component(s) of the plant reproductive tissues that are fed upon. We begin with nectar (nectarivory) and pollen (palynivory) feeders, which we functionally lumped together with pollinators because they typically consume nectar and/or pollen rewards. We then consider florivory, which we define as consumption of whole or partial flowers before seed coat formation. When pollination is successful, a range of organisms feed on seed (spermivory) and fruit (frugivory), both pre- and post-dispersal. These categories are helpful in understanding the animal taxa involved, but species often feed on multiple stages of plant reproductive tissues. In discussing the taxa potentially impacted, we highlight economically and culturally important species, as well as those that are legally protected.

Nectarivory and palynivory

Many consumers of nectar and pollen also are pollinators, typically defined as animal pollen vectors (i.e. they move pollen between different flowers). Pollinators represent a substantial component of animals that feed on plant reproductive tissues, however, not all nectarivores and palynivores are pollinators. There are cases where flower visitors consume pollen and/or nectar without transferring pollen between flowers. One example of this is nectar robbery where a flower visitor circumvents flower opening to remove nectar without contacting anthers or stigmas. Pollination as an interaction likely evolved from palynivory by beetles although palynivory without at least some incidental pollen transfer may be rare given that most palynivores will visit multiple flowers and likely carry some pollen on their bodies. The exception is in monoecious or dioecious plants where palynivores may be able to focus exclusively on male flowers or plants. Three of our focal genera – *Pinus*, *Populus* and *Pseudotsuga* – are wind-pollinated and thus do not produce floral nectar (Fig. 4). These taxa likely have a community of palynivores, although we could find no studies characterizing the pollen-feeding assemblages of these taxa.

The primary tree genus that is animal-pollinated and also the focus of transgenic developments is *Eucalyptus*, which has a diverse and substantial community of flower visitors that vary depending on geographic location and species. This list includes marsupials, birds and a diverse assemblage of insects. In plantations without active pollination management, seed production is often low and pollen-limited, which may result in reduced seed consumer abundance and/or richness compared with other land uses. Most *Eucalyptus* species provide copious and sugar-rich nectar resources, and beekeepers in many parts of the world will bring honey bee colonies to plantations during flowering to promote honey production.

However, honey bee colonies apparently cannot be sustained on *Eucalyptus* floral resources alone, likely due to the relatively low protein content of pollen. Among flower visitors are several species that are of conservation concern in some regions, including the Swift Parrot (*Lathamus discolor*) and the Blue-Eyed Cockatoo (*Cacatua ophthalmica*), which are endangered and vulnerable species that frequently feed on nectar, flowers and pollen. Two mammals, the Lesser Short-Tailed Bat (*Mystancina tuberculate*) and the Black Capuchin (*Sapajus nigritus*), also have been shown to consume *Eucalyptus* fruit and pollen.

Florivory

Although florivory includes consumption of, or damage to, pollen and ovules (or immature seeds) in addition to other floral structures (e.g. bracts, petals, pistils, stamens), here we exclude strict pollen/ovule consumption from florivory as already discussed.

Florivory has been relatively understudied and there is little information on how florivory impacts the tree genera that are the focus of this review. One exception is the Douglas Squirrel (*Tamiasciurus douglasii*), which is known to consume entire mature pollen cones of Douglas-fir (*Pseudotsuga menziesii*). Another exception is *Populus* catkins, as several bird species use them for food. Cedar Waxwings (*Bombycilla cedrorum*) consume male catkins of Eastern Cottonwood (*Populus deltoides*). In experimental feeding trials, researchers showed that captive waxwings lost body mass on a diet of *Viburnum opulus* fruits unless their diet was supplemented by *P. deltoides* catkins, whose pollen was shown to be an important source of supplemental protein. Similarly, quaking aspen (*Populus tremuloides*) flower buds and catkins comprise a significant portion (66–100%) of the winter and spring diet of the ruffed grouse (*Bonasa umbellus*), although its high amounts of secondary metabolites also reduce its nutritional value. Various bird species may also use *Populus* catkins indirectly as a source of invertebrate food. For example, long-snouted weevil (*Dorytomus* spp.) larvae feed on pollen grains within staminate catkins and may provide a supplemental food source for birds during energy-intensive migration.

Spermivory

Seeds and plant ovules represent a rich resource utilized by various animals. We treat this separately from frugivory, or fruit consumption, which often also involves seed consumption, and which is covered below. Seed and ovule feeding (hereafter referred to as 'seed' for simplicity) can take place in various floral developmental stages and ecological contexts. Seeds can be consumed pre- or post-dispersal, by insect larvae that develop exclusively within a single seed and by vertebrates that consume many seeds over the course of a single day. In general, the mode of seed dispersal is of ecological importance for spermivory, given that wind- (or gravity-)dispersed seeds (including *Pseudotsuga*, *Populus* and *Eucalyptus*) tend to be of low nutritive value relative to most animal-dispersed seeds (including those of *Pinus*).

Two of our focal tree genera – *Populus* and *Eucalyptus* – are not considered animal dispersed and have relatively non-nutritious seeds that do not serve as a substantial resource for animal communities. For example, *Eucalyptus globulus* is considered gravity-dispersed and most seeds are estimated to fall within 20 m of the parent plant. A study of this species in Spain, where it is an exotic, found no evidence of any seed predation or animal-mediated seed movement. Similarly, *Populus* has small wind-dispersed fruits with single seeds that are unlikely to be an important food source for vertebrates.

However, the dichotomy between animal dispersal and wind/gravity dispersal is not perfect, and of one our focal genera, *Pseudotsuga*, is a counterexample. Its seeds are winged and thought to be dispersed primarily by wind and gravity, but are food items for: a number of mammals including mice, voles, shrews and chipmunks; birds including the Winter Wren (*Troglodytes hiemalis*), Pine Siskin (*Carduelis pinus*), Song Sparrow (*Melospiza melodia*), Golden-Crowned Sparrow (*Zonotrichia atricapilla*), White-

Crowned Sparrow (*Zonotrichia leucophrys*), Dark-Eyed Junco (*Junco hyemalis*) and Purple Finch (*Haemorhous purpureus*); and a range of insect seed predators. Relatively specialized feeders on *Pseudotsuga* seeds include the aforementioned Douglas Squirrel and potential subspecies of red crossbill, including the Douglas-fir Red Crossbill (*Loxia curvirostra neogaea*).

Of our focal tree genera, spermivory is likely the most important in *Pinus*. Pine seeds are considered to be wind- or gravity-dispersed in *c*. 85 of the 110 species in the genus, with the remaining 25 or so considered to be primarily animal dispersed, including the well-known Clark's Nutracker (*Nucifraga columbiana*)–Whitebark Pine (*Pinus albicaulis*) symbiosis. A rich community of animals feed on pine seeds ('pine nuts' to human consumers), including from species that are considered wind-dispersed.

This includes several species of conservation concern including the Tenerife Blue Chaffinch (*Fringilla teydea*), a species endemic to Spain that primarily eats *Pinus sylvestris* seeds. Similarly, the Maroon-fronted Parrot (*Rhynchopsitta terrisi*) and Thick-billed Parrot (*Rhynchopsitta pachyrhyncha*) are spermivores native to the southwest USA and Mexico that feed primarily on seeds from pines such as *Pinus ponderosa*, *P. menziesii*, *Pinus engelmanii* and *Pinus arizonica*.

Frugivory

Frugivory, or consumption of fruit, technically only occurs in angiosperms because gymnosperms do not produce fruit (they are named for their naked ('gymno') seed ('sperm')). Thus, seed consumption in gymnosperms, including *Pinus* and *Pseudotsuga*, is considered spermivory. Of our target species, frugivory in the angiosperms – *Eucalyptus* and *Populus* – is not substantial, as discussed above under spermivory.

Specialization on resources

The degree that species specialize on reproductive resources will influence the extent that altering reproduction will impact populations and communities. In general, it is frequently acknowledged that highly specialist pollinators and frugivores are rare. This is particularly the case in temperate systems where interannual and seasonal variation in flower and fruit availability makes specialization an evolutionarily risky strategy. For those species that do specialize on these resources, flower modification may have negative impacts. We expect that specialization will be less likely to occur when plantations use non-native tree species, but this will depend partially on evolutionary similarity between tree species and native species in the area, and the length of cultivation in the new geography. In addition, specialists may be uncommon in plantation forests, such that flower modification may have few impacts because these species may already be absent.

LANDSCAPE-LEVEL IMPACTS

Management practices often have effects beyond those observed at local scales; composition and configuration of habitats can have a considerable influence on biodiversity at regional and global scales. Such effects become particularly pronounced as landscapes become more homogeneous (i.e. beta-diversity is reduced). Given uncertainty about the degree that GE-induced changes to tree reproduction may influence habitat quality across taxa, here we consider five potential scenarios for landscape-scale effects: (1) creation of low-quality or nonhabitat, (2) plantations as permeable matrix, (3) the emergence of ecological traps, (4) changes in nutrient cycling; and (5) impacts on local and regional climate.

GE plantations as low-quality or nonhabitat

Removal of reproductive structures is most likely to render habitats of reduced quality, or nonhabitat, for pollinators and seedeaters. Much has been written on the effects of habitat loss and fragmentation on native biodiversity over the past 20 yr. Ultimately, it is the spatial extent of habitat loss due to GE plantations across the landscape that will determine compatibility with biodiversity. If GE plantations result in only minor habitat loss at the landscape scale (e.g. < 5–20%), we consider population-level effects to be highly unlikely. At these levels of habitat loss, population declines should be marginal and remaining habitat would tend to be highly connected, which facilitates inter-patch dispersal.

However, both theoretical and empirical work have indicated that thresholds exist where habitat decline results in disproportionate population declines. Mechanisms for such thresholds are many and include reduced dispersal capacity once landscapes become highly fragmented by nonhabitat, Allee effects (i.e. positive density dependence), and increased risks of predation and parasitism. The amount of GE plantation development that can take place at the landscape level before it compromises gamma diversity (i.e. the regional species pool), ecosystem functioning and population viability of native species depends on the dispersal capacity of organisms to recolonize managed stands. In turn, this will be a function of the spatial configuration of plantations and the degree to which they impede animal movement.

GE plantations as permeable matrix

Although reproductively modified GE plantations may not serve as breeding habitat for pollen- and seed-dependent animals, for some species the cover provided by tree canopy may facilitate movement among patches. In such cases, a matrix dominated by plantations could be preferable to other types of habitat loss that would occur through urbanization or agricultural development. We predict that mature forest-associated species are more likely to benefit from this positive 'matrix' effect. Conversely, for species that preferentially breed and disperse though early seral habitats (e.g. butterflies and many species of pollinators), closed-canopy GE plantations are more likely to constitute a barrier, thereby potentially reducing population viability.

GE plantations as ecological traps

Organisms regularly depend on environmental cues to make habitat selection decisions. However, in anthropogenically altered environments, previously reliable cues may become maladaptive. In such cases, organisms can become 'trapped' by their behavioral responses, resulting in reduced survival or reproduction. Given the novelty of GE reproductive modification, and the lack of studies about how organisms that feed on reproductive structures use them as cues, little is known about the degree that such plantations could constitute ecological traps.

If native species select habitat based on structural cues associated with the forest rather than direct cues such as the pollen or seeds (as food), the potential exists for GE plantations to act as traps. In such cases, even small amounts of GE plantations could potentially have negative impacts on populations because individuals could be drawn from otherwise appropriate habitat toward suboptimal locations with reduced fitness. Simulated population models could assist in understanding the degree of 'trap' effect required in relation to other demographic parameters before substantial risks to populations are encountered. However, the biological knowledge on which such models must be built is presently lacking, in no small part due to the rarity of all types of GE plantations, reproduction-modified or otherwise (*c.* 0.00033% of plantation area: less than *c.* 500 ha, mostly poplars in China), in relation to a global plantation estate of *c.* 150–200 million ha (of total forest area of *c.* 4 billion ha; R. Sedjo, pers. comm., May 18, 2016).

Role of pollen in nutrient cycling

Enriched with vegetation growth-limiting elements such as nitrogen, tree pollen may provide a significant supplement of macronutrients to forest ecosystems early in the growing season. For example, researchers found that a species of ectomychorrhizal fungus (*Paxillus involutus*) was able to consume nitrogen and phosphorus from the extensive pollen deposited by a member of our focal genera (*P. sylvestris*) and return a significant proportion of these nutrients to its host tree *Betula pendula*.

They demonstrated experimentally that *Betula* individuals grown with pollen and mychorrizal symbionts contained significantly more nitrogen and phosphorus than those grown without pollen, resulting in increased seedling growth. However, compared to litterfall, overall nutrient inputs into forest ecosystems via pollen appear to be small.

For example, researchers found pollen deposition in temperate pine forests in Korea to be *c.* 0.3% that of total litterfall by weight, with nitrogen inputs corresponding to 1% and 6% of total nitrogen and phosphorus, respectively. Pine pollen may nonetheless be a significant source of nutrients during the critical early growing period. In single-species plantations of *Pinus* and other wind-pollinated species, however, such inputs would likely be the same or less than the loss of nutrients from pollen release.

Tree pollen may provide an important source of nutrients for freshwater ecosystems. Researchers, for example, found that *Pinus* pollen enhanced productivity in a small (0.27 km^2) boreal lake system by increasing phosphorus and carbon availability. From May to

June, an estimated 7 kg km^{-2} of pollen (0.5% P by weight) was deposited onto the lake, providing a potential subsidy of 10 kg of phosphorus. The increased levels of phosphorus (and correlated increases of carbon) were associated with significant increases in diatom, phytoplankton and zooplankton abundance around the time of pollen deposition.

Role of pollen in regional climates

Changes in the abundance and types of tree pollen may significantly influence local and regional climate. Bioaerosols, including pollen grains present in large quantities, can alter cloud properties and influence atmospheric radiative forcing by scattering and absorbing solar and planetary radiation. Pollen also has been shown to be an efficient source of cloud condensation nuclei, but its relative importance to cloud formation is unclear given that it is often at low density and of short duration compared to other natural and anthropogenic sources of such nuclei. As with most other reproductive modifications, the climatic impacts of modified pollen production would be highly scale- and context-dependent.

SOCIAL DIMENSIONS

Support and opposition for GE

In addition to biological considerations, societal responses toward GE and related technologies can impact their adoption and regulation. Although specific studies on societal responses to reproductive modification in trees are scarce, research associated with more general use of GE in trees provides insight into whether people might reject or accept reproductive modification.

Concepts including attitudes, risk, knowledge and trust are related to acceptance of GE (Fig. 5). Attitudes involve evaluating an entity such as GE with some degree of favor or disfavor. Compared to agriculture, attitudes toward GE in trees and forests are less understood and there are only a few studies of attitudes about reproductive modification in forest trees.

In a sample of students in multiple countries (mostly European), researchers found generally positive attitudes toward using GE of trees in plantations, but not in wild natural forests. In a sample of the public in the USA, researchers found that attitudes were less favorable toward using GE in wild forests compared to other forestry techniques (e.g. breeding, planting, thinning), but GE was more favorable for addressing tree diseases (e.g. chestnut blight) compared to other forest impacts (e.g. climate change, increasing growth for timber; Fig. 6). GE also was more acceptable in native species compared to adding genes from exotics or distantly related species. Attitudes toward GE also vary between experts (e.g. forest scientists, forestry agencies) and the public, with experts often having more favorable attitudes. Researchers found that, compared to the public, experts (scientists, agencies, companies and nongovernmental organizations involved in forest issues) had more positive attitudes toward GE in forestry (Fig. 6).

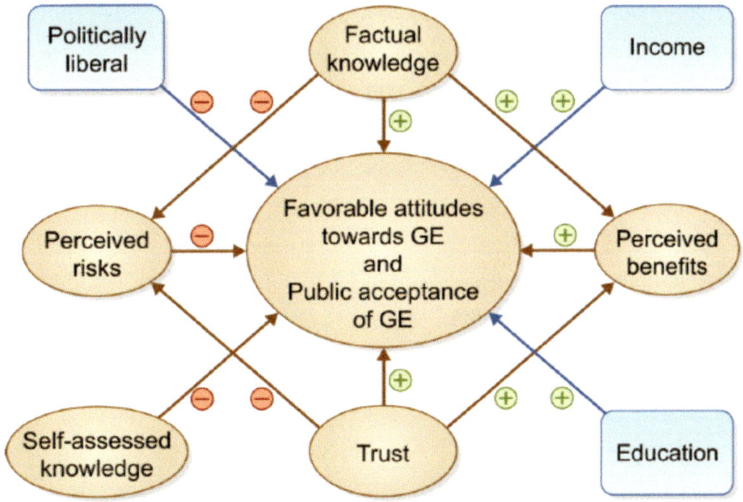

Fig. 5. Conceptual model of proposed relationships among favorable attitudes and public acceptance of genetic engineering (GE). +, positive relationship between concepts; –, negative association. Blue boxes, observed/directly measured single variables; brown ovals, latent concepts measured with multiple variables.

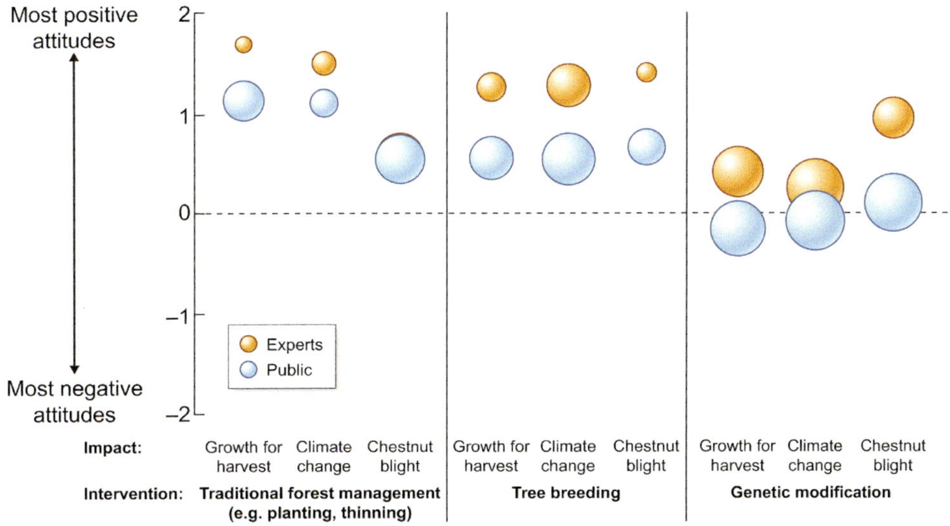

Fig. 6. Public and expert attitudes to forest health interventions in the USA. Survey results of public (n = 278) and expert (n = 195; scientists, agencies, companies and non-governmental organizations involved in forest issues) attitudes toward using different interventions for addressing various forest health impacts.

RISKS AND BENEFITS OF GE

Most research on public responses to GE has focused on risk perceptions, which are subjective evaluations of threats posed by hazards. In general, people express concern and skepticism toward new and unknown hazards that are not well-understood. Given its novelty, reproductive modification in trees is likely to be associated with risk factors such as newness, harm, dread, unnaturalness and unfamiliarity. Research examining public responses toward approaches for addressing forest impacts, for example, showed that biotechnologies (e.g. GE) were perceived as riskier than more well-known and familiar approaches such as planting and thinning. Interestingly, cross-breeding with related species (e.g. American chestnut with non-native Asian chestnut) was considered riskier than transgenic approaches (i.e. inserting genes from unrelated species such as from bread wheat to the American chestnut), suggesting that perceptions of naturalness or familiarity may elicit different concerns because Asian chestnut species, although more closely related to the American chestnut than wheat, are not as familiar to the public as wheat (i.e. source of bread). Researchers compared perceptions toward various GE applications and found more concern when interspecies boundaries were crossed ('transgenics').

Researchers identified interfering with nature and severity of consequences as two of the most important dimensions of risk perceptions related to GE. Reproductive modification of forest trees might be viewed similarly, especially given the potential biodiversity impacts outlined earlier. Reproductive modification may be viewed as particularly unnatural because it represents a fundamental change to the natural process of procreation. This viewpoint has been emphasized by monikers such as 'terminator genes' in activism campaigns against reproductive modification technologies and their potential impacts on ownership of these technologies.

There have been only a few studies examining perceived risks associated with GE in trees. Researchers compared consumer behaviors associated with GE paper products, woody biomass energy products and other wood products, and found that public safety issues and environmental impacts were among the largest concerns. Researchers examined perceptions toward transgenic forest plantations and found that potential for gene escape causing contamination of wild forests (i.e. gene flow between transgenic plantation trees and wild trees) was the greatest perceived risk. Other concerns included susceptibility to disease from lignin content modification and impacts from higher herbicide inputs (i.e. from use of herbicide resistance traits).

Researchers examined risks of eucalypt plantations and also found invasion potential and associated negative impacts on natural forests as primary concerns. These findings are consistent with Friedman & Foster who surveyed forest agency employees and found that loss of adaptation and genetic diversity, and changes in ecosystem components, were the largest perceived risks toward artificial regeneration in trees. Likewise, researchers found that members of the public were most likely to agree that changing existing genes in trees and adding transgenes (from wheat or other distant species) will change the genetics of wild or native trees and cause long-term negative impacts that are currently unknown. Gene-flow risks, however, could be mitigated or avoided by the genetic containment mechanisms that are the focus of this paper. Thus, information and education about this new capability might affect risk perceptions.

Although risk perceptions dominate the literature, acceptance of reproductive modification also may be related to the extent that people view these applications as beneficial (e.g. high growth rates, control invasive species). Researchers investigated perceptions of students (mostly European) toward GE trees in plantations and found that respondents perceived several benefits, including reductions in pesticide inputs, restoration of contaminated soils and higher tree productivity. These benefits, however, may differ based on the scale of production (e.g. plantation owners vs smaller community forests).

Knowledge and trust associated with GE

Given the complexity of reproductive modification, most people are likely to be uninformed and lack knowledge about this topic. Yet, knowledge and awareness are often related to positive attitudes and acceptance of GE. In the context of transgenic trees, researchers examined self-assessed knowledge and found that over 60% of students (mostly European) said they understood the meaning of transgenic trees and were willing to purchase GE products. However, fewer than half were aware of whether these trees could be deployed in plantation forestry (e.g. grown commercially, sold on the market). The investigators also believed there to be 'a serious perceived lack of knowledge about potential benefits and risks of the cultivation of transgenic forest trees'.

Given the technical nature of reproductive modification and likelihood that public knowledge about this topic is low, trust in knowledgeable experts (e.g. forest scientists, agencies) is an important consideration for understanding perceptions and acceptance. Trust is the willingness to rely on those responsible for making decisions or taking actions affecting public wellbeing. The public relies on trusted sources (e.g. scientists, agencies) to assess complex and unknown issues, and greater trust generally is associated with lower perceived risks, higher perceived benefits, more positive attitudes and higher acceptance. Sources such as agencies, companies and interest groups that are trusted by some people also can use information campaigns to influence perceptions and persuade people to support or oppose GE in various contexts (e.g. on private vs public land, plantations vs wild forests).

SYNTHESIS

Although reproductive modification using GE may have benefits for productivity or containment of exotic species on its own, we consider that its primary use will be to facilitate the use of GE for other traits that would increase productivity, reduce losses to stressors, or improve wood-processing efficiency. Thus, an important question is whether benefits of GE are sufficient to offset the added costs to breeding of having sterile or partly sterile trees, and if economic costs of technology, regulatory approval and market limitations are sufficient to justify use of GE in general and reproductive modification in particular. Technology costs are considerable barriers, as many species and genotypes are difficult and costly to transform under present technology, further emphasizing the need for large benefits to justify GE.

However, most published studies of the benefits of GE have only been of small-scale field studies of a few genotypes under highly controlled conditions, and the strong market-based preclusions to field studies make it unlikely there will be significant new, large-scale public research in the near future. GE, therefore, is likely to remain very limited in scale, potentially enabling considerable research on the ecological impacts of reproductive modification and other GE traits before their possible large-scale use.

Land sharing vs sparing

Conservation biologists are increasingly debating the benefits of land sparing (high-yielding agriculture on small land area) or land sharing (low-yielding wildlife-friendly agriculture covering more land). GE tree plantations have the potential to provide considerable production benefits, and such innovations are akin to developments that have occurred since the beginning of the agricultural revolution, which enabled high levels of food production on small land areas. However, those innovations resulted in large changes to the genetic composition of wild relatives of crops in many areas, as well as large reductions of general biological and genetic diversity within what became farms; similar potential changes via GE appear to be less acceptable today, presumably due to it being viewed as a more artificial process or the lack of knowledge on how prior innovations arose and have benefitted humanity.

Central to the debate about whether GE plantations will provide indirect conservation benefits is whether growing more trees per unit area will ultimately result in more land set aside for conservation. Interestingly, there appears to be only a weak tendency for increased crop productivity to translate into higher rates of nature conservation in agricultural systems. Improved crop-growing efficiencies likely decrease per unit production costs, which could boost demand and provide further incentives for agricultural expansion into previously undeveloped areas. Thus, intensification needs to be coupled with 'strong governance' or zoning that requires parallel land conservation efforts.

Zoning approaches are not new to forest management. Researchers proposed 'triad' methods where a forested land base is divided into three zones: intensively managed, extensively managed and protected areas. Several modeling exercises have explored the 'optimal' relative proportions of these zones, but clearly such decisions will depend greatly on the forest type and biodiversity goals under consideration, as well as the relevant political, social and economic contexts.

With respect to GE plantations, optimal amounts of different management types will depend greatly on the amount and configuration of habitat required by native species, as well as whether they are capable of moving through and/or breeding in GE plantations with reproductive modification. In addition, the large structural and species changes within the intensively modified plantation systems where GE is likely to be most used are, in general, already complete or underway, and have already had large impacts on reproductive resources – with or without GE added.

To the extent that such systems are already part of intentional or unintentional 'triad' systems to support biodiversity will to a large extent determine whether the added impacts of reproductive or other GE modifications are judged to have significant additional biological impacts. Such 'zoning' systems are commonly in place in many areas due to government controls on land uses or market-based controls such as from forest certification systems.

How should the Precautionary Principle be applied?

At first glance, following the 'Precautionary Principle' to biodiversity conservation would seem to necessarily require limited implementation of GE plantations, including those with modified reproduction. However, if this technology results in reduced land area required for wood production, or increased economic efficiency with spin-off benefits for society and environment – as has been argued forest plantations have done in many places, and that field trial experience with GE trees suggest could likely be delivered – application of the Precautionary Principle is by no means straightforward. As demand for wood products continues to expand, the risk of not using GE may come in form of increased rates of harvest in primary forest, conversion of native forests into non-GE plantations, and increased impacts to wild and planted forests from climate change-associated abiotic and biotic stressors. Thus failing to develop GE options could be viewed as being in opposition to the Precautionary Principle.

It is therefore critical that future policy decisions on GE be informed by the full suite of trade-offs between traditional forestry practices and implementation of GE plantations. Understanding the trade-offs, of course, requires continued research, both of GE benefits and biodiversity impacts. Unfortunately, both types of research have become difficult as a result of regulation and market obstacles (primarily forest certification systems) to field plantings. This preclusion is unfortunate as certification systems, such as PEFC, point to a lack of adequate research on GE trees as a reason for their exclusion: '… as the scientific evidence of potential benefits and dangers of genetically modified organisms (GMOs) and its impact on biodiversity remains insufficient and the society has not completed its debate, the PEFC General Assembly has determined that GMO cannot be considered as part of PEFC certified material'.

RESEARCH PRIORITIES

There are a number of research avenues to improve understanding of the possible paths and impacts of reproductive modification in plantation trees.

GE reproductive modification

Both the science and technology underlying GE forestry is in its infancy; basic genetic and genomic science on which GE is based continue to expand rapidly. Perhaps the most promising path to new and more effective methods for reproductive modification is through direct gene editing, such as by the CRISPR-Cas system, which can be used for mutating genes required for male and/or female reproduction, and can be reversed if needed for further breeding. Such mutations should be highly if not perfectly stable, and after initial research to establish efficacy, can be identified and selected (based on rapid DNA analysis) years before natural flowering occurs (i.e. as soon as GE shoots are produced in the laboratory). Trees with desired changes can undergo accelerated flowering studies of their effects in the laboratory (e.g. using *FT* gene overexpression technology) followed by field studies of their impacts on reproductive and vegetative development. These studies should test for possible off-target effects and instability, improve delivery and removal methods, and monitor for pleiotropic effects on tree growth and development. Among the target genes should be those that are active late in reproduction, such as those required only during pollen germination and embryo development. In theory, knock-outs of the latter types may produce flowers, and even fruits and seeds, that can serve as normal resources for organisms that feed on them.

Biodiversity

Our literature searches have revealed an absence of studies directly addressing the effects of reproductive modification on plantation biodiversity. Thus, the most urgent need is for field sampling and experimental perturbations to analyze the extent of use, and dependence of plantation biota, on reproductive structures. This knowledge is needed for informing the design of hypothesis-driven stand and landscape-level experiments, and models of what landscape impacts might be and how they could be mitigated.

It is also needed to help interpret the biological and social significance, and trade-offs, associated with reproductive modification compared to the many other human impacts on biological diversity. By understanding effects on biodiversity, this information also will provide insight into other ecosystem services that may be altered. Given there are so few GE plantations established around the world – and no species commercialized with reproductive modification – there appears to be ample time for surveys and experiments to inform management should social and political conditions become conducive to wider use of GE plantations.

Social dimensions

If reproductive modification is to be used for managing gene flow from GE and exotic plantations, it will be essential to understand what drives public attitudes and perceptions of benefits and risks. It also will be important to understand how reproductive modification is perceived by different stakeholders (e.g. public, experts, NGOs), as gene flow is itself a major concern voiced by opponents of GE in forestry.

An important element of acceptance is likely to be familiarity. Given scientific uncertainty around reproductive modification, fear of unknown potential impacts will likely continue until science can shed light on these impacts and effectively communicate them in the context of other impacts on biodiversity (e.g. from agriculture, urbanization, plantations) and similar types of common technological interventions (e.g. neutering pets, birth control, seedless fruits, GE foods).

Investigating benefits and costs of GE in forest plantations also will require ethical and political considerations such as how much biotechnology is needed, who makes final decisions and who takes responsibility.

Conclusion

Existing and developing genetic engineering (GE) technologies appear to be able to provide solutions to the problem of gene flow from GE and exotic plantation forests. However, there is little biological or social science research to inform about its impacts, the scale(s) at which it may become a concern, its significance in relation to other major sources of biodiversity perturbations, or the options for mitigation in the diverse species and coupled forest–human ecosystems where they might be applied. There are numerous opportunities for research to improve and extend GE technologies, and to fill gaps in knowledge of biodiversity impacts, but constraints in the form of market and regulatory restrictions to field research – and associated disinvestments in research from companies and governments – may make it difficult to fill these gaps in the foreseeable future.

11

FOREST AS HABITAT

INTRODUCTION

Forest is a habitat for many plants and animals because it provides a suitable environment for them. For example, neem, Sheesham, Palash, tiger, porcupine, elephants, jackal etc. The ground beneath the tree of the forest is the forest floor. It consists of a root, soil, decomposers and decomposing organic matter.

Long, woody trees in the forest also provide shelter creepers and also the climbers. Millions of people live in forests and all of them including living outside the forests depend on the products we get from the forest for their livelihoods. Forests are home to 80% of the world's terrestrial biodiversity. To look at, a forest habitat is a collection of trees that cover a specific area. There are lots of different levels to a forest:

- **The forest floor** - when leaves, branches, and fruit fall from the tops of trees, this is where they end up. It's also called the 'recycling layer' because when the debris breaks down, it all becomes part of the forest again. You can find lots of insect species in this layer of the forest too.
- **The herb layer** - this layer is made up of ferns, wildflowers, grasses, and tree seedlings. This grows thickest in forests that have a more open canopy and understory.
- **The shrub layer** - just above the herb layer, the shrub layer's plants are larger. This is usually made up of bushes and shrubs with twigs and branches.
- **The understory** - this is a layer of younger trees; they're still growing taller and aren't as tall as the oldest trees in the forest yet.
- **The canopy** - when the leaves and branches of the trees and plants intertwine, this is called the canopy. It covers the top of the forest and can vary between being thick with little of the sky visible, and more open which means you can see lots of the sky.

- **The emergent layer** - this is above the canopy and consists of trees that have grown taller than the rest; they're poking out above everything else.

Forests are home to80% of the world's plant and animal species(many of which are still waiting to be discovered), andthree hundred million people live in forests across the globe! Animal and plant species live in every layer of the forest, so it's easy to see why forest habitats are so rich with wildlife.

Not only are forest habitats essential for biodiversity, but they also hugelycontribute to the world's oxygen levels. Trees absorb carbon dioxide and other greenhouses gases and release oxygen in their place, which helps to combat climate change. Without forests and woodlands, the world wouldn't have as many different plants and animals, and we'd have much less oxygen to breathe.

COARSE WOODY DEBRIS

The term coarse woody debris (CWD) refers to a range of different sizes and types of woody material that can be found in forests, including logs, snags, chunks of wood, large branches, and coarse roots. The definition of what constitutes coarse, as oppose to fine, woody debris varies between researchers, but a typical minimum diameter is 2.5 cm. CWD is a very important habitat feature, particularly for fungi, invertebrates, mosses, and lichens, and may also provide sites for establishment of tree seedlings and sites for storage of carbon, water, and nutrients. Snags are used by many animal species, particularly birds, as sites for nesting, perching, and roosting. The ecology of CWD in temperate forests is reviewed by researchers. Methods for assessing CWD are described by researchers, on which this account is partly based. The simplest method to assess the input of CWD from living trees is to determine rates of tree mortality by using permanent sample plots or tagged trees. This may underestimate inputs, because large branches and broken tops of trees are not included. Input to CWD can also be measured on cleared plots, or by marking or mapping pieces present at the beginning of the period of observations. Alternatively, historical reconstructions of forest disturbance history can be used to estimate CWD inputs from catastrophic disturbance events.

When surveying, a distinction is generally made between standing dead trees (snags) and lying dead trees (or downed logs). The species, volume, biomass, decay status, number of woody fragments, surface area, and ground cover (projected area of fragments) are variables that are commonly recorded. Size class distributions and decay class distributions of wood pieces may be reported, along with descriptions of spatial patterns of distribution (such as accumulation in riparian environments) and orientation (for example, alignment in relation to slopes and direction of prevailing wind, an issue that should be borne in mind when sampling). Information on CWD is now routinely collected in many national or other largescale forest inventories, stimulated by the common inclusion of CWD volume as an indicator of sustainable forest management. Simple, subjective assessments can be made by visually assessing CWD at locations within a forest stand. However, if there is a need to compare different areas or to monitor changes over time, then more objective methods are prefereable. Note that when measuring CWD, variances tend to be high and therefore a large number of samples is typically required.

FOREST FRAGMENTATION

Introduction

During the second half of the 20th century, the rate of species extinction reached an almost unprecedented level in earth's history and some authors have suggested that the world is experiencing its sixth major extinction event. Four major human-induced processes are responsible for the present biodiversity crisis: (1) over-exploitation of species by hunting and fishing; (2) environmental deterioration, leading to decreased habitat quality (e.g. through eutrophication or salinization); (3) introduction of exotic species; and (4) fragmentation of habitats. The relative contribution of each of these processes to biodiversity loss is not clear, but it has been suggested that habitat fragmentation may be one of the most important threats to species diversity.

The habitat fragmentation process encompasses essentially three components: (1) pure loss of habitat; (2) reduction of patch size; and (3) increasing spatial isolation of the remnant habitats. Pure loss of habitat directly leads to species extinction. The role of decreasing patch area and increasing isolation on species diversity, however, is much more complicated and elucidating it has become a major topic in ecological research. Small habitat fragments can only contain small populations, whereas isolated habitat patches do not receive or send migrants to other fragments.

Species become extremely vulnerable to extinction below a certain population size, even when the habitat quality of the fragment where they reside is optimal. Extinction probabilities are expected to increase further when these small populations are isolated from each other, preventing them being 'rescued' from extinction by colonizing individuals from other fragments and hampering recolonization of empty fragments after extinction. Thus, populations are expected to become extinct when a certain threshold value of patch size and isolation is reached.

A second important consequence of decreasing fragment area is not directly related to the population size of the occurring species, but to the relative increase of the edge length of the habitat fragment, which allows potentially negative edge effects to penetrate into the fragment and affect the viability of the occurring species. Edge effects can be defined as the flux of matter, energy and species directed from the landscape matrix into the habitat fragment, and their extent increases significantly in highly fragmented landscapes.

It is important to note that in reality all landscapes and habitats are naturally heterogeneous and in a certain sense fragmented. Windthrow in a forest or the flooding of a wetland, for example, can create naturally fragmented habitats. Natural landscape heterogeneity, however, generally differs from human-induced fragmentation in terms of the smaller temporal and spatial scale and the higher degree of connectivity between the remaining fragments. Therefore, it can be expected that many species are more or less adapted to a certain degree of natural habitat fragmentation.

Forest fragmentation can be induced by natural causes like lava flows and other natural calamities or by man-made causes like the conversion of forest to farmland, and infrastructure development in Protected Areas. The anthropogenic fragmentation of pristine terrestrial habitats is occurring at an accelerating rate globally.

Implications

Fragmentation affects biodiversity by decreasing the habitat, and because of the opening up of the once extant habitat, now we have changes in physical and biological properties, and the disruption of movement and dispersal patterns of the native wildlife.

Habitat Loss

Habitat destruction, defined as the elimination of the conditions necessary for animals and plants to survive, not only impacts individual species but the health of the global ecosystem.

Loss of habitat is one of the biggest ramifications of forest fragmentation. When a forest area is fragmented, the plant and animal species in that area are rendered shelter less. This could endanger many endemic species and can even lead to its extinction. This has far-reaching impacts on the planet's ability to sustain life.

For example, if we consider the Himalayan ranges, a study by WWF shows that habitat loss is extensive in the region with - over 75% of the original habitat destroyed or degraded.

The pressures are ever increasing due to unsustainable extraction of resources and the incursion of linear infrastructure such as roads in this geologically and ecologically fragile region.

Forest Edges

Fragmentation leads to increased light levels, higher daytime temperatures, higher wind speeds, and lower humidity. It changes and disrupts the microenvironment at the fragment edge. Species which are sensitive to these changes such as temperature and humidity can be wiped out from these fragments.

Moreover, these factors also contribute to forest fires, which is a grievous threat to biodiversity. Fire can also spread to these fragments from nearby agriculture fields or other human settlements. In this process, many species will be eliminated.

These patches also tend to be vulnerable towards invasive species, leading to altered ecosystems.

Disruption of Movement

Forest fragmentation divides a larger forest into two estranged forest patches. This would obstruct the free movement of animals from one region in the forest to another. When there is no other route for movement between landscapes the animals try to move through human-dominated landscapes which often leads to conflict with the animal getting killed, loss of life, livestock or property.

Due to lack of habitat for wildlife to move, human-wildlife conflict with tigers, leopards, elephants, wild boars and other ungulates are on the rise. From the perception of wild animals, areas of human settlement or agricultural lands are nothing but part of the larger habitat that existed in the past.

Wildlife Corridors

Retaining or restoring connectivity between fragmented forests is of greatest importance. Ensuring connectivity between forests and Protected Areas is what we call corridors. As mentioned earlier, loss of connectivity between forests results in human-wildlife conflict as wildlife especially long-ranging animals such as elephants and tigers come out of the forest into human-dominated landscapes that result in crop-raiding, damage to life, property and live-stock causing animosity towards wildlife on part of communities residing adjacent to the forests. Lack of connectivity also often results in local extinctions of species, inbreeding, lack of genetic diversity and non-adaptation to climate change.

These corridors provide a safe travel route for animals moving from one region int the forest to another. Thus, animals are not forced to wander into the human-settlements, thereby mitigating conflict.

Wildlife corridors may be natural or artificial. Natural corridors are usually thin strips or a series of small clumps of high-quality habitat that connect the isolated patches. Artificial corridors are generally constructed where there is a lot of human activity. 'Land bridges' or underground tunnels are artificial wildlife corridors that provide a safe way for animals to cross a road throughout the day.

EDGE CHARACTERISTICS AND EFFECTS

Forested habitats are often distributed as patches within a landscape. The characteristics of the edges of such patches, where they meet other habitat or land-cover types, often differ from those of patch interiors. Characterization of habitat edges has become a major theme in ecological research and, in response to the widespread concern about the impacts of forest fragmentation, has also become important for practical forest conservation and management. Most investigations of edges focus on describing ecological patterns, such as changes in abundance of species with respect to the edge, but edge characteristics may influence a number of ecological processes such as dispersal, migration, predation, competition, and gene flow.

Remote sensing methods such as aerial photographs can also be used to provide descriptions of edge characteristics. Forest edges can be described by various attributes such as length, width, shape, vertical, and horizontal structure, density, or interior to edge boundary. Researchers describe an assessment procedure for the forest edge that was operationally applied in the second Swiss national forest inventory (NFI). In the Swiss NFI, sample plots are distributed in a systematic grid. Whenever a forest margin lies within 25 m of the centre of a field plot, an assessment of the forest margin is conducted. A 50 m line forms the basis for the forest edge assessment. Along the transect, floristic diversity, habitat features (especially for birds and insects), and the aesthetic value for recreational purposes are assessed.

To provide an example of a typical research investigation, the approach used by Lopez-Barrera and Newton for their work in the montane forests of Chiapas, Mexico, is described here in some detail. To describe edge characteristics, transects 80 m long and 10 m wide were established in each of six study sites. The transects were established perpendicular to the forest edge, running from the forest (60 m) through the edge into the forest exterior (20 m into the neighbouring grassland). The choice of transect length was influenced by the scale of heterogeneity observed within forest fragments and neighbouring forest-free areas, and to avoid the proximity of other nearby edges influencing the results. The edge was defined as the line coinciding with the base of bordering mature (> 30 cm dbh) tree stems (following Oosterhoorn and Kappelle 2000). Within each transect, all the woody plants were counted and measured, to provide an assessment of stand structure and composition.

This investigation was also designed to examine the effects of edge contrast: the difference between 'hard' forest edges, defined as those where adjacent vegetation has a simple homogeneous structure, thus creating a sharp contrast with adjacent forest cover, and 'soft' edges, which where characterized by more structurally complex successional vegetation maintained for 15-20 m from forest edges (Fig. 1). Particular care was taken in this study to establish genuinely independent replicates, for example by establishing transects across edges of six different forest fragments. Lack of replication, or pseudoreplication, is a feature of many edge studies.

Once the edge characteristics have been described, typically an investigation also examines the distribution or behaviour of species across the edges, or some particular ecological process (Fig. 2). For example, Researchers describe the effects of different forest edge types on seed dispersal, predation, and germination and seedling establishment in relation to plant-animal interactions and the dynamics of forest patches within the landscape.

A thorough review of recent ecological research into habitat edges is provided by researchers, who highlight the following methodological points:

- Edges are generally defined as boundaries between distinct patch types, so the identification of edges depends on how patches are defined within a landscape.

Patch definition can occur at a variety of scales, and obviously varies with the characteristics of the study area.

Fig. 1 Vegetation may differ in mean height and density across edges, leading to edges with different characteristics. For example, 'soft' (low-contrast) edges (a) may be differentiated from 'hard' (high-contrast) edges (b). The degree of edge contrast may affect the permeability of the edge to flows of energy, material and organisms.

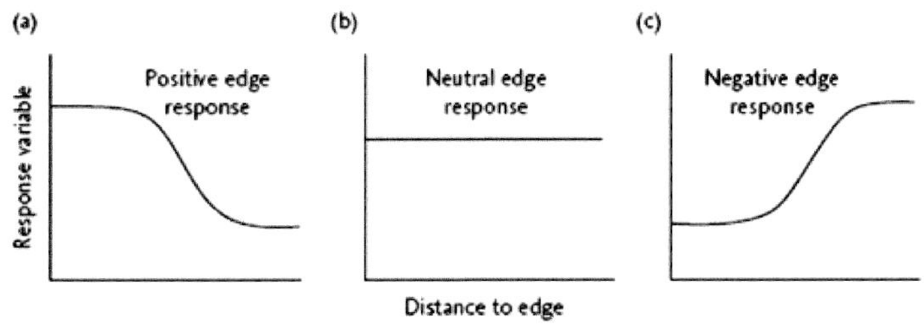

Fig. 2 Three classes of ecological edge responses with respect to distance from the closest habitat edge. Responses are generally categorized as (a) positive edge responses, where the variable of interest increases near the edge; (b) neutral responses, where there is no pattern with respect to the edge; and (c) negative responses, where the variable decreases near the edge.

- Studies of edge effects in which the variables that determine organism abundance are assessed are much more powerful than purely correlational studies, but they are relatively rare in the scientific literature.
- Most edge studies have low replication, few distance categories, and little penetration into patches, and this limits their ability to detect responses.

Little progress has been made in extrapolating responses measured at local scales to larger scales, despite the value of the latter to management and conservation strategies. Two models are available for 'scaling up' edge responses to the landscape scale: the core area model and the effective area model. Core area models use an estimated distance of edge influence to determine the amount of habitat in a patch that is not affected by edges (the core area); this approach is often used when considering reserve designs. In situations where most core habitat has already disappeared, the effective area model may be preferred. This method describes density (or other variables) as a function of distance from edge, allowing quantitative predictions of distributions to be made for an entire landscape.

It is important to determine the distance that edge effects extend into habitat patches. This value is often referred to as the depth ofedge influence (DEI). In many investigations, DEI is determined by visual inspection, and results are influenced by study design (for example, the length of survey transects and the number of distance categories). Robust methods for analysis of DEI including use of appropriate statistical analyses are described by researchers. Abiotic and plant responses are generally reported to extend up to 50 m into patches, invertebrate responses up to 100 m, and bird responses 50-200 m.

Most studies ignore the potential effect of being near more than one edge. Linear distance to the closest edge is generally used as the main explanatory variable, and therefore care should be taken to avoid placing plots near corners or other converging edge types, to limit their potential influence. It is possible that edge responses are different near multiple edges, or where edges converge, but these aspects have rarely been examined.

Ideally, the effect of edges on population dynamics should be examined by assessing the key parameters that ultimately determine distributions (birth, death, immigration and emigration rates), but few studies have achieved this to date. Many studies have measured community changes near edges, with the most common result being an increase in species diversity, but the causes and implications of such changes in diversity near edges often remain unclear.

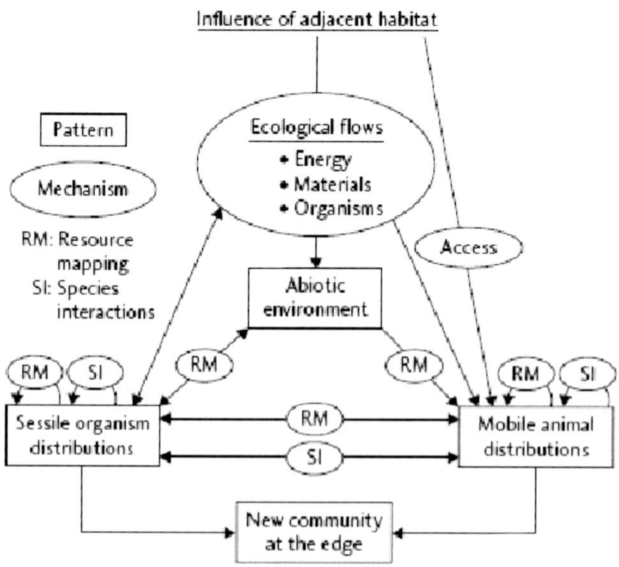

Fig. 3. A mechanistic model describing how the distributions of organisms are altered near habitat edges. Patterns in the abiotic environment, organism distributions, and

community structure (boxes) are influenced by four principal mechanisms (ovals). Ecological flows of energy, material, and organisms across the edge influence the abiotic environment as well as organism distributions. Organisms map onto changes in the distribution of their resources. Changes in species' distributions near edges can lead to novel species interactions that can further influence abundance and distributions. All these changes in species distribution lead to altered community structure near edges.

Researchers also provide a mechanistic model to provide a framework for edge studies (see Fig. 3).

KINDS OF FOREST HABITATS

There are generally three main types of forests:

Tropical

Tropical forests are also known as 'rainforests' because they occur in areas where there are high levels of rainfall and no dry season. These kinds of forests have an extremely thick and closed canopy and they're the most biologically diverse in the world. The most famous rainforest is the Amazon Rainforest in Brazil.

Boreal

Boreal forests grow where temperatures are freezing for 6-8 months of the year. Because of these harsh conditions, the forests usually consist of coniferous trees, including pine and birch trees. These forests are found at the north of the globe, most commonly inRussia, Canada, Alaska and Scandinavian countries.

Temperate

Temperate forests are found in-between tropical and boreal regions, located in what's called the 'temperate zone'. They have high levels of precipitation, humidity, and the most common trees are deciduous. There are two types of temperate forest: a deciduous forest and an evergreen forest. You can find these in the Eastern United States, Canada, Europe, China, Japan and parts of Russia.

The two types of trees that make up most forests are deciduous and coniferous.

Deciduous trees have leaves flat leaves that change from green to orange or golden brown in the autumn, fall off in the winter and grow back in the spring. Maple and oak trees are deciduous trees.

Coniferous trees have sharp pines in place of leaves and they grow cones, which is where their seeds are stored. Most conifers are evergreen trees (which means they stay green all year long and their leaves don't fall) but some aren't. 'Conifer' simply means 'cone bearer'.

PLANTS LIVE IN A FOREST HABITAT

Tropical Forests

Did you know that two-thirds of plant species in the world are found in tropical forests? They thrive in the humid environment and provide food and shelter for the many animal species that live there too. These plants include:

- Orchids
- Ferns
- Kapok Trees
- Banana Trees
- Rubber Trees
- Avocado Trees
- Bamboo
- Cacao
- Goliath Water Lilies
- Passionfruit Flower

The forest floor is where most of the plants grow, but this also means that there's higher competition for nutrients and light. This is why some plants have adapted to grow on branches, and there are even 'aerial plants' which collect nutrients from the air!

Boreal Forests

Boreal forests have long, freezing winters, short, dry summers and often experience wildfires. This means that the plants that grow there have had to adapt to harsh conditions. As a result, the main species of plants in these forests are coniferous. The coniferous trees that grow in boreal forests are:

- Pine
- Fir
- Larch
- Birch
- Spruce
- Aspen

Shrubs also grow in boreal forests, including alder, mountain ash and willow shrubs. Some even grow fruits such as blueberries and cranberries, which are a source of food for the animals there, as well as aboriginal people. There are also different species of herbs, lichens, mosses and fungi.

Since the soils in the boreal forest are acidic and difficult to collect nutrients from, some plants have evolved a unique way of gaining nourishment - from animal protein! The Sarracenia pitcher plant, for example, lures its prey (usually insects) into a pool of digestive liquid. Once the insect is broken down, the carnivorous plant takes in all the nutrients.

Temperate Forests

Unlike tropical and boreal forests, a temperate forest's environment is always changing with the seasons. It distinctly experiences each season, which means winters are cold, and summers are warm. The soil in these forests is also very fertile, which means it's the perfect place for a variety of plant species to grow.

On the forest floor, you can typically find lichen, moss, wildflowers, ferns and other small plants. In the understory, you can find shrubs such as the rhododendron and ones with fruits such as gooseberries and blueberries. There are flowering plants, such as the White Trillium. High up in the trees' branches, you might also see Mistletoe, which steals water and nutrients from inside branches.

There's a mixture of trees in the temperate forest, thanks to its changing environment. There are hardwood trees, such as:

- Maple
- Birch
- Oak
- Magnolia
- Sweetgum
- Beech

And there are also conifer trees, like spruce, fir and pine trees.

ANIMALS LIVE IN A FOREST HABITAT

Just like plants, forest habitats are full of a huge variety of animal species, and they can differ for each forest type too. Forests also provide shelter for forest habitat animals, whether it's within the tree roots or trunks, or up in the branches.

Tropical forests

In tropical rainforests, the largest group of animal species found there are insects. What's more, almost 40-75% of all species on Earth live in these forests! They're hugely responsible for our planet's diverse wildlife.

The forest floor receives very little sunlight (about 2%) and it's home to the larger animals of the forest, including the tapir, the okapi (also known as the forest giraffe), and the Sumatran rhinoceros. There are also lots of reptiles (snakes, lizards, turtles, tortoise), amphibians (frogs, toads, salamanders, newts), and insects (African Weaver Ant, Amazon Caterpillar, Rhinoceros Beetle, Heliconius Butterfly) in this layer.

The understory layer receives a little more light than the forest floor at around 5%, but it's still covered by the dense canopy. Here, there are small mammals, birds, and even more reptiles. The jaguar prowls this layer as a predator, hunting the smaller mammals and birds as its prey. Other forest habitat animals in this layer include:

- Boa Constrictor
- Poison Dart Frog
- South-American Coati

Animals live in trees

On the canopy, there are more agile animals such as birds and monkeys, but there are also reptiles and invertebrates too. Animals that live in the trees are known as arboreal animals, and they spend most of their time eating, sleeping, playing and rearing their young in the trees.

Geographically, most arboreal animals are concentrated in tropical forests, but you can also find them in all forest habitats across the globe. Some examples of animals that live in the trees include insects, arachnids, amphibians, reptiles, birds and mammals. The emergent layer is home to species such as the King Colobus, a black-and-white species of monkey, as well as the Crowned Eagle and the Large Flying Fox, which is actually a bat.

We've compiled a list of some other familiar animals you'll find living in the trees:

- **Three-toed Sloth.** Sloths are the world's slowest mammal, in fact,they're so slow that algae grows on their furry coats!
- **African Grey Parrot.** These talkative fellas are known for their extreme intelligence, earning them the title of "The Einsteins of the Bird World."
- **Toucan.** The word "toucan" comes from the sound the bird makes and their songs often sound like croaking frogs.
- **Hornbill.** These guys sure can eat; the greater hornbill can consume as many as 150 figs in one meal!meal!
- **Kinkajou.** These small mammals can turn their feet backwards to run easily in either direction along branches or up and down trunks.

HOW DO ANIMALS THAT LIVE IN THE TREES SURVIVE?

Tree-top living sure isn't easy, but animals that live in the trees have learnt to adapt to a life among the clouds (the emergent layer). Thanks to evolution, many tree-top animals have gained specialised adaptations in their bodies which helps them survive in their habitat. Below is a list of some of these fascinating adaptations:

- **Prehensile tails.** These are specially adapted tails which are able to grasp onto things. They help with a variety of tasks such as gathering food, climbing and holding onto objects (such as tree branches, food etc.)
- **Gliding membranes.** Stretchy membranes between legs help arboreal animals such as flying squirrels to glide easily from tree to tree.
- Low centre of gravity. Having a low centre of mass minimises the chances of falling out of trees!
- **Swinging arms.** The ability to swing from tree to tree using just the arms is a special adaptation called "brachiation".
- **Feet of grit.** Having feet which are able to grip tightly onto branches is imperative for a life in the trees. Claws, adhesive pads and flexible ankle joints that can turn forwards and backwards (such as the Kinkajou's) are all adaptions useful for gripping tightly.

Boreal Forest

Boreal forests are home to more than 85 species of mammals, from rodents to marine mammals. There are numerous species of deer, including:

- Western Roe Deer
- Elk
- Reindeer
- Siberian Musk Deer
- Mule Deer

There is also a range of **predators** that hunt in the boreal forests, such as:

- Arctic Fox
- Tundra Wolf
- Polar Bear
- Amur Leopard
- Siberian Tiger
- Eurasian River Otter
- Eurasian Badger

Down on the forest floor, there are plenty of small mammals like hedgehogs, red squirrels, flying squirrels and even the Siberian Chipmunk and Arctic Hare.

In boreal forests near oceans and rivers, you can find:

- Northern Fur Seal
- Walrus
- Humpback Whale
- Narwhal
- Orca
- Pacific White-sided Dolphin

Temperate Forests

Since the vegetation in temperate forests is seasonal, there are fewer mammals in these forests than in tropical ones. The mammals that do inhabit these forest habitats will hibernate in the winter, and the birds that live there will often migrate to somewhere warmer, and then come back in the spring.

In temperate forests in the US, you can find Brown Bears. They live in the woodland during the spring and the autumn season but, come winter, they prefer rocky areas where they can find caves to stave off the cold weather and snow.

Other mammals you can find in the temperate forest include:

- Red Foxes
- White-tailed Deer
- Raccoons
- Porcupines
- Opossums

In temperate forests in the UK, it's possible to see some of these animals:

- Grey Squirrel
- Fallow Deer
- Vole
- Sparrowhawk
- Brown Long-eared Bat
- Hazel Dormouse
- Pine Marten
- Purple Emperor Butterfly
- Wild Boar
- Grass Snake
- Long-eared Owl

LIFE OF ANIMALS IN FORESTS: EXAMPLES

Life of animals in a forest habitat is quite different from other biome. Diverse forest animals require varying kinds of habitat, conditional to the type of food and shelter they need. Some select the vegetation found in young, open forests.

Others necessitate features like snags or fallen logs, which are usually found in older forests. Some animals live in forests of any age. As forests age and the habitat configuration alter, the occupying wildlife will alter as well.

Life among the trees can be rough. However, owing to some quite special adaptations, a large quantity of marvelous animals thrives in our forests, such as squirrels, bears, and the white tailed deer of Central America.

Evolution is at the origin of all life on earth, making animals acclimate to live on. After evolving for millennia, the species below are all seamlessly adjusted to forest life.

Goshawk

The goshawk is one of the UK's most mysterious birds and a mystic forest animal, but it's anything but reluctant and reserved. These raptors are heat-seeking bullets prepared with a lethal range of adaptations.

Fig. 4. Goshawk

Goshawks thrust themselves through holes in the overgrowth by pushing up and launching their talons forward. They fold up their wings to soar between openings in the trees, all the while their tail fans out like a third wing to sustain lift.

In the meantime, semi-transparent additional eyelids close to shield their eyes from thorns. With this amazing agility, goshawks are capable to capture squirrels, wood.

The grouse is also native to the UK and is a game bird found in pine forests.

Pine Marten

The pine marten's habitat is, regrettably, part of what makes this animal so uncommon. It prefers woodland, but is generally limited to Scotland because of habitat loss and past maltreatment.

Fig. 5. Pine Marten

Strong semi-retractable claws make them brilliant climbers. Also, they have flexible ankle joints, which mean they can scurry up and down trees when chasing prey.

Treecreeper

Treecreepers scuttle up tree trunks effort-lessly. They are impeccably adjusted for life in the forest, with powerful feet and elongated, curved claws.

Fig. 6. Treecreeper

These come in use when searching for insects in the gaps of tree bark. A delicately curved bill captures prey while steadiness is kept with a firm tail.

It is further believed that the bill and claw length of treecreepers may differ rendering to the seasons. This lets them make the most of scavenging prospects throughout the year.

Stag Beetle

The large mandibles are remarkable. However, they aren't what characterizes the stag beetle a forest expert. That is all down to their larvae. Stag beetles spend anywhere from three to seven years in their larval phase.

Fig. 7. Stag Beetle

Concealed beneath the soil, they feed on that woodland specialty: rotten wood. Grown beetles live for just a few months, expending up their fat stores, but will also eat fruit that fallen on the ground and drink tree sap.

Red Squirrel

It's no surprise that red squirrels are perfect for forest survival. Their extended tail and sharp claws assist them to dart up and down soaring trees and jump from branch to branch. Similar to pine martens, their flexible ankles allow for incredible maneuverability.

Fig. 8. Red Squirrel

But red squirrels have an extra adaptation. With four toes on their front paws but five at the back, they have even better control when up in the trees.

Wild Boar

Nonexistent for centuries, wild boar have come back to some of the UK's woods. This is because of absconding from farms and perhaps unlawful releases.

Their powerful snouts are adapted to digging through woodland soil, while thin brown hair aids them to blend in among the trees. Piglets are even stripy like humbugs for further camouflage. What makes the wild animal most triumphant, though, is their extremely diverse diet. They will feast on almost anything found in the forest, from bird eggs and worms to acorns and carrion.

Purple Emperor Butterfly

The purple emperor is one of the most obscure creatures in the UK. It spends most of its life up above the canopy, flapping around in quest of aphid honeydew and tree sap. But this isn't their only food source.

Fig. 9. Purple Emperor Butterfly

Purple emperor butterflies have bizarre palates. They will come down from their large realm above the oaks to feed on urine, animal droppings and carrion. They can't get enough of the salts in this strange butterfly diet: a vital part of their nutrition.

Brown Long-Eared Bat

Majority bat species grab their insect target from mid-air. The brown long-eared, on the other hand, chooses to grab its food from the trees. They have even been identified to capture spiders from their webs. It is all because of those amazing ears. Brown long-eared bats have outstandingly sensitive audible range. Joined with their silent form of echolocation, this makes them deadly predators.

Woodlouse Spider

Tiny but impeccably formed. The business end of a woodlouse spider makes it a daunting forest hunter. They might just be 3cm long at the most, however these small spiders come with remarkable mandibles. This spider's powerful jaws are designed to pierce the tough armour of woodlice, their only prey. Instead of spinning webs, they hunt at night by searching warm, damp places for roaming woodlice, capturing them in their forward-facing jaws. During the day they hole up in special hiding places lined with silk.

These spiders guard their woodland habitat, stalking woodlice beneath logs and stones. Their strong fangs come into play when they hunt: penetrating the louse's hard exoskeleton without difficulty.

Fig. 10. Woodlouse Spider

Tree Kangaroo

Tree kangaroos exist in low-lying and hilly rainforests in Papua New Guinea, Indonesia, and the far north of Queensland, Australia. They have adapted to life in the trees, with smaller legs and tougher forelimbs for climbing.

Fig. 11. Tree Kangaroo

They are the biggest arboreal mammal in Australia. Like all macropods, as an alternative of sweating, tree kangaroos lick their forearms and let the evaporation to help cool their bodies when hot.

Many tree kangaroo species are very uncommon and most are declining in number. They undergo habitat loss because of deforestation.

Giant Panda

Pandas live mostly in bamboo forests up in the highlands of western China, where they live virtually completely on bamboo. They play a vital part in the bamboo forests by scattering seeds and helping growing of vegetation.

Panda's undergo habitat loss because of the construction of roads and railroads, which piece apart the forest, sequestering panda populations and stopping breeding trends, and decreasing pandas' access to the bamboo they need to subsist.

Fig. 12. Giant Panda

Saola

Saolas are one of the most rare and most endangered species on Earth. They are a counterpart of cattle but look like antelope. They are extremely threatened and are found only in the Annamite Mountains of Vietnam and Laos.

As forests vanish under the chainsaw to make way for farming, estates and infrastructure, saola are being forced into smaller spaces. Fast and large-scale infrastructure in the area is also dividing saola habitation

Orangutan

He name orangutan is translated to "man of the forest" from the Malay language, and they are the world's biggest tree-climbing mammal. They make nests of trees of vegetation to slumber in at night and relax in during the day.

Fig. 13. Orangutan

They are "gardeners" of the forest, playing a crucial part in seed dispersion in their habitats. Their habitat is rapidly vanishing to make way for oil palm farms and other agricultural farms. Currently, more than 50% of orangutans are found outside secure areas in forests under administration by wood, palm oil, and mining corporations.

Orangutans exist on the islands of Borneo and Sumatra. The island of Sumatra has lost 85% of its woodlands and an analogous amount of devastation is happening on the island of Borneo.

African Forest Elephant

African forest elephants live in the thick rainforests of west and central Africa. They are comparatively smaller than African savanna elephants, with more oval-shaped ears and less curved tusks that point downward.

By consuming more fruits from more tree species than possibly any other big vertebrate, an African elephant is important for the spreading and germination of several rain forest trees. For some of these species, their seeds will only germinate once they have passed through and with the elephant's stool.

In brief, elephants are actual forest gardeners. Their range has been decreasing quickly from 3 million square miles in 1979 to less than 1 million square miles at present. Poaching has additionally destroyed their left over populations by over 65% from 2002 to 2013.

These elephants are being moved into smaller islands of sheltered areas, further hampering their freedom to wander and long term existence.

Bald Eagle

The Bald Eagle is only found in North America and it has been the national emblem of U.S since 1782. Before that, it was considered a spiritual symbol for indigenous people.

Fig. 14. Bald Eagle

The rare animal species was endangered at first, but slowly grew in population and is no longer threatened.

Bald eagles are important to the environment. They prey on dead animals, that that helps with the clean-up process.

They are voracious hunters, keeping the overall population of animals strong. They kill the weak, slower, and old animals, and leave the healthier ones to survive.

Gray Squirrel

The mammal is lives in a variety of forest settings but prefers old hardwood trees. It is considered a minor nuisance in some areas of Northern America.

Fig. 15. Gray Squirrel

The Gray Squirrel has also been called the official state mammal of North Carolina.

They are integral to the forest ecosystem. They consume a lot of seeds from plants and trees. When they catch these seeds, it helps distribute the seeds, hence maintaining the tree's growth.

They are small, yet fast creatures that love to eat truffles. Due to their eating activity, they also disperse truffle fungal spores.

RISKS TO FOREST HABITATS

The biggest risk that forest habitats around the world face today is deforestation. Trees in forests are being cut down at an incredibly fast rate (around 3.5 billion to 7 billion every year) which means there are fewer resources and less shelter for the species that live and grow there. Trees are often cleared to make room for agriculture and cattle farms, but they're also cut down for timber and paper.

Forests are also at risk from forest fires, which can devastate the habitat in a short amount of time.

The Amazon rainforest (the world's largest forest) has lost at least 17% of its forest cover in the past 50 years because of human activity. The island of Sumatra, in Indonesia, has lost 85% of its forests. This leaves many animal species without habitats and without access to their normal source of food, which puts them at risk of endangerment or even extinction.

HABITAT MODELS

Introduction

Biodiversity issues in sustainable forest management (SFM) are changing, which brings along the need for new analytical tools. A major change is that sustainability is increasingly defined through connected ecological and social complex systems at multiple scales; this adds adaptive capacity and resilience among key qualities of SFM, along with traditional expectations to sustain a supply of specific forest goods and services.

Thus, biodiversity concerns have transcended traditional nature conservation to become an integrative issue that underpins ecological resilience, adaptive capacity of ecosystems, and many ecosystem services. It is yet unclear how such a perspective will be put into practice. However, given the schism between broad political acceptance of SFM and of forest protection versus the continuing loss of forest biodiversity, there is an unprecedented need for clear biodiversity targets and tools.

An obvious goal in sustaining biodiversity is to manage harmful environmental pressures and threats rapidly, proactively, and effectively. Geospatial models have long been used to anticipate futures by spatial planning of forests, including biodiversity targets. Many modeling approaches and techniques have been developed for depicting and accounting for biodiversity across forest landscapes.

Recent technological progress enables mass recording of biodiversity variables, e.g., by combining molecular sampling, observations, and remote sensing. However, such advances in biodiversity modeling are not easily picked up by forestry and conservation planners to specify general ecological guidance. Thus, biodiversity assessment practices for SFM or in protected forest habitats are based mostly on convenient landscape metrics and woody vegetation proxies.

We highlight a biodiversity response variable as a critical issue for useful geospatial models in SFM. The large and diffuse literature on such variables indicates a narrow disciplinary focus of most spatial models. We identify at least four 'interdisciplinary gaps' to be considered by the biodiversity modeling community.

- Biodiversity-representation gap—attempts to describe biodiversity comprehensively are common when analyzing current management situations, while most scenario analyses do not address representation beyond woody vegetation.
- Goal-setting gap—biodiversity distribution across the landscape does not tell managers how to set management priorities and goals without highly technical, data-rich decision-support computing.
- Scale-relevance gap—typical units of forest management and conservation decisions are either single trees, forest stands, or mosaics of stands (landscapes), while most biodiversity data are collected or modeled in other units (plots; pixels; etc.) that cannot be easily combined for decision support. Local biodiversity patterns, in turn, result from wider and longer-term ecological processes, which are difficult to explicitly incorporate in the models.
- Feedback gap—realistic biodiversity models tend to become very complicated, which undermines their updating, reduces advantages over adaptive management, and limits communication and uptake by the wider public. Infrequent or one-sided communication, in turn, reduces the ability to mobilize knowledge for action.

In brief, geospatial predictive models can be irreplaceable tools for biodiversity issues in SFM, given the large area of forests, long temporal scales of forest development, and the vast nature of biodiversity. However, there are apparently communication boundaries between the modeling community and other stakeholders (biodiversity researchers; policy-makers; forest and conservation managers; wider public). For example, the first two interdisciplinary gaps above together reflect 'biodiversity concerns' that have been recognized as difficult to model. Social and governance studies suggest that the failure to manage communication boundaries by meaningful simplification makes biodiversity difficult to grasp even for professionals. The question is how to build models that clarify biodiversity issues and help to plan a meaningful and understandable future.

Here, we revisit the conservation concept of 'focal species' as proposed by Lambeck, who proposed setting environmental standards in a specific context according to the most sensitive species to each threatening process in the environment. Managing for a full set of such species might then encapsulate the biodiversity conservation aim of a landscape.

We elaborate this concept in a spatial modeling perspective to demonstrate how it can—mostly through the binding element of 'threatening process'—operationalize multiple issues of ecological sustainability and bridge the interdisciplinary gaps listed above. We retain Lambeck's original term for 'focal species', while acknowledging that it has been loosely used in the literature and must be routinely rechecked against the original definition.

HABITAT QUALITY MODELS

Habitat index as a surrogate for biodiversity value

With the rise of new powerful statistical techniques and GIS tools, the development of predictive habitat distribution models has rapidly increased in ecology. Such models are static and probabilistic in nature, since they statistically relate the geographical distribution of species or communities to their present environment. The analysis of species–environment relationship has always been a central issue in ecology.

The quantification of such species–environment relationships represents the core of predictive geographical modelling in ecology. These models are generally based on various hypotheses as to how environmental factors control the distribution of species and communities. We depart from the concept of the species' habitat, considering the "optimal" habitat, the area where the presence of a species is due to suitable conditions for its survival.

This "quality" concept conceived for this modelling approach represents the optimal environmental conditions for a particular type of forests known to sustain an important numbers of species considered as a surrogate for biodiversity value. The biodiversity value we use to support the forest habitat quality models is based upon the concept of "naturalness" developed by several authors.

The problem we still have is that in the case of biodiversity conservation, empirical evaluation models based on real field data for all species of interest cannot be expected to become available. Up to date habitat models based on species census data are too restricted in area and time consuming. Then, the challenge is to develop methods and practices of locating and evaluating suitable sites for threatened species.

The main interest within the framework of this research is to improve the existing forest management planning, in this sense it is essential that the decision alternatives be assessed with respect to a combination of expert knowledge and habitat models. The one way of dealing with this problem, as proposed in this work, is to use habitat quality indices that reflect the quality of the forest by identifying possible causal relationships between forest structure, environmental data, and ecological conditions. We depart from the hypothesis that all species have specific habitat requirements, which can be described by habitat factors. These factors are connected to the critical characteristics of the habitat, e.g. to those of vegetation and soil, but also areas surrounding the habitat can influence the habitat quality (e.g. spatial structure of landscape elements).

Habitat factors can also be classified according to the particularity of the factor: a deterministic habitat factor has to be present in a high-quality habitat, but a non-deterministic factor has a trade-off with some other factor. Thus, a deterministic factor can be taken as a noncompensatory habitat characteristic whereas a decrease in a non-deterministic factor can be compensated by an increase in the value of another non-deterministic factor, as expressed in the habitat-evaluation model.

The habitat index reflects the value and importance that an area potentially possesses in terms of biodiversity. The calculated habitat index for the different forest habitats were used as the sole ecological variable in optimizing the site selection for conservation. The first step in assessing the quality is to determine the forest habitat factors on the basis of an analysis of existing studies and expert knowledge and forest inventory data. Here, judgements made by experts, in particular in terms of key species requirements of forest habitat suitability, were applied to develop the tresholds for the different habitat models.

Habitat quality assessment for conservation

Habitat quality assessment and habitat suitability maps constitute an useful approach in the design of management plans that seek to expand or create new protected areas. This approach helps to satisfy a number of conservation goals based on habitat characterisitics for certain species of particular importance. In this sense, habitat modelling generated using spatial statistics and GIS can help in the characterizations of habitat requirements and the localization of suitable habitats.

Our proposal is to develop habitat quality indices as a surrogate for biodiversity values. The method helps to decide where to protect forest biodiversity based on the habitat value of the forest. The work relies and builds upon data from the Finnish National Forest Inventory (NFI) and other related databases from permanent inventories. These indices should reflect the quality of the forest by identifying possible causal relationships between forest structure, environmental data, and ecological conditions.

This departs from the hypothesis that all species have specific habitat requirements, which can be described by habitat factors. These factors are connected to the critical characteristics of the habitat (e.g., vegetation or soil, but also areas surrounding that can influence the habitat quality (e.g., spatial structure of landscape elements).

A SPATIAL MODELING PERSPECTIVE ON FOCAL SPECIES

Theoretical Background

Simply put, the practical question of environmental management is where to do what to sustain environmental quality. For the SFM principle of maintaining and restoring biodiversity, it translates to locating and managing threats and opportunities for biodiversity on dynamic landscapes. There are four classes of spatial management decisions involved: stand-scale management for single or multiple goals; landscape design by combining stand-scale goals for landscape functions (including setting aside protected areas); regulating forest benefits and values in time; and managing for uncertainty at multiple scales.

In our view, the strength of Lambeck's concept of focal species is that it integrates these strategic aspects in a way that is understandable to the wider public, thus serving the stakeholder participation principle of SFM. Specifically:

- Selecting well-defined sensitive species to represent a full set of threats to biodiversity simplifies practical biodiversity concerns. A useful input is the red-listing of species based on the International Union for Conservation of Nature (IUCN) framework, which also considers 'projected declines' based on potential threats and changes in habitat area and quality.
- Managing (avoiding, mitigating, or reversing) each threat to sustain focal species in actual landscapes links the concern with management responses and implicitly addresses some uncertainty (e.g., maintaining population 'at the safe side'). A key simplification is that focal species serve simultaneously as biodiversity indicators and management targets, while such a link is unspecified in other biodiversity schemes for SFM.
- The uncertainty component can be further scrutinized by scanning for future threats that emerge from changes in the environment or production forestry, and by explicitly incorporating adaptive management and precautionary measures.
- Lists of focal species can be suited to environmental, social, and cultural contexts—depending on local species pools, knowledge, tradition to survey particular species groups, and priorities set by legal protection of species or acceptable costs.

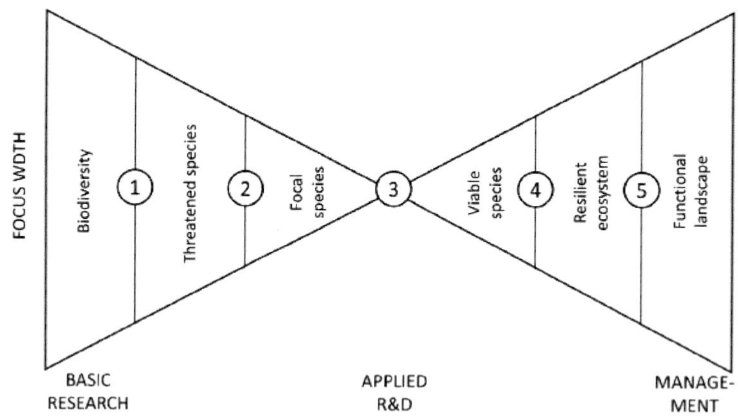

Fig. 16. A framework of targets and activities that link basic biodiversity knowledge (left triangle) and sustainable forest management (SFM) (right triangle) through the nexus of focal species habitat modeling. The activities indicated by numbers: 1, Red-listing of threatened species; 2, distinguishing focal species by listing major threats; 3, focal species habitat modeling; 4, habitat conservation; 5, landscape design.

The opportunity for spatial models in this framework is to predict focal species' distribution or performance in real landscapes subject to expected or designed change. Such spatial predictions are derived from species–habitat relationships, where specific 'threats' refer to limiting factors or population processes of the focal species. Note that, for modeling these links, ecological niches of the species must be understood beyond correlative patterns in current distributions; thus, so-called black-box modeling approaches that fit environmental parameters without understanding their ecological meaning are discouraged.

Due to the underlying logic that 'sensitive species illustrate a general threat', the maps derived from properly parameterized spatial models might then help to depict, analyze, and communicate broader 'where' and 'how' of sustaining forest biodiversity in the spatial and temporal scales for which the parameter values are available (Fig. 16). Depending on how closely the species' distributions follow environmental threats in time, such maps may also reveal past or present spatial extent, severity, and reversibility of the threats.

The methods for predictive mapping of species distributions can be divided into correlative (inductive) and mechanistic (deductive) models. The inductive methods, where predictions are derived from statistically linking empirical observations with habitat characteristics, include many algorithms and programming tools available. The algorithms basically differ depending on the species data (presence-only, presence–absence, or finer scales) and shapes of its habitat function. The deductive methods are based on prior insight into the species' requirements, with a wide range of more and less formal approaches, including procedures for systematizing expert knowledge. For either class of models to enable spatially explicit management guidance, they should be able to depict landscape change (including alternative management scenarios) in terms of the factors that indicate threats.

There are several basic caveats in interpreting predicted distributions of focal species directly for broader biodiversity management. First, each species can be limited by multiple factors, and its distribution is affected by stochastic events and population processes. This implies that both realized and potential (habitat) distributions of species affected by the same threat only overlap partly and to an extent that varies in time. Hence, increasing model prediction accuracy for a specific species—a major technical aim of distribution modeling—can paradoxically reduce the insight obtained from the model about wider biodiversity. Second, uncertainty of most environmental parameters increases when predicting the future, and data quality is usually reduced toward the past as well.

This, too, means general problems with predictions, specifically for complex models. The issue is to find simple and robust habitat characteristics that change predictably in time. Third, most sensitive species may be very rare or even extirpated in the degraded landscapes where habitat improvement is most needed; thus, their habitat prediction may not be practical or reliable for the remaining biodiversity. Instead of completely ignoring such species, a possibility is to add species less vulnerable to the same threat to be able to cover a broad range of environmental change.

An alternative to habitat modeling of threatened species is to map threats (hazards) directly. Such 'exposure maps' have been created by remote sensing of whole landscapes (e.g., for fire frequency, deforestation, night-time lights or modeling point observations based on landscape characteristics (for example, poaching threat maps from camera trapping of poached animals). Threat maps, though, are not explicit about likely biodiversity responses, which in turn limits objective-setting and cost-effective spatial analysis of conservation actions. In a structured decision-making process, focal species thus serve as a multi-purpose tool to set the objectives, choose among actions, and to measure the success.

Published Spatial Models of Focal Species Performance

Modern techniques of creating species habitat maps for forest management prescriptions originate from the rapid development of spatial analysis in environmental protection, wildlife ecology, and threatened species research in the 1980s. Notably, the United States (US) Forest Service developed Habitat Evaluation Procedures in the 1970s, which became increasingly formalized, supported by guidelines of use and computer programs. Along with the appearance and acceptance of GIS-techniques and data, such procedures transformed from individual assessment to automated landscape analysis. In the high-profile conservation case of the spotted owl (*Strix occidentalis*), GIS models were linked with population models in real landscapes first in 1992. In parallel, there was a development from prescribing forestry activities from the perspective of a single subjectively selected species toward comprehensive sets of species to represent different niche dimensions of habitat specialists or species requiring large areas. In retrospect, these US approaches appear closer to SFM than Lambeck's Australian view that emphasized protection and restoration. The new aspect brought up by the latter was, however, that species should be used to analyze habitat futures, not just to maintain the present values.

To characterize the field's development since then, we performed a search of modeling studies that spatially predicted the performance (incidence, trends, or demography) of representative species in real forest landscapes and through time, and in response to threats that could be mitigated by forest management or conservation.

Forestry perspective *s. lat.* (i.e., including also policies, planning of landscapes and set-asides) and met at least three of the following four criteria:

A. Addressed environmental threats that also affect wider, at least partly known range of forest taxa (note that, for practical purposes, we here restricted the focal species surrogacy assessment to species diversity only);

B. Included management options or approaches that also affect wider, known species diversity;

C. Were a part of a legitimate planning process;

D. Described how the focal taxa were selected based on specific threats and their surrogate value to represent wider species' diversity.

If these criteria were supported by references only, we also checked the original publications. Finally, we integrated a series of papers by the same research group in the same study system, and assessed potential gaps in the search string for additional searches on specific topics (metapopulation models; forest water bodies).

In fact, many field surveys have addressed potential non-vertebrate surrogate taxa. For example, forest fungal surrogates have been explored in many studies, including the matching of selected wood-inhabiting species with threats. Specialized lichens appear suitable for guiding multiple management dimensions, but spatial models for that remain scarce. Modeling for decision-making may have thus contributed to the taxonomic bias in SFM, which is usually attributed to insufficient stakeholder knowledge.

A similar gap appears in ecosystem coverage regarding the management of small freshwater bodies, notably headwater streams in forests. Again, there is well-established literature on the indicator value of many aquatic or semi-aquatic taxa, including suggestions to use some invertebrates, fish, amphibians, or birds as broader management targets. Relevant spatial models are, however, rare and tend to focus solely on the species' indicator value or its conservation perspectives.

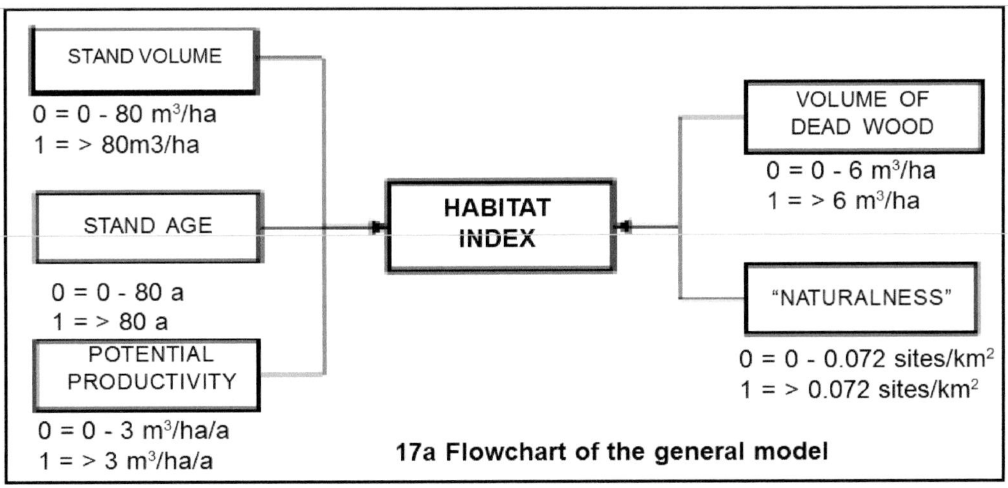

17a Flowchart of the general model

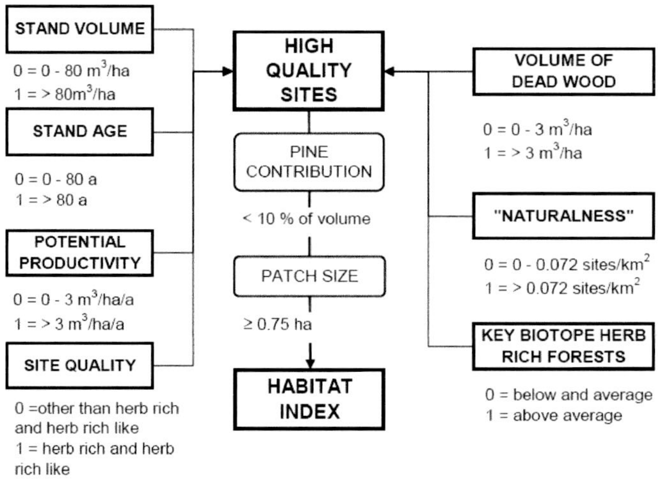

17b Flowchart of the herb-rich habitat quality model

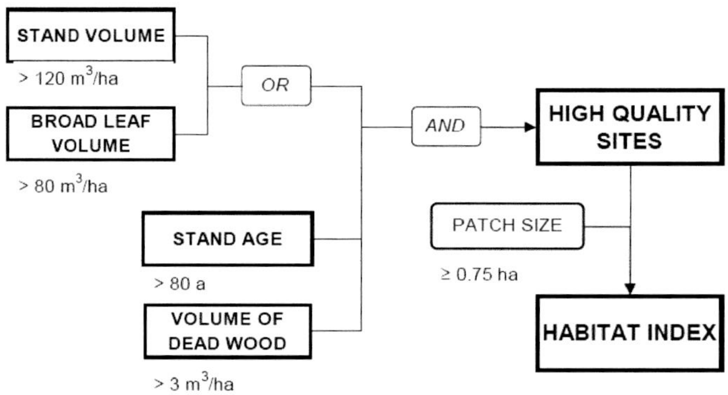

Fig. 17c. Flowchart of the general model for high biodiversity forests presenting the model inputs, their thresholds and the operators and conditions applied.

HABITAT QUALITY MODEL

The models include two types of input layers derived from the NFI: i) multisource thematic maps representing estimates for the volume of growing stock and of individual tree species, stand age, productivity of the site and site quality and ii) interpolation layers derived directly, without any auxiliary information, from the field plot data representing estimates for the volume of dead wood (CWD), area contribution of key biotopes and density of sites with no forest management actions for the last 30 years or more. All data were integrated into a GIS platform using Geostatistical Analyst and Spatial Analyst in ArcGIS.

All input layers were reclassified according to specific thresholds values based on literature review, expert knowledge, forest characteristics based on NFI data and MS-NFI and on landscape patterns analysis derived from the input maps (Fig. 17).

The conditions for each of the models are presented in the flowcharts (figure 17a,b,c). The flowcharts for the three conditions represent the logical analysis performed once the input layers were produced as explained before. All input maps have the same extent, the same resolution (50 m) and values for forestry land. However, the thresholds and conditions differ according to the type of habitat modelled (Fig. 17 and Fig 18).

It must be considered that since no coherent species data are available for this extensive area, we are not aiming to predict habitat suitability to any particular species, but rather to produce a spatially-based habitat quality model at a regional level as aforementioned.

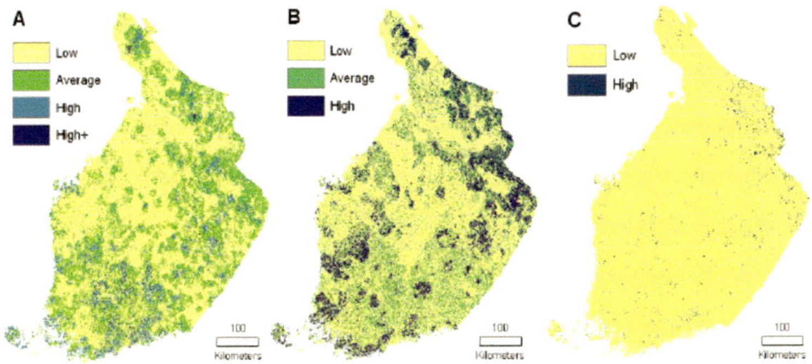

Fig. 18. Maps of habitat indices resulting from the habitat models: A. herb-rich habitat model, B. general habitat model, C. constrained habitat index. Light colors show areas of low habitat quality while dark areas are related to a high quality

Some sites were deemed high quality sites by all or two of the models, and the overlapping was removed so that the high quality sites according to the constrained model were kept as they are and those sites were excluded from the sites pin-pointed by the other models. This way, the sites selected according to the herb-rich model rules were excluded from the sites established by the general model. Pixels in an individual input layer meeting or exceeding the threshold value were assigned the value 1; pixels not meeting the threshold value were assigned value 0 (i.e., a Boolean approach). All input layers were then added resulting in a map with values ranging from 0 to 7.

The resulting layer was reclassified so that pixels whose value was below the average were assigned a value of "Low" habitat quality, pixels whose value was between the average and one standard deviation from the average were assigned a value of "Average" quality, and pixels whose value exceeded the average by at least 1 standard deviation were assigned a "High" value (Fig 18).

The models produced aim at representing forest biodiversity and hence the modelling was confined to forested biotopes. The habitats to depict and the variables used were chosen in accordance with the habitats and criteria mentioned in accordance to the Forest biodiversity programme for Southern Finland (Ministry of Environment) as explained before.

We produced three habitat quality situations: the habitat quality models produced (Fig. 18) will be called: Herb-rich Habitat Quality Model ("Herb-rich"), General Model for High Biodiversity Value Forests ("General") and Constrained Habitat Quality Model (also "Constrained") (Fig. 18). With the constrained model we aim at depicting forests in natural or semi-natural state with large amounts of coarse woody debris, dominant volumes of broad leaf and a patch size of equal or more of 0.75 ha.

Considering the dissected forest habitat the patch size considered constitutes an important threshold. Those habitats are regarded as the most important heath forest habitats for biodiversity in the Finnish forests (Ministry of Environment). The herb-rich model depicts herb-rich forests which are the most fertile forest habitats in Finland with the highest species richness. For this habitat type, class 1, showing low forest habitat quality covered 52.0 % of forestry land, class 2 for an intermediate quality accounted for 32.1 % of forestry land and class 3 showing the highest quality of forest habitat based on the constraints considered for the model accounted for only 15.9 % in the result layer. After this, a 4th class was separated from class 3 by first setting a further constraint for the pine contribution to be a maximum of 10 % of total volume. This provided a four class with a higher constraint setting a minimum patch size of 0.75 ha for areas being classified into that 4th class of high habitat quality (Fig. 18 A-High+). The general model aims at depicting the same habitat type as the constrained model but is less restrictive thus focusing on slightly different habitats that present a minimum of spatial continuity, we considered then a minimum forest patch area.

HERB-RICH HABITAT QUALITY MODEL

The result layer for the herb-rich habitat quality model showed the probability of an area to be herb-rich forest. This figure is derived from the results obtained of the 4 classes calculated as explained in the methods section. In this way, we obtained the forest area that represented larger patches of dominant herb rich forest from low to high habitat quality. Figure 19 provides a good communication tool for policy makers to show the potential for further conservation actions. The details showed in the windows allow the comparison of the red polygons boundaries of conservation areas with the dark blue shaded areas representing high values of "Herb Rich Habitat Quality". Considering the low percentage of protected areas of herb rich habitat, it may be important to expand existing conservation areas and/or create corridors to enlarge and connect existing protected areas. The Herb Rich model allows a rapid evaluation of good candidates' areas to target enlargement and protection, take a look at all the blue areas (High quality habitat) that are not receiving any protection (Fig. 19). Furthermore, when examining 2 km wide buffer areas around conservation areas it can be noted that some types of conservation areas have more valuable herb rich forests in their vicinity than others (Fig. 19).

We assessed the connectivity between protected areas and areas with high probability of containing herb rich forests patches and high quality mineral soil forest patches through a multiple distance ring buffer analysis (Fig 20).

A more connected area can be highlighted to the northeast of the study area (Fig 20). While a good candidate area to improve conservation of new areas and expand the conservation network can be identified towards the south (Fig 20)

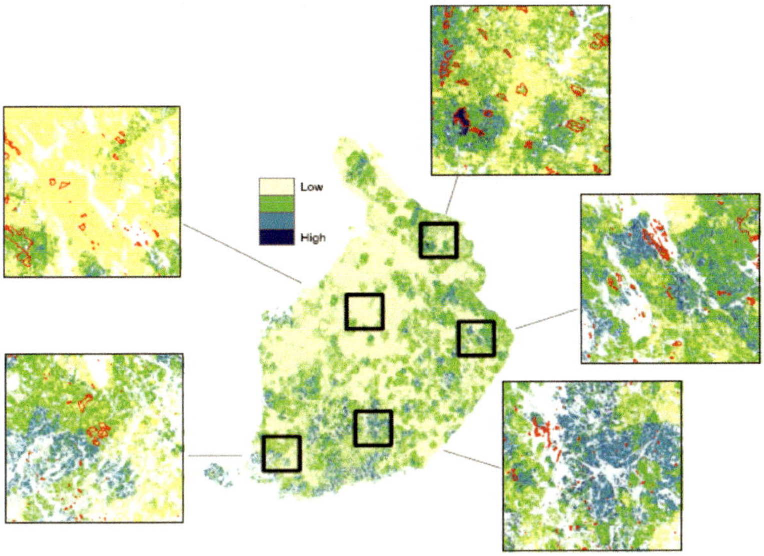

Fig. 19: The result layer of the herb-rich habitat quality model. Blue areas denote high habitat quality according to the variables and criteria used. In red the boundaries of protected areas to be compared against the habitat quality delineated areas.

A recent report advocates that herb rich forests should be conserved especially in know high-density herb rich forest areas independent of size. In this sense, the spatial model produced shows regions of special interest that should be considered in future conservation planning efforts.

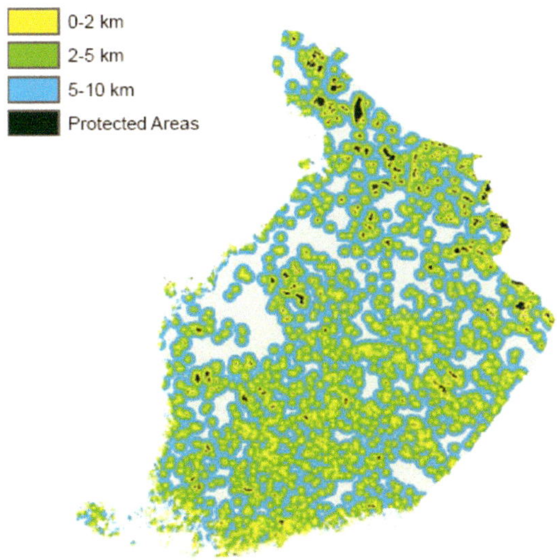

Fig 20. Map of protected areas and multiple distance buffers

12

TOWARDS EFFECTIVE FOREST CONSERVATION

INTRODUCTION

Before understanding the conservation of forest and wildlife it is important to understand the general definition of the forest. There are several definitions of the forest; the article encompasses the definition provided by FAO.

A forest is defined as a land area of more than 0.5 hectares with a tree canopy cover of more than 10% that is not predominantly used for agriculture as well as other quasi purposes. An important requirement is trees should be capable of attaining a length of 5 meters in situ and achieving the canopy cover requirement in young forests or locations where tree development is climatically inhibited.

In simpler words, forests are areas where most of the species of flora and fauna are found per square kilometre. It is the lungs of the planet that take up the carbon dioxide released by the animals and give oxygen to the environment. Forests are defined as the areas of lands that grow various species of trees, plants, shrubs, herbs. The importance of forest and its conservation can be understood by the commodities it provides. The forest serves as a natural habitat of numerous animal species, it helps in balancing the impact of carbon emission, it helps in global temperature regulation. Apart from it forest also maintains the various biogeochemical cycle.

Forests are the primary source of many medicinal plants, woods, food, raw materials, and fabrics. About 31% of the earth's surface is covered by forest, but as the modern age arises, the rate at which the trees are being cut-down is increasing gradually. Hence, there is an importance in conserving the woods and their associated wildlife.

Environmental activists consider forests as one of the top 5 natural resources on earth. Today we shall look at how wonderful our forests are to us and why we should immediately end its destruction.

There is more to forests than just a massive collection of trees. It is a natural, complex ecosystem made up of a wide variety of trees that support a massive range of life forms.

WORLD DISTRIBUTION OF FORESTS

Apart from trees, forests also include the soils that support the trees, the water bodies that run through them, and even the atmosphere (air) around them. Forests of the world are a natural wonder that humans have sadly taken for granted.

Forests come in many sizes and forms. For example, the piece of land with huge trees and many animals, birds, and water bodies running through it in a part can be called a forest. In the same way, the large belt of thick, evergreen trees running from Peru to Brazil *(called the Amazon Rain Forest)* is also a forest.

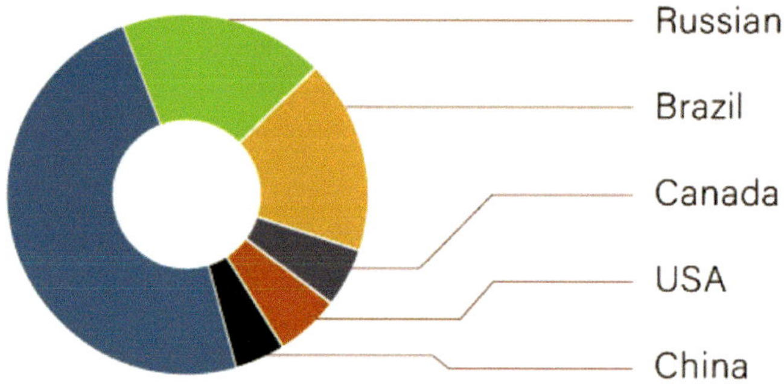

Fig. 1. More than half of the world's forests are found in only five countries (FAO, 2020)

Between 2015 and 2020, the rate of deforestation was estimated at 10 million hectares per year, down from 16 million hectares per year in the 1990s. The area of primary forest worldwide has decreased by over 80 million hectares.

Forests are hugely important for life on earth. This is because it serves as an ecosystem and sustains life for millions of animals, birds, and animals that live in the rivers and streams running through these forests. It also does a lot of good to the atmosphere in climate control, as well as supplying oxygen for human sustenance.

About a third of the world's land area, around 4 billion hectares, is covered with forests. Forests are home to the vast majority of terrestrial species, particularly due to the high biodiversity in the large tracts of tropical forests located in the Congo Basin, the Amazon Basin and South-East Asia.

Forests also play a major role in climate change mitigation and the regulation of rainfall and global freshwater cycles. Natural forests and wildlands also provide, on average, 28 per cent of total household income in communities in and around forests – nearly as much as agricultural crops. However, the future of forests is under threat, mostly due to commodity-driven deforestation. Most agricultural expansion is now taking place in forested areas of the developing and emerging worlds; the land is being converted on a large scale to meet food-, energy- and fibre-related needs.

Reduction of deforestation and forest degradation can be achieved in many ways, and diverse policies and programmes have been tested across the tropics. The case of protected areas represents a multifaceted tool that has been applied on a very large scale with varying degrees of protection. Integrated conservation and development projects (ICDPs) have also been a favoured approach for many donors, due to their alleged capacity to deal with problematic trade-offs by providing lasting solutions of productive change.

Market-based approaches and payments for environmental services (PES) have attracted much attention for their focus on agents' self-interests that can be made compatible with conservation purposes. In a similar vein, the private sector has signed zero-deforestation commitments under the threat of boycott campaigns and risks of reputational damage. Briefly, these interventions may be summarized into three main categories: command-and control/disincentives (e.g. protected areas), incentives (e.g. PES) and enabling measures (e.g. capacity-building).

It is clear, however, that the creativity used to design conservation policies and programmes has not been matched by thorough efforts to evaluate their capacity to deliver impacts. Any such attempt should preferably be based on the application of scientific methods – ideally, experimental methods where many sources of bias are controlled for. That said, other methods such as descriptive research with case studies may also yield essential information about local contexts and challenges.

Whatever the methods used for the evaluation of outcomes, a robust definition of the interventions is required. However, conservation interventions commonly combine elements in "intervention/policy mixes" that make them hybrid across categories, and thus harder to evaluate (e.g. PES implemented inside a protected area, or capacity-building as part of an incentive mechanism). For this reason, we devote special attention to the definition of categories of interventions, which is a first and necessary step towards an assessment of existing evidence of what works and what does not. We subsequently discuss the gaps that we believe remain to be filled.

In 2016, the International Initiative for Impact Evaluation (3ie) published an evidence gap map (EGM) of forest conservation interventions to identify existing and missing evidence of the relationship between given intervention types and their outcomes and impacts on the ground. Our current study continues that work, and is supported by the Independent Evaluation Unit (IEU) of the Green Climate Fund (GCF) and the Center for International Forestry Research (CIFOR). The GCF aims to support forest and land-use projects that have the potential to reduce greenhouse gas (GHG) emissions within the framework of REDD+, towards achieving the GCF vision of paradigm shift/ transformational change. This strategy involves a diversity of interventions that go beyond the forest sector to promote conservation and sustainable management in the landscape. Therefore, an assessment of the effectiveness of different types of conservation interventions is critical to help support evidence-based decision-making by the GCF and other organizations that invest in forest-based climate change initiatives, such as the UN-REDD Programme, the Climate Investment Funds (CIF), the United Nations Environment Programme (UNEP) and the Global Environment Facility (GEF).

REASONS WHY FORESTS MATTER

We have briefly discussed the importance of forest which leads to the requirement of conservation of forest. Let us look into some of the important reasons for conserving the forest.

- The primary reason to conserve forests and wildlife is to provide oxygen to breathe. Without the existence of trees, oxygen cannot be formed, and carbon dioxide may be accumulated in the environment.
- The products of the forests, like medicinal plants, fabrics, and raw materials, give employment to millions of people. Without the woods, all of these people lose their jobs and cannot earn a livelihood.
- Forests provide shelter to a wide range of animals. If the trees are destroyed, these animals also die as they do not have their natural habitat. This destruction makes many of the endangered species of plants and animals extinct.
- Forests play a role in the carbon cycle by absorbing carbon dioxide from animals.
- They bring about a moderate atmospheric temperature and prevent global warming and also reduces the greenhouse effect. Global warming causes an increase in the sea level by melting the glaciers and ice in the north and south poles.
- Forests can absorb most of the water during a flood and hence prevent soil erosion and rescue us from the natural calamities.
- A significant economy can be generated by converting the forest zones into tourist spots and viewing the wildlife as a first-hand experience.

FRAMEWORK, SEARCH PROTOCOL AND DATA EXTRACTION

The PICO Framework

Several steps were followed to construct the EGM. It is structured with what is commonly labelled the Population, Intervention, Comparator, and Outcome (PICO) framework:

- Population refers to the "subject" (including ecosystems) upon whom an intervention is applied or who is ultimately affected by the intervention.
- Intervention refers to the action, policy, programme or project to which the subject is exposed.
- Comparator refers to control groups that have not been exposed to the same intervention.

- Outcome refers to all relevant outcomes that result from the relevant population being exposed to a relevant intervention based on a causal chain (in theory).

A tentatively exhaustive set of categories and subcategories is provided in the PICO framework. Building off the previous EGM report and guidance from GCF documents, but also relying on our experience in the field, we present a broad set of categories for forest conservation interventions in Table 1.

Table 1. PICO framework

Population		**Population contains two components:**
	•	Local households, communities and companies directly targeted by a forest conservation intervention
	•	Forest ecosystems targeted by a forest conservation intervention in a low- or middleincome country (using the World Bank definition). Included are terrestrial forest ecosystems, mangroves, and agroforests
Intervention		Interventions are policies, programmes or projects, that include the following:
	1.	Incentives: PES-like with clear conditionalities
	1.1.	Private and non-governmental organization interventions such as Coasean agreements ("user-financed PES")
	1.2.	Subsidy programmes established by broader public sector institutions or user representatives, and often tax financed ("government-financed PES")
	2.	Protected areas, broadly speaking: various categories as defined by the International Union for Conservation of Nature (IUCN), based on the degree of protection/tolerance of human presence and activities within their borders
	3.	Conservation based on local sustainable practices involving local communities: promotion of non-timber forest product production under conditions such as "extractive reserves"; community management generally speaking (community-based, co-management...)
	4.	Intra-governmental deforestation-curbing regulations and incentives:
	4.1.	Transfers from central to local government based on conservation performance (e.g. the Brazilian ecological value added tax [ICMS-Ecológico])
	4.2.	Central government publicly exposing high-deforesting municipalities ("blacklisting") and cutting their access to credits and subsidies while listed
	5.	Conservation based on product markets:
	5.1.	Environmental certification, with consumer-financed sustainability premiums (for forest products, such as the Forest Stewardship Council for timber, or crops, such as the Roundtable on Sustainable Palm Oil for palm oil)
	5.2.	Moratorium (barring non-sustainable providers in the value chain)
	5.3.	Zero-deforestation commitments (reaping publicity gains for consumer recognition of sustainability)

	6.	Indirect conservation based on improved technologies and/or substitution effects:
	6.1.	Increased agricultural yields, Borlaug hypothesis of "land sparing"
	6.2.	Improved cooking stoves
	6.3.	Plantations with clear conservation purposes (e.g. for fuelwood supplies)
	6.4.	Agroforestry with clear conservation purposes
	7.	Indirect conservation based on enabling conditions:
	7.1.	Environmental education/awareness building
	7.2.	Capacity-building
	7.3.	Improving the rule of law
	7.4.	ICDPs
	8.	Land tenure reforms that include conservation objectives (including indigenous/local land demarcation, titling enforcement processes [e.g. the Brazilian CAR – Environmental Rural Register])
	9.	Land swaps such as moving a concession from a High Conservation Value area to a degraded area or an area without forest cover.
		Comparator Comparable populations (forest ecosystems or human populations or companies – see above) at sites without the implementation of a forest conservation intervention as defined above.
		Outcome Relevant environmental and socioeconomic outcomes (or impacts) may include those that are faced by the population (as defined above) directly as a result of the implementation of a forest conservation intervention (as defined above), including maintenance or improvement of forest ecosystems, reduced GHG emissions, livelihoods and employment.
		In addition, we consider related characteristics that define the quality of the impacts: costeffectiveness (how much positive outcome against the cost of the intervention) and leakage (outcomes taking place outside of the boundaries of the intervention). The list is not closed and other outcomes such as participation, equity and others might also be considered.

APPROACHES TO FOREST CONSERVATION

The forest resources are valuable as an integral part of the ecosystem, from the commercial point of view, and as providers of shelter to wildlife. Today forests provide the raw materials for over 5000 products worth about 23 million dollars.

They support industry which employs 1.3 million people. In fact, forests are still the natural habitats of several species of plants and animals, as well as of several tribal groups of the world. But, the most unfortunate setback came in the form of commercial exploitation, which resulted in mass destruction of forest cover year after year.

Originally, over two-fifth of the land area of the earth, exclusive of the Polar Regions, or about 1,200 million hectares was covered with natural forests. But, now more than one-third of this area has been robbed by man of its natural protective cover and has been turned into barren land.

The history of the exploitation of forests is as old as man himself, but during earlier times it was balanced through a natural growth process because at that time forest cutting was done for personal or community use only. But with the expansion of agriculture, forest lands have been cleared.

More destruction has been done after industrial revolution and urbanisation. During the colonial period commercial exploitation began and this was the main cause of the depletion of forests.

The commercial use of forests nowadays has reached such an extent that it has become a threat to the environment in the form of:

i. Increase in temperature,
ii. Lesser precipitation,
iii. Increased rate of soil erosion,
iv. Increase in frequency and volume of floods,
v. Loss of soil productivity,
vi. Extinction of several species,
vii. Non-availability of several essential forest products, and
viii. Imbalance in ecosystem.

The harmful effects of deforestation are so much that all over the world people and authorities have realised that forest resources must be conserved properly in order to protect the ecosystem.

The forest is a national resource and a social asset. It yields a great social profit which lies wholly outside the realm of business.

But, at present, most of the forests of the world are so over-used that experts predict dire calamities in the not-too-distant future and irreparable damage on a catastrophic scale. If properly used and put on a sustained yield basis, it will be one of man's greatest resources and for this; conservation of forest is the only alternative.

METHODS OF FOREST CONSERVATION

The following steps should be taken for the conservation of forests:

Regulated and Planned Cutting of Trees:

One of the main reasons of deforestation is commercial felling of trees. According to an estimate, about 1,600 million cubic metres of wood have been used for various purposes in the world. Although trees are considered as perennial resource, when exploited on a very large scale, their revival cannot be possible.

Therefore, cutting should be regulated by adopting methods like:

i. Clear cutting,
ii. Selective cutting, and
iii. Shelter wood cutting.

The clear cutting method is useful for those areas where the same types of trees are available over a large area. In that case, trees of same age group can be cut down in a selected area and then marked for replantation.

In selective cutting only mature trees are selected for cutting. This process is to be followed in rotation. Shelter wood cutting is where first of all useless trees having been cut down followed by medium and best quality timber trees.

The time gap between these cuttings is helpful in re-growth of trees. In regulated cutting only one-tenth of the forest area is selected for use and rotational system is always followed for their protection.

The forest can be managed in such a way that a timber crop may be harvested indefinitely year after year without being depleted. This technique is called the 'sustained yield' method adopted by many countries of the world.

Control over Forest Fire

Destruction or loss of forest by fire is fairly common; because trees are highly exposed to fire and once started it becomes difficult to control. Sometimes, the fire starts by natural process, i.e., by lightning or by friction between trees during speedy winds, while in most cases it is started by man either intentionally or unintentionally.

According to an estimate, during the period from 1940 to 1950, in the US alone, fires consumed an average of 21.5 million acres of timber yearly and as many as 1,175,664 cases of forest fires occurred during 1955 to 1964 period.

Throughout the world, forest fire is common and in most cases they were begun by man. As John D. Guthrie, former fire inspector of US Forest Service has written: "To stage a Forest Fire you need only few things – a forest, the right atmospheric conditions, and a spark either from a lightning bolt or a match in the hands of a fool or a knave. The formula is simple – the larger the forest, the drier the air, the bigger the fool, the bigger the fire you will have".

In order to save forests from fire it is necessary to adopt latest techniques of firefighting. Some of the fire suppression techniques are to develop three metre wide five lanes around the periphery of the fire, back fires, arrangement of water spray, fire retardant chemicals should be sprayed from back tank and if possible by helicopters. There must be a trained staff of firefighters to control the fire.

Reforestation and Afforestation

The sustained yield concept dictates that whenever timber is removed, either by block cutting or by selective cutting, the denuded area must be reforested. This may be done by natural or artificial methods. Similarly, any forested land which has been destroyed by fire or mining activities should be reforested. In rugged terrain aerial seeding is the method of choice.

Besides all this, fresh afforestation programmes should be started. New plantations will not only increase the forest cover but also help in making up the eco-balance. For afforestation, selection of trees should be done according to local geographical conditions and care must be taken during initial growth of the trees.

Check over Forest Clearance for Agricultural and Flabitation Purposes

Most of the present-day agricultural land was once forested and then cleared for the use of agriculture. But now it has reached the stage where further clearance will be dangerous for the entire ecosystem.

There are tribals in some parts of Asia, Africa and South America, where shifting cultivation is still a part of their system of land procurement. According to an estimate, about 40 million sq km of land is used for this purpose by 200 million tribals of the world.

For the conservation of forest, this should be checked and an alternative system should be suggested to them. Similarly, for the development of villages, towns and cities, forest lands have been cleared and this process continues to this day causing loss of forest cover. This also should be checked and green belts around cities should be developed.

Protection of Forest

The existing forests should be protected. Apart from commercial cutting, unorganised grazing is also one of the reasons. There are several forest diseases resulting from parasitic fungi, rusts, mistletoes, viruses and nematodes which cause the destruction of trees. The forests should be protected either by use of chemical spray, antibiotics or by development of disease resistant strains of trees.

Proper Utilisation of Forest Products and Forests

Generally, forests have been cut for logs and rest of the tree – stump, limbs, branches and foliage, etc., are left out in the forest as worthless debris. Further waste occurs at the sawmill. There is a need to use all this waste material. Now several uses have been developed and products like waterproof glues, board, etc., can be obtained.

Similarly, forests can easily be used or developed as tourist centres. By using them as tourist centres the country can earn substantial foreign exchange. This practice has been adopted by many countries, both developed and developing.

The concepts of 'national park' and 'game sanctuary' have now become popular and every country has developed its unique forest area as a 'national park'. In India alone, there are as many as 21 national parks. This scheme is a good method of forest conservation.

SUSTAINABLE FOREST MANAGEMENT

Forest management is a set of improved techniques that make it possible to use natural resources aiming at their maximum conservation. In other words, it is the rational and environmentally appropriate use of nature's resources.

This way, the extraction of wood and vegetation will always be balanced. In addition, due to the reforestation and conservation of the area, the site is preserved, preventing the region from being negatively affected and rendered useless.

Thus, these practices are supported by the principle of **continuity** and ensure that natural resources are always replenished — in order to prolong biodiversity and sustain a circular economy for companies and the supply chain.

It is worth emphasizing that forest management is a mandatory practice by certain country laws. Therefore, organizations that do not comply with the law are subject to fines and legal proceedings. In addition to putting the natural environment, the company's productivity and reliability at risk.

Sustainable Forest Management Practices

Forest conservation in modern forestry utilizes multiple techniques:

- Afforestation and reforestation enlarge forest areas on our planet.
- The ability to distinguish and treat tree diseases or pest infestations empowers forestry managers to save their farms and mitigate losses.
- Replanting forests after harvesting contributes to ecologically sustainable forestry.
- Selective logging and thinning prevent from felling the entire stand.
- Pruning saves from logging the whole trees for wood and stops pathogens' spreading.
- Clear-cutting or removal of mature trees contributes to forest health and stimulates offspring growth.
- Controlled burn naturally revives forests – on condition the process does not go beyond control.
- Specific training boosts foresters' proficiency in sustainable forestry techniques.
- Weighted planning facilitates the solutions for more sustainable forestry.
- Satellite monitoring enables remote control of the forest state and timely response to deviations.

There is **no universal recipe for every forest**, and the list goes far beyond the above-mentioned practices. Each forestry management technique should consider commercial value, forest specifics, environmental aspects, communities' needs, and stakeholders' interests. Besides, the collaboration of forestry managers with local businesses, authorities, NGOs promotes more effective decision-making.

Continuous Monitoring of Forest Stands as a Sustainable Forestry Practice

Regular inspection determines an effective forest farm run. Yet, this is not always feasible, especially when it comes to vast, distant, or hard-to-reach locations. It is where remote sensing is particularly handy. Satellite monitoring is among the most effective forest management practices allowing foresters to greatly improve their businesses.

EOSDA has developed the EOS Forest Monitoring software to facilitate foresters' efforts – a smart tool for remote forest control and forestry management. In particular, EOSDA EOS Forest Monitoring users can enjoy the following:

- Monitor forest stands remotely;
- Assess overall forest health;
- Track changes and receive regular problem area alerts;
- Detect and report illegal logging with satellite-based deforestation maps;
- Compare planned and actual tree felling scope;
- Control forest regrowth;
- Analyze aois before and after deforestation, and more – all in one place.

IMPORTANCE OF SUSTAINABLE FORESTRY

The significance of sustainable forestry is explained with the forest benefits for all living beings on our planet. Thus, sustainable forest management benefits are the same as forest benefits as such, with the important difference that people can enjoy them for a long time.

Despite all sustainable forestry advantages, global deforestation rates are alarming, especially in the tropical latitudes, and the biggest threat to forestlands is agriculture. Around 80% of deforestation in the world occurs for the sake of farming.

The impact is the most severe in Latin America, Africa, and Asia due to palm oil, soybean, or beef production. It is why sustainable forest management in developing countries summons particular attention.

MODERN SUSTAINABLE FORESTRY IMPLEMENTATION

Permanent availability of forest benefits essentially depends on sustainable forestry management facilitated with remote sensing. However, much depends on regular people, too. So, raising public awareness encourages a sparing and responsible attitude to forests, contributing to nature protection. Paid employees and volunteers replant billions of trees worldwide in the U.S., Pakistan, China, Africa, and other countries.

Joint efforts of forestry managers, governments, ecologists, and communities can remarkably improve forest conservation. Satellite monitoring is capable of promoting this noble mission by guarding the forest heritage and tracking tree growth.

BENEFITS OF FOREST MANAGEMENT

Forest management carries a series of benefits that extend to all links in the chain that use products derived from vegetation and native areas. Thus, **everyone benefits** — the natural environment, companies, suppliers, customers and society.

Below, see what are the main advantages of this practice.

Production continuity

Continuity governs sustainable forest management, working with **ma**ximum preservation of the natural environment and encouraging practices that save resources, both for the ecosystem and for the company itself.

Thus, the management and annual production planning are safer, as there is a schedule of uses and replacement of resources, always aiming for the least interference in natural and industrial cycles.

Increases profitability

With the guarantee that resources and production are better managed, business profitability also increases. The reason is that less losses and damages occur, especially in regard to environmental laws and the organization's sector.

With the recurrent use of raw materials and their constant replacement, the finance company is also able to work in a more planned way, knowing when and on which fronts to invest, considering the best periods — minimizing the risk of losses.

Meets standards and laws

Environmental laws are often created, especially due to studies that assess the impacts of unregulated extraction of raw materials on the planet. Thus, new standards are created based on research that indicate techniques and standards to reduce impacts on the natural environment.

By strictly following this criterion, the company that adopts forest management as a corporate policy is able to produce better results in every way. In addition to maintaining its suitability before the courts and supervisory bodies of the sector.

Maintains forest conservation

Forests and woods are preserved through forest management, as nothing is removed beyond what is scheduled to be extracted from the area. This way, species have their life cycles respected, and the soil is not damaged to the point of being unusable and useless.

With forest conservation, animal species, insects and all the biodiversity of natural areas is protected. It is noteworthy that these beings and the local vegetation exert influence on conservation beyond deforestation and the regional climate, interfering even with the health of the local community.

Greater job security

As forest management is not done illegally, all the rules and laws that compete for the legal exploitation of the area are respected. Thus, work safety in the use of machinery, vehicles and other parts of the chain is guaranteed, and the employees' health is also preserved.

This security generates legal support for employees. Thus, in the event of an accident at work, they have all the support to resort to the competent labor bodies.

Therefore, the company is also safer to meet financial and legal needs, in case any problems occur during operations.

Greater reliability for the company's chain

Working based on forest management generates greater confidence for everyone involved in the organization's production chain. Knowing that the company has an environmental management and follows the forest management law, suppliers, customers and employees are more engaged with the company. In this sense, the company increases its credibility, becoming a high quality reference for the market and talents who want to join the company.

Opens up greater market opportunities

Today, the market is more likely to invest and operate in partnership with companies that are linked to sustainability.

For this reason, the organization that invests in forest management expands the chances of being seen and pursued for excellence, **increasing business** possibilities.

REVITALIZES THE REGION

The region where forest management is carried out is revitalized, because with the vegetation preserved, the **area tends to gain value** — more specifically, cultural, environmental and real estate value, attracting people and investments to the region.

Sustainable use of tree species

"Sustainable" means to maintain, continue, and keep, while "forestry" is the science and art of managing forests. Thus, sustainable forestry is about caring for and managing forests to provide the natural resources, such as wood and clean water, we need now and in the future. It also means sustaining other things we value from the forest like wildlife habitat and beautiful landscapes. Sustainable forestry is concerned with all parts of the forest—trees, smaller plants, soils, wildlife, and water. It involves protecting forests from wildfire, pests, and diseases, and preserving forests that are unique or special.

Sustainable forestry also puts an emphasis on people. People need forests for the resources they provide, and as a place to live or to make a living. Sustainable forestry ensures that future generations will have forests to meet their needs and values. As you can see, sustainable forestry is complex and can involve many things. Let's look at a few of these things more closely. You will also see how sustainable forestry can mean different things to different people.

FOREST MANAGEMENT

To care for and manage a forest in a sustainable way it is necessary to use responsible management practices. These are often specifically adapted to each site. One of the most important practices is to look at whether the forest has enough natural seeds, seedlings, and tree sprouts (all called *regeneration*) to make a future forest. Excessive populations of white-tailed deer in a forest greatly reduce regeneration by eating young trees. Too many ferns or too little sunlight can also play a role. Many sustainable forestry practices can protect or encourage forest regeneration. They include putting up a fence to exclude deer, controlling weeds and other plants, and removing some trees to allow more sunlight to reach down into the forest.

Fig. 2. Tree seedling in forest

When it comes to removing trees, or *timber*, from the forest, many practices can assure the forest's future. It is a good practice to think as much about which trees to leave as about which trees to cut. The trees left in the forest will continue to grow, occupying and regenerating the forest for many years to come. If they are chosen properly, the remaining trees can provide many of the same values and resources, and, perhaps, new ones in the future.

Fig. 3. Sustainable forestry involves forward thinking. It's just as important to think about which trees will be left after a harvest as which trees will be cut

Fig. 5. Assuring that a forest has enough new trees to regenerate after a harvest is an important practice. Tree tubes can protect seedlings and stump sprouts from deer browsing

Other sustainable forestry practices include protecting forest streams and wet areas. Harvesting trees can disturb and expose soil in small areas. This is especially true on roads built for driving the machinery used to remove trees. The flow of rain or other water across, and under, roads must be carefully managed with *culverts* (large pipes) and proper road design. This helps prevent soil from washing into a stream or wet area. Too much soil in water, or *sediment*, is harmful to aquatic life. Trees and other vegetation left undisturbed adjacent to streams or wet areas can also prevent soil from entering streams. These areas are called *buffer strips.*

BROWSING CONCERNS

Forest Health

Sustainable forestry is also about maintaining forest health. Many living and nonliving things make a forest unhealthy. These include insect pests, diseases, wildfires, and extreme weather. Normally, the loss of individual or small groups of trees is not a big concern. When larger numbers of trees die, it may signal a forest health problem. Sustainable forestry seeks to promote forest health by reducing things which seriously damage the forest.

Caring for Special Sites

All forests are special, but some areas are more unique than others. It is a principle of sustainable forestry to protect unique areas that are not found elsewhere or are uncommon. One example is very old forests. Almost all of the 17 million acres of forests in Pennsylvania were clear-cut at least once by the early 1900s. However, a few small areas were never cut. Most of these old forests are currently preserved.

Different Approaches

People view sustainable forestry in many different ways. To a forest landowner, sustainable forestry might mean selling timber and yet passing the land to an heir in good condition. To a logger (a person who cuts trees for a living) sustainable forestry might mean protecting the trees left in a forest, and constructing roads properly. To a professional forester, sustainable forestry might mean selecting appropriate forestry practices for a given site. All of these people want to sustain forests, even though each views the forest in different ways. Sustainable forestry is a broad but important concept. It requires concern and commitment on everyone's part.

Fig. 6. Pennsylvania has very few old growth forests left. Most of these are protected for future generations to enjoy and study. Some are over 500 years old!

MODERN TECHNOLOGY FOR SUSTAINABLE FOREST MANAGEMENT

The importance of forests is rapidly growing not only for 'traditional' resources like food, timber, and non-timber forest products, but also for a range of other ecosystem services like water purification, climate regulation, carbon removals and carbon storage, necessary to ensure a harmonic balance for the existence of life. This growing importance makes sustainable management of forests even more critical for dependent populations, economic growth, and climate action.

In this regard, data from forests is becoming vital. Forest managers need to track more and more parameters to assess the health of forests. Having access to data helps forest managers monitor threats, prioritise interventions based on the location's needs, and identify areas at risk so that advanced measures to address these challenges can be taken. Forests offer nature-based solutions to address climate change; reliable and transparent data is becoming mandatory to access national and international climate finance via mechanisms of carbon markets and REDD+. Central and South American countries like Costa Rica, Argentina and Colombia have already received funding from the *Green Climate Fund's REDD+ Results-Based Payments* pilot for their demonstration of emissions reduction through effective monitoring processes. Monitoring and reporting are key for assessing interventions made and supporting sustainable adaptive management. Today, large volumes of reliable and transparent data translate to better forest management.

Several pilots across the world, in countries like the United States, China, Sweden and Finland, validate this.

The challenge? Data collection from forests is demanding

Forest data collection is complicated due to the dynamic nature of forest ecosystems. There are different species of trees, as well as trees of different ages. Forest growth is highly dependent on microclimatic and geophysical conditions, which change over the course of a year. Forests cover large areas, making manual data collection cumbersome. Even though remote sensing and satellite imagery have made data collection easier, they come with their own challenges. Satellite data is heavy (terabytes of data across different spectrums), coarse (often >15 m resolution), and periodic (data can only be collected when the satellite is above the forest patch in question).

Emerging technologies like *Machine Learning (ML)*, *Artificial Intelligence (AI)*, single-window digital platforms, and *Internet of Things (IoT)* hold much promise in addressing these challenges by reducing redundancy, increasing scale (finer data at more frequent intervals), and improving accuracy in the data collected from forests. For example, real-time data on forest soil health can be collected through ground-installed devices connected via IoT. ML and AI techniques can be employed to analyse the collected data and detect areas with deteriorating soil health, prompting targeted remedial action. Forest managers get insights at a granularity previously unimaginable.

Emerging technologies in forest management

For example, the US-based start-up NCX is using ML to sift through field data collected by the *US Forest Service* and train their software to identify the size and species composition of 0.2-hectare grids of US forests from satellite imagery. ML algorithms are adept at activities like identification of anomalies in images, identifying tree species, counting trees, etc., thus reducing human input.

A combination of ML and AI can be powerful. While ML helps sift through large troves of data, AI can help inform decision making by computing and delivering insights remarkably quickly. The start-up 20tree.ai is applying ML/AI on high-quality satellite data to assess forest health and provide managers with detailed feedback. A task that would otherwise take months, is being completed within minutes.

Realtime data at scales of <10 meters can transform how we monitor threats like fires and illegal logging, where each second is crucial. Pilots around implementing emerging technologies for such purposes are already underway, for example by Dryad Networks for real-time forest fire monitoring and the Hitachi – Institute of Wood Science & Technology, Bengaluru collaboration to prevent illegal logging.

This sector can greatly benefit from technologies like IoT and single-window platforms. While the interconnected devices in the IoT collect data on several parameters of the agroforestry patch, dedicated single-window platforms can map, tag, and register agroforestry resources and monitor progress from the planting stage to the felling stage. This provides a detailed and holistic insight into agroforestry in India and reduces transaction costs by removing barriers like excessive paperwork. The Government of India already has a single-window platform called e-Green Watch, which automates and streamlines processes related to plantation and forestry activities. Emerging technologies can be linked to existing digital infrastructure like *e-Green Watch* to accelerate adoption.

Attempts in Indian policy to adopt emerging technologies

The *Department of Electronics and Information Technology, Ministry of Electronics and Information Technology*, already has a draft IoT policy that outlines different applications of IoT in the country. Given the potential benefits of these technologies in the forestry sector, robust pilots are necessary to inform the policy and design measures to integrate smart technologies into Indian forest management. At the same time, the capacities of the Indian forest management system must be improved to apply these technologies. Decades of field data collected by Indian foresters can be harnessed to scale the use of these technologies.

Indo-German development cooperation is making preliminary attempts to support stakeholders in applying emerging technologies for forest and biodiversity management. For example, the Indo-German Biodiversity Programme, commissioned by the *German Federal Ministry for Economic Cooperation and Development (BMZ)* and implemented by *Deutsche Gesellschaft für Internationale Zusammenarbeit (GIZ) GmbH* together with the *Himachal Pradesh Forest Department (HPFD)*, is developing a *Long-Term Ecological Monitoring (LTEM)* system. The LTEM is an online database system that aims at understanding the dynamics of forest ecosystem and developing appropriate management strategies. For instance, the system helps forest managers identify areas of grazing pressures and assesses *Non-Timber Forest Products (NTFP)* potential and use. Over the next few years, Indo-German development cooperation will focus on developing dedicated pilots where the use of emerging technologies for forest landscape restoration, sustainable forest management, and trees outside forests will be demonstrated.

The improved accuracy and transparency from using advanced digital tools will make it more attractive for private-sector players and landowners to participate in national forest management. This will raise the value of standing trees and forests, increase finance flows, and usher in a new era of forestry in India, thereby supporting India in meeting its national and international commitments on climate change and forestry.

GIS (Geographic Information System)

A geographic information system (GIS) is a computer system for capturing, storing, checking, and displaying data related to positions on Earth's surface. GIS can show many different kinds of data on one map, such as streets, buildings, and vegetation. This enables people to more easily see, analyze, and understand patterns and relationships.

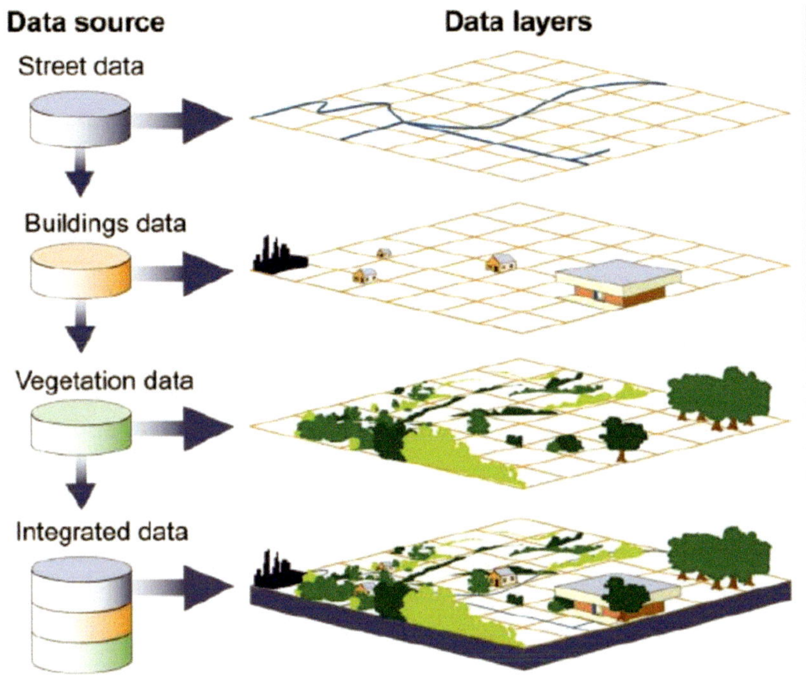

Fig. 7. Geographic Information System

A geographic information system (GIS) is a computer system for capturing, storing, checking, and displaying data related to positions on Earth's surface. By relating seemingly unrelated data, GIS can help individuals and organizations better understand spatial patterns and relationships.

GIS technology is a crucial part of spatial data infrastructure, which the White House defines as "the technology, policies, standards, human resources, and related activities necessary to acquire, process, distribute, use, maintain, and preserve spatial data."

GIS can use any information that includes location. The location can be expressed in many different ways, such as latitude and longitude, address, or ZIP code.

Many different types of information can be compared and contrasted using GIS. The system can include data about people, such as population, income, or education level. It can include information about the landscape, such as the location of streams, different kinds of vegetation, and different kinds of soil. It can include information about the sites of factories, farms, and schools, or storm drains, roads, and electric power lines.

With GIS technology, people can compare the locations of different things in order to discover how they relate to each other. For example, using GIS, a single map could include sites that produce pollution, such as factories, and sites that are sensitive to pollution, such as wetlands and rivers. Such a map would help people determine where water supplies are most at risk.

Data Capture

Data Formats

GIS applications include both hardware and software systems. These applications may include cartographic data, photographic data, digital data, or data in spreadsheets.

Cartographic data are already in map form, and may include such information as the location of rivers, roads, hills, and valleys. Cartographic data may also include survey data and mapping information that can be directly entered into a GIS.

Photographic interpretation is a major part of GIS. Photo interpretation involves analyzing aerial photographs and assessing the features that appear.

Digital data can also be entered into GIS. An example of this kind of information is computer data collected by satellites that show land use—the location of farms, towns, and forests.

Remote sensing provides another tool that can be integrated into a GIS. Remote sensing includes imagery and other data collected from satellites, balloons, and drones.

Finally, GIS can also include data in table or spreadsheet form, such as population demographics. Demographics can range from age, income, and ethnicity to recent purchases and internet browsing preferences.

GIS technology allows all these different types of information, no matter their source or original format, to be overlaid on top of one another on a single map. GIS uses location as the key index variable to relate these seemingly unrelated data.

Putting information into GIS is called data capture. Data that are already in digital form, such as most tables and images taken by satellites, can simply be uploaded into GIS. Maps, however, must first be scanned, or converted to digital format.

The two major types of GIS file formats are raster and vector. Raster formats are grids of cells or pixels. Raster formats are useful for storing GIS data that vary, such as elevation or satellite imagery. Vector formats are polygons that use points (called nodes) and lines. Vector formats are useful for storing GIS data with firm borders, such as school districts or streets.

Spatial Relationships

GIS technology can be used to display spatial relationships and linear networks. Spatial relationships may display topography, such as agricultural fields and streams. They may also display land-use patterns, such as the location of parks and housing complexes.

Linear networks, sometimes called geometric networks, are often represented by roads, rivers, and public utility grids in a GIS. A line on a map may indicate a road or highway. With GIS layers, however, that road may indicate the boundary of a school district, public park, or other demographic or land-use area. Using diverse data capture, the linear network of a river may be mapped on a GIS to indicate the stream flow of different tributaries.

GIS must make the information from all the various maps and sources align, so they fit together on the same scale. A scale is the relationship between the distance on a map and the actual distance on Earth.

A projection is the method of transferring information from Earth's curved surface to a flat piece of paper or computer screen. Different types of projections accomplish this task in different ways, but all result in some distortion. To transfer a curved, three-dimensional shape onto a flat surface inevitably requires stretching some parts and squeezing others.

A world map can show either the correct sizes of countries or their correct shapes, but it can't do both. GIS takes data from maps that were made using different projections and combines them so all the information can be displayed using one common projection.

GIS Maps

Once all the desired data have been entered into a GIS system, they can be combined to produce a wide variety of individual maps, depending on which data layers are included. One of the most common uses of GIS technology involves comparing natural features with human activity.

For instance, GIS maps can display what man-made features are near certain natural features, such as which homes and businesses are in areas prone to flooding.

GIS technology also allows users to "dig deep" in a specific area with many kinds of information. Maps of a single city or neighborhood can relate such information as average income, book sales, or voting patterns. Any GIS data layer can be added or subtracted to the same map.

GIS maps can be used to show information about numbers and density. For example, GIS can show how many doctors there are in a neighborhood compared with the area's population.

With GIS technology, researchers can also look at change over time. They can use satellite data to study topics such as the advance and retreat of ice cover in polar regions, and how that coverage has changed through time. A police precinct might study changes in crime data to help determine where to assign officers.

One important use of time-based GIS technology involves creating time-lapse photography that shows processes occurring over large areas and long periods of time. For example, data showing the movement of fluid in ocean or air currents help scientists better understand how moisture and heat energy move around the globe.

GIS technology sometimes allows users to access further information about specific areas on a map. A person can point to a spot on a digital map to find other information stored in the GIS about that location. For example, a user might click on a school to find how many students are enrolled, how many students there are per teacher, or what sports facilities the school has.

GIS systems are often used to produce three-dimensional images. This is useful, for example, to geologists studying earthquake faults.

GIS technology makes updating maps much easier than updating maps created manually. Updated data can simply be added to the existing GIS program. A new map can then be printed or displayed on screen. This skips the traditional process of drawing a map, which can be time-consuming and expensive.

FOREST RESTORATION

Traditional approaches to ecological restoration

The community approach

A biological community is a group of organisms that interact and share an environment. Within a community, organisms may compete for the same resources (competition), profit from the presence of other organisms (facilitation) or use other organisms as a food source (trophic interaction). In stable communities, these interactions lead to predictable, directional changes in community structure known as ecological succession. Succession is an important guiding principle in the community approach to ecological restoration. The restoring forest is a dynamic ecosystem, with changing species composition and forest structure, but interventions and management steer the forest towards a desired climax or pre-disturbance community structure.

These interventions are usually designed to accelerate natural succession or to bypass intermediate successional phases. Basically, the community approach is focussing on restoring forest biodiversity per se. The many studies that apply facilitation as a restoration tool of woody communities are typical examples of the community approach to forest restoration. Planting late-successional tree species (protégé species) under early-successional shrubs (nurse species) has been shown to be an effective means of restoring forests under high abiotic stress.

Insights from alternative stable states theory have also been useful to guide restoration practices that focus on community structure. In severely degraded systems, alternative stable states may make efforts to restore pre-disturbance communities difficult, if not impossible. In such cases, a single intervention may not suffice to induce forest regrowth: succession fails and the community is blocked in a low diversity/low biomass state.

Exclusion of grazing animals may be an effective means for woodland restoration in degraded drylands, but only when soil moisture conditions also improve. Wet pulses caused by climatic oscillations such as the El Niño Southern Oscillation may provide such necessary additional impulse to induce a regime shift that leads to forest restoration. Similarly, planting and sowing of late successional tree species (an intervention to overcome seed limitation) has been found effective for the restoration of highly complex forest on bauxite mined sites, but only after careful site preparation and topsoil handling or replacement (interventions to overcome survival limitation caused by soil compaction, decreased soil porosity and infiltration capacity, and the loss of soil biota).

The ecosystem approach

Restoration of species richness and community structure over time implies increasing ecosystem complexity and functionality. In the ecosystem approach, restoration of ecosystem functions such as primary production, energy flows and nutrient cycles, is the guiding principle on which restoration efforts are based. Basically, this approach aims at restoring suitable abiotic conditions that allow (passive) recolonization of species.

The ecosystem perspective typically starts from a landscape point of view, building on spatial heterogeneity and broad spatial scales. The connections or barriers between neighbouring ecosystems have an effect on the resource balances and set limits on the communities that can be restored. Reforestation of degraded sites with trees that alter the physical and chemical characteristics of the soil and that affect the biochemical cycles through litter fall or root activity presents a typical example of the ecosystem approach to forest restoration.

The biodiversity - ecosystem function approach to ecological forest restoration

The study of the relation between biodiversity and ecosystem functioning is a rapidly growing field. The traditional view that has dominated ecology until the 1990s started from the idea that species distribution patterns resulted directly from the abiotic and biotic (species interactions) components determining the environment. In the early 1990s, however, this view was challenged, when one started to realize that species diversity also affects the abiotic environment, and even the functioning of ecosystems. The functioning of an ecosystem incorporates processes such as decomposition of organic matter, fixation of carbon, nutrient and water cycling and degradation of toxic compounds. Meta-analyses of the results of mainly small-scale biodiversity experiments have shown that, on average, ecosystem functions increase with increasing species number. The success of the idea that biodiversity affects ecosystem properties and functions - some have called it a paradigm shift in ecology - can be explained by the fact that it offers a comprehensive framework to evaluate the consequences of biodiversity loss caused by human activities, and at the same time provides a powerful incentive for biodiversity conservation and ecological restoration.

Researchers was the first to propose that restoration ecology may benefit from insights from the BEF framework, and this idea has been further elaborated by researchers. Here we build on these ideas and put them in a forest restoration context. In contrast to more traditional approaches, restoration based on the BEF perspective strongly focuses on restoring the relationship between biodiversity and ecosystem functioning. In what follows we list some important considerations regarding forest restoration that can be inferred from the BEF framework.

We are aware that foresters have already adopted the BEF framework in setting up large experiments where the effects of tree species richness on ecosystem functions are evaluated. Nevertheless, we believe that forest restoration efforts may benefit from such an overview, in particular since ecosystem functioning and functional (bio)diversity has received very little attention in a forest restoration context so far.

RESTORING MULTIPLE FOREST FUNCTIONS REQUIRES MULTIPLE SPECIES

One of the major functions of forest ecosystems is carbon fixation, which is directly related to the ecosystem services carbon sequestration and the provision of fire and construction wood. There is evidence that tree diversity has a positive effect on ecosystem production. Based on the largest data set ever analysed in this context to date (12,000 permanent forest plots in eastern Canada), researchers reported that, after controlling for environmental and climate differences between plots, tree productivity was positively related to stand biodiversity. These results confirm earlier work in 5000 permanent plots in Mediterranean forests across Catalonia (NE Spain). In a reforestation context, researchers found that mixed plantations in Costa Rica performed better than monocultures for all growth variables considered, including height, diameter at breast height, volume, and above-ground biomass.

Also in natural stands of tropical forest with high environmental and spatial variation, positive effects of tree species diversity on tree carbon storage were found. Positive effects of tree diversity on above-ground productivity are certainly not an universal pattern, however, and above-ground biomass production and soil carbon fixation may also respond differently to tree diversity in plantation forests. This corroborates the result of a meta-analysis of BEF experiments where it was found that high biodiversity treatments do not always outperform the best performing monoculture. In a forest restoration context, where fast growing tree species with strong global timber markets are readily available, this may suggest that monocultures are an option. However, evidence is accumulating that focussing on one single ecosystem function often overlooks an important aspect of biodiversity: the possibility of one species to contribute to different ecosystem functions at the same time.

Because different species often influence different ecosystem functions, focussing on one function in isolation will strongly underestimate the biodiversity required for maintaining an ecosystem with multiple functions, at multiple times and places in a changing environment. Although the evidence only comes from grasslands and aquatic environments so far, it convincingly shows that species redundancy is unlikely to occur when several ecosystem functions and services are considered in combination.

Therefore, it is highly unlikely that species poor plantations will outperform species diverse tree assemblages for a combination of forest ecosystem functions, including above-ground biomass production, disease resistance, carbon fixation, nectar provision, erosion control, water capitation, N_2-fixation and fruit production. It is therefore of special importance that reforestation efforts clearly define the ecosystem services and functions that the restored forest is intended to deliver. Also, it is important to realize that ecosystem functions of restoring forests may change over time because of changes in tree sizes, forest structure and relative importance of functional groups, even if there are no changes in tree species composition.

Finally, it should be noticed that although there is already some knowledge on the effects of tree diversity on forest productivity, it is not known how understory shrub diversity, and even herbaceous species, affect forest productivity or other ecosystem functions. This may, for example, happen through these species' impacts on litter decomposition, on water capture and on the diversity of soil biota.

Restoring stable forest functions requires multiple species

The hypothesis that larger species diversity leads to higher stability of ecosystem functioning has been a point of debate for half a century, and it has re-emerged within the BEF framework. The main ideas behind the biodiversity *vs.* ecosystem stability concept are functional response diversity and functional compensation. This occurs when positive changes in the level of functioning of one species (a species becoming functionally dominant) are associated with negative changes in the functioning of other species. This compensation drives the stabilization of ecosystem properties such as biomass production. Basically, the stability of the functioning of an ecosystem can be measured in three ways: i) the long term variability of an ecosystem property through time in relation to background environmental variation (*variance*); ii) the impact (*resistance*); and iii) the recovery (*resilience*) of ecosystem properties to discrete disturbances. As it is expected that these discrete and extreme disturbances such as extreme climate events and pest and disease outbreaks will become more frequent under the predicted climate change, it is very important to incorporate insights from the relation between biodiversity and stability of ecosystem functioning into forest restoration projects. It is crucial to realize that, just as the degree of species redundancy decreases when multiple ecosystem functions are considered (see earlier), there is currently strong experimental evidence that in changing environments, more species are required to guarantee ecosystem functioning than in constant environments.

Evidence for the latter comes from studies that have related forest tree diversity with measures of stability of forest ecosystem functioning. Researchers used satellite imagery to estimate the impact of the extreme 2003 summer drought on canopy greenness of different forests types in Spain, by quantifying the NDVI (normalized difference vegetation index). NDVI correlates with ecosystem CO_2 fluxes. These authors reported a positive relation between woody species diversity and resistance of canopy greenness against drought in forests on dry locations, whereas no such relation was discovered in more moist forests. Similarly, researchers related stability in stand productivity across 64 years with conifer diversity in the Sierra Nevada, USA.

They found a significant relation between species richness and resilience of stand productivity after recurrent severe droughts. Resistance to drought was, however, not related to species diversity. These studies partly support positive biodiversity effects on stability of biomass production, but they also show that patterns may be complex, vary across ecosystem types, and depend on the measures that are used to quantify stability. In any case, temporal stability of ecosystem functioning is an important consideration for projects aiming at forest restoration, especially under the current global change scenario. Again, it is not known whether understory shrubs and herbaceous species contribute to the stability of forest ecosystem functioning.

FOCUS ON FUNCTIONAL DIVERSITY RATHER THAN ON TAXONOMIC DIVERSITY

Whereas general biodiversity measures are based on taxonomy in the first place (species presence or absence), functional diversity measures relate to what organisms effectively do in an ecosystem, quantify the distribution of traits in a community or measure the relative magnitude of species similarities and differences. How to best measure functional diversity is a much debated question, but researchers summarize five useful multivariate functional diversity measures. Some authors have suggested that functional diversity measures are particularly suitable or even better to predict the interactions between biodiversity and ecosystem processes. Using a tree diversity index based on among-species variation in seed mass, wood density and maximum height Paquet and Messier showed that this measure outperformed a taxonomically based diversity index in explaining tree productivity. Researchers demonstrated that removing certain functional groups from a tropical forest had more important effects on the above-ground carbon pool than randomly removing species. Researchers, on the contrary, reported that functional group richness performed worse than tree species richness, but this was likely due to a rather rudimentary functional group delineation.

Thus, when selecting tree species for forest restoration, these findings suggest focusing on functional groups based on relevant plant traits. While these traits are readily available for species from temperate regions by now, the establishment of plant trait databases for tropical tree species and the centralisation of all available data in a general database are important works in progress. Maximizing functional diversity can be achieved by quantifying the functional diversity of the species mix used for restoration. This can be done by delineating emergent or functional groups (assemblages of species performing similar functional roles), or by using more complex, continuous or non-grouping measures of functional diversity. The selection of relevant plant traits remains, however, crucial with respect to the forest ecosystem functions to be restored. Researchers provide a comprehensive list of species traits that can be used to quantify functional diversity of tree mixtures used for reforestation of European temperate forests. Selected traits included nominal (e.g. leaf type, crown architecture), ordinal (e.g. adult light requirements, height growth vigour) and scale variables (e.g. leaf N concentration, litter C:N ratio).

A better mechanistic understanding of how species traits and their interactions affect ecosystem functioning is also important, however, to be able to proactively analyse different reforestation scenarios and their impact on forest functioning. In this context, it is important to realize that relationships between functional traits and ecosystem functions such as carbon storage in natural populations are not always transferable to tree plantations and *vice versa*.

EFFECTS OF GENETIC DIVERSITY EXTEND UP TO THE ECOSYSTEM LEVEL

Whereas conservation biologists have acknowledged the negative fitness consequences of reduced genetic diversity for decades, forest restoration projects may still incorporate very few genotypes. There is evidence, however, that monoclonal populations are more vulnerable to pathogens than genetically diverse assemblages. The point that we want to make here, however, is that the effects of stand genetic diversity can be expected to extend far beyond the fitness of the individual trees or stands. It is only recently that it has become clear that variation in population genetic diversity or in genotype composition can have far-reaching ecological effects. The ecological consequences of genetic diversity (coined 'community genetics') have been demonstrated at different levels of organization, from the population over the community to the ecosystem. For example, plant genotypic diversity and genotype identity have been shown to affect biomass production and community invasibility, and also the invertebrate diversity of the higher trophic levels. It was also shown that litter decomposition and nutrient release differed between different *Populus* genotypes, indicating that selection of tree genotypes may have profound and long lasting effects on ecosystem functioning of restored forests. Although a discipline as community genetics is in its infancy, there is already some evidence to suggest that there are extended consequences of plant genetic variation, up to the level of the ecosystem properties. The selection of specific genotypes, and the genotypic diversity of tree assemblages, may therefore have major implications for the functioning and the resilience of forests.

Synchronize above- and below-ground biodiversity

The above-ground biodiversity of forests also comprises fauna with important ecosystem services that include pollination, pest control and seed dispersal. The ecosystem services of birds, for instance, have been well documented and in the light of forest restoration birds have been shown essential for dispersing tree seeds into restoring areas and overcoming seed dispersal and germination limitation. Far less is known about the role of below-ground biota and the linkages between trees and these biota. The study of soil microbial community structure and functioning has traditionally received little attention in ecology.

But as with above-ground biodiversity, there is evidence that below-ground diversity has a significant impact on ecosystem functioning. In a series of simplified tropical forests, researchers found significant positive relationships between the diversity of arbuscular mycorrhizal fungi (AMF) and ecosystem net primary productivity, and between AM fungal community evenness and ecosystem phosphorus-use efficiency. The quick development and availability of molecular tools such as t-RFLP and next-generation sequencing to quantify microbial diversity, together with the strong focus of the BEF approach on the functionality of ecosystems, has resulted in an increased interest in the role of microbial soil community diversity in driving processes such as organic matter decomposition and plant nutrient uptake.

Because ecological restoration is usually occurring on highly disturbed or degraded sites, it is important that above-ground-below-ground species linkages are permanently considered during the restoration process, and more specific, that there is a synchronisation between above- and below-ground species associations. Clearly, the crucial question is whether below-ground microbial community simply follows the introduced tree and shrub species, or whether some kind of inoculation is required. Among the relevant soil micro-organisms, arbuscular mycorrhizal fungi (AMF) and ectomycorrhizal fungi (ECMF) can be expected to play a major role during restoration of degraded sites. Many tree and shrub species associate with AMF and ECMF, which provide nutrients in exchange for plant carbohydrates. Recent evidence has shown that at least ECMF are dispersal limited, and are less abundant on isolated trees.

This finding may urge for some kind of active inoculation of degraded restoration sites. How to successfully apply soil microorganisms in particular restoration projects, however, is an almost empty research field. Whereas fundamental insights in the role of AMF in structuring grassland communities are growing, it remains largely unknown how these fungi contribute to successful restoration and the few available reports on the effects of large scale inoculations in grasslands did result in contradictory conclusions. Also, the inoculation of tree roots with mycorrhiza has received some attention in forest restoration projects, but the results are not straightforward. This leads to the conclusion that at present much remains to be understood about how below-ground microbial diversity contributes to successful restoration of forest functions. Newly available molecular tools to quantify microbial diversity combined with detailed measurements of forest functioning are likely to increase our insights in how to apply below-ground biodiversity for restoration purposes.

Restored forests are often novel ecosystems

While restored forests may deliver similar ecosystem services and conserve levels of biodiversity comparable to the pre-disturbance vegetation, restored forests rarely match the composition and structure of the original forest cover. Large changes in ecosystems will usually result in novel systems, comprising different species, interactions and functions. In this context, it is important to realize that both the recent tendency towards accepting perennial, global change driven changes to the environment and the increasing application of the BEF framework to ecological restoration may facilitate the acceptance of using non-native species in forest restoration. While many ecologists still consider autochthony of species a prerequisite for their use in ecological restoration, a focus on species' functions rather than on species' origins is already advocated by others as being a "more dynamic and pragmatic approach to the conservation and management of species". In this sense, the BEF approach may be at the source of a paradigm shift in restoration ecology.

ADAPTIVE MANAGEMENT

AFM (Adaptive forest management) is forward-looking and aims to preserve and develop the functionality of specific forests as a prerequisite for fulfilling the future need for forest ecosystem services. This is dedicated to all measures that adapt intact forests to changing growth and management conditions due to environmental setting, but also, e.g., due to diverse economic perspectives. Researchers introduced three pillars of AFM: (1) *knowledge* of both environmental settings including uncertainties, but also of perception changes among decision-makers; (2) *options* to identify forest adaptive capacity, to protect forest performance and to apply AFM strategies; and (3) *decisions* to repeatedly optimize AFM according to significant evaluation outcomes. Thus, the AFM concept produces feedback loops of silvicultural interventions and management aims against the background of changing environments and varying owners' perspectives. With this, AFM represents a flexible forest management concept, but with distinct local reference considering small-scale variations of climate and site conditions.

To be clear, AFM is not the same as adaptive management although AFM may be usefully applied within an adaptive management framework. Contrarily, adaptive management may not be useful in guiding adaptation under rapidly changing climatic conditions as it relies on information gained from management experiments under current conditions to guide future actions that may be conducted under quite different conditions of novel climate. Several strategies for adaptation under global change have been described in relation to tolerance of ecological novelty, or how different the future ecosystem is relative to the historic past.

Forest landscape restoration—the regional concept

FLR, in contrast, is the process of regaining ecological functionality and enhancing human well-being across deforested or degraded forest landscapes. The FLR approach seeks to balance different values/functions at the landscape scale such as water regulation, wildlife habitat, and biodiversity or carbon storage. Most of the many relevant restoration techniques for FLR are not new, but the new is that FLR requires the involvement of a wide range of stakeholders and competences to fulfill the landscape approach in populated regions. Nevertheless, a central element in restoration management is the use of ecological key concepts such as succession, disturbance, functional characteristics of species, or safe sites.

The contexts of FLR projects vary according to biome, landscape history, and social factors such as governance, tenure, and technical capacity. FLR projects are highly heterogeneous also because they begin with different initial objectives such as offering employment in economically constrained areas, reduction of soil erosion in agricultural land (China), landscape rehabilitation in abandoned farmland (Eastern Europe), reduction of natural hazards (human-populated mountain areas), carbon sequestration (Ireland), or reconstruction of fragmented habitats in degraded landscape (Italy). Too often, FLR is backward-looking and aims to return to historical conditions of species composition, stand structure, or both but this is not inherent in the FLR approach.

Nevertheless, the FLR concept is still being refined to accommodate new perspectives, such as technical problems (lack of large-scale experience), goal conflicts between the various stakeholders involved, the inclusion of non-forest land use within the landscape including agroforestry measures, or even the appropriate measure of success.

Integrative adaptive measures concept—helpful for adaptation and restoration success?

In the following, we discuss how the AM concept could serve as the link between FLR and AFM. In addition, two essential ambits of stand and landscape-related concepts of AM and their contribution to maintain or restore forest functionality are looked at in more detail.

AM interaction with biodiversity issues

There is strong evidence that tree species richness and high genetic diversity positively affect the adaptive capacity of forests against climate change. But can species-rich and genetically diverse forests better restore basic functions in forest landscapes? One of the emerging research questions is how biodiversity affects the functionality of a forest ecosystem (functional biodiversity research. Forests rich in woody species often contain plants with different "strategies" concerning establishment and competitiveness (plant functional types). Therefore, resources such as light, water, and nutrients can be spatially and temporally used by different species, which in some cases lead to a superior productivity of diverse compared to mono-specific forests.

Moreover, does tree species diversity positively affect resistance/resilience of forests to disturbances or stresses? That is, could pre-disturbance functionality be better restored in a diverse stand, because tree species with different response patterns to these stresses can compensate for losses of more vulnerable species? In this respect, there is increasing agreement on the role of non-native species in the provision of important ecosystem services such as desired products or habitat for other species in the future. The difficulty of removing all non-native species from ecosystems contradicts the still dominant goal to push ecosystems back to historical composition and function.

Another important measure to enhance the restoration capacity of a forest after disturbance is to retain a significant amount of ecosystem legacies (e.g., seed trees, deadwood, stand remnants), thus increasing the structural diversity of stands. Legacies provide seed dispersal, nutrient translocation, water storage, and the maintenance of genetic information in the recovery phase of an ecosystem after disturbance. Moreover, stand-level legacies contribute as important habitat to faunal species richness, e.g., as antagonist species which can curb biotic disturbances. Therefore, legacies increase the number of potential pathways for ecosystem restoration after disturbance. This fits well with the general goal to manage forests for resilience.

Advantageous hereby are multi-aged stands with structural diversity—they have the potential to increase the resistance and resilience of both stands and forests. There are several ways and approaches to achieve this. In central Europe, even-aged mono-specific forests are currently being converted into uneven-aged mixed forests for multiple purposes. Moreover, the integration of disturbance into forest management can be a means to achieve this goal. Here, researchers give an overview of concrete measures to promote uneven ageness and structural diversity: emulation of disturbances and carefully designed salvage operations, emphasizing the important role of retained elements of the stand and variable treatment intervals or intensities. Most of these measures are more feasible at the landscape scale because uneven-aged stands with high structural diversity gradually lose their stand compartment structure. With the integration of stand and landscape perspectives in the AM concept, adaptive features like tree species richness, structural diversity, and enhanced gene flow can be managed more effectively both at the stand and landscape levels.

AM contribution to reduce vulnerability and increase resistance and/or resilience

Vulnerability can be described as the probability with which an environmental system can be damaged through changes in the environment, society, or both after taking into account reduction of its adaptive capacity. A variety of measures that reduce vulnerability in a forest stand or landscape play a positive role in restoring the resilience potential of a forest ecosystem after disturbance.

Site preparation can be a central measure to enhance the regenerative capacity of an ecosystem by removing inhibiting factors for seedling's growth or by increasing the variability of site conditions. Furthermore, the use of stress-tolerant plant material (improved seedlings) or stress-tolerant provenances is essential to overcome the vulnerable juvenile growth phase of trees or to reclaim a disturbed or degraded ecosystem.

Another important tool to reduce vulnerability of forest ecosystems is assisted migration of more adapted or adaptive species whereby species (often non-natives) are intentionally transferred to regions outside of their natural range that are characterized by a matching climate which represents the artificial extension of the range distribution of a more resilient species. Many forest conversion activities in Europe in the past decades have already applied this approach to replace endangered species.

Moreover, in regions where past land use led to significant degradation (e.g., Denmark), the transfer of suitable mostly non-native species proved to be successful in establishing ecosystems with a high degree of novelty (transformational restoration). Assisted migration also encompasses the choice of appropriate stress-tolerant provenances, e.g., towards extreme weather events like heat and drought. This goes along with the assisted gene flow concept to translocate pre-adapted individuals to facilitate adaptation of planted forests to climate change. In particular, marginal provenances originating near the drought-induced range limits can provide tree species ecotypes more tolerant to drought. Yet, there is evidence that these ecotypes often maintain their stress tolerance at the expense of growth.

Restoring forest landscapes can be accompanied by systematically combining plant (tree) species with specific adaptive traits that make the ecosystem more resilient against climate change. During the course of stand management, the reduction of stand density most likely increases the individual stability and stress resistance of the remaining trees in the stand, especially if management is concentrated on previously selected superior future crop trees.

At the same time, in specific cases, the increase of stand density with different tree species could be beneficial in terms of resilience potential and provision of ecosystem services. The application of silvicultural systems, maintaining—on a long term—low or moderate stocking levels or smaller target diameters, does not contribute that much to increase the forest's resilience but certainly decreases the risk for disturbances, such as storm and fire.

A comprehensive vulnerability reduction needs to account for biotic stress. Climate change modifies the population dynamics of pests and pathogens and needs to be considered by AFM as well as FLR. Moreover, more advanced forest adaptation and restoration measures need to acknowledge the overarching challenges that reduce their effectiveness such as loss or degradation of forest land.

ASSESSING THREATS AND VULNERABILITY

Effective conservation depends strongly on a full appreciation of the causes of biodiversity loss. Yet identifying such threats or 'threatening processes' has received remarkably little attention from researchers. Surprisingly, the issue is ignored by many conservation biology textbooks. There is enormous scope for improving methods for assessing threats and diagnosing their impacts, and improving the quality of information available to conservation managers. However, identifying the precise causes of decline in the abundance of a particular species can often be surprisingly difficult.

Different types of threat may be identified. Direct threats are those that are directly responsible for loss or degradation of forests, or their associated biodiversity. Indirect threats are the underlying causes of such direct threats. For example, an underlying threat such as a government policy may be responsible for the direct threat of forest conversion to agriculture. Other terms used to describe threats include 'drivers' or 'pressures'; these terms may be preferred because they imply that effects on biodiversity can be either negative or positive, whereas the term 'threat' implies only negative impacts. However, it should be remembered that different authors interpret these terms in different ways.

Researcher identifies the following main methods of assessing threats:

- Conceptual modelling, used to illustrate the relation between threats and their impacts, and for providing a strategic framework for identifying appropriate management interventions.
- Threat matrices. Matrices can vary from simple tables to complex logical frameworks linking different threats and interventions to conservation targets. Matrices are relatively simple to implement and can readily be updated, but their dependence on subjective information is a weakness.
- Participatory threat mapping, which can involve use of pictorial maps or three dimensional models to elicit information about changes in forest habitat quality or quantity, when working with community groups.
- GIS-based mapping, incorporating quantitative spatial data. Direct threats, such as habitat fragmentation, can be assessed and displayed by using GIS. Spatial models of forest dynamics can be used to explore and illustrate the potential impacts of different threats on forest extent, structure and composition.

Researchers provide a detailed review of the concept of vulnerability in conservation planning, noting that information on threatening processes and the relative vulnerability of areas and features to these threats pervades the process of conservation planning. Researchers defined vulnerability as 'the likelihood or imminence of biodiversity loss to current or impending threatening processes'. They extend this definition by distinguishing three dimensions of vulnerability, *exposure*, *intensity*, and *impact*, and provide the following information regarding their measurement:

- Exposure can be measured either as the probability of a threatening process affecting an area over a specified time, or the expected time until an area is affected. Exposure is commonly measured categorically, for example as 'high', 'medium', or 'low' suitability for agriculture, but has also been measured on continuous scales by some authors. Maps can be produced illustrating the relative exposure of different areas to a particular threat.
- Intensity measures might include magnitude, frequency, and duration of the threat. Examples include livestock density, volume of timber extracted per hectare

of a forest type, or the density of an invasive plant species. Intensity can also be estimated categorically, and can be mapped across whole planning areas.

- Impact refers to the effects of a threatening process on particular features such as the distribution, abundance, or likelihood of persistence of a species of interest. For example, logging may have much greater impact on animal species dependent on old growth forests than on species that inhabit a variety of post-logging stages. Impact might also depend on the spatial pattern of the threatening process, for example on the degree of connectivity between old growth patches retained after logging operations.

As conservation planning is generally spatial, a key issue is whether vulnerability can be mapped, and therefore integrated with other spatial information such as forest cover, boundaries of management units etc. According to researchers, this requires mapping of spatial predictions of the future distribution of threatening processes. Maps of exposure can be based on the current distributions of threats and knowledge of variables that could predispose areas or features to those threats. For example, the likelihood of forest conversion to agriculture is often related to the suitability of the soil for agricultural crops, topography, and proximity to infrastructure or population centres. Spatial predictions of intensity are less common in the literature than predictions of exposure; for some threats, such as forest clearing, intensity is considered as binary (either cleared or not).

Impact is the most difficult dimension of vulnerability to map, as this may require feature-specific information on the effects of different levels of intensity, spatial information on features relative to variations in intensity, and ways of integrating this information across assemblages of species, sets of vegetation types, or other groups of features.

Researchers also reviewed the different methods that have been used by previous researchers to assess the three dimensions of vulnerability. The methods were grouped into four groups based on the types of data used. Little information is available in the literature to indicate which approach would be preferable in a given situation; it is likely that a combination of approaches will give the best overall result.

As noted by these authors, a comprehensive assessment of vulnerability would consider all of the threats affecting an area and also include the dynamic responses of threats to conservation actions. Combining vulnerability scores for multiple threats is analytically tractable and can be achieved by differentially weighting threats to reflect their relative importance, ideally informed by their respective impacts. An example of the latter approach is provided by researchers, in their global assessment of the conservation status of tropical dry forests. Here, GIS techniques were used to analyse and combine maps of different threats derived from remote sensing data to identify those forest areas vulnerable to multiple threats. Such methods, coupled with spatial analysis and statistical modelling techniques, offer the potential for powerful and sophisticated analyses of threats at a range of different scales.

It should be remembered that risk analysis and hazard assessment are widely practiced by foresters, and a substantial literature exists on this topic. This literature has been little used by the conservation community to date. Such measurements can be used as a basis for assessing risk, as described briefly below for some of the most important types of disturbance.

- Fire. Forest fire risk is generally assessed by identifying the potentially contributing variables and integrating them into a mathematical expression or index. The index is used to indicate the level of risk, and can be mapped. A wide variety of different approaches are used to produce such indices, which vary particularly in terms of the timescales involved. Estimates of the probability of fire occurrence are typically based on variables such as the amount and type of fuel available for burning, topographic variables, vegetation characteristics, and meteorological variables.
- Wind. A number of different approaches to analysing and modelling windthrow risk have been developed. Statistical models use empirical information on damage collated over a number of years in selected areas, whereas deterministic models consider tree or stand characteristics and the windiness of a site or the critical wind speed. An example of a widely used mechanistic model is GALES, developed by the Forestry Commission in the UK, which estimates the threshold wind speeds required for overturning and breaking the mean tree of a stand. The resulting decision support system (DSS), ForestGALES, can be used to predict the wind damage risk for 19 different conifer species, growing in homogenous, even-aged stands in Europe. Spatially explicit models of forest dynamics such as LANDIS also offer tools for exploring the potential impacts of different wind regimes.
- Herbivory. The determination of browsing damage on forest regeneration at a given time can be used to forecast impacts in the future, for example when the timber is harvested or when the function of the forest is seen to have been compromised. The effects of browsing and the resulting damage can be decades apart. In order to estimate the long-term impacts of browsing, indicators may need to be specified for young forest stands at the time when top-twig browsing is no longer possible. Ultimately the risk of damage by herbivores is a function of animal behaviour, something that is difficult to predict but is also an active area of research.
- Deforestation. Research has indicated that deforestation can be related to a range of factors, such as population density, population growth, agricultural expansion, income levels, and amount of timber harvesting. Typically, deforestation is assessed by comparing the forest cover at different times by means of remote sensing images.

Modelling of land-use changes such as deforestation requires combining spatially explicit ecological data with information on socio-economic as well as biophysical variables. A range of models of land-use change have been developed, based on a variety of analytical approaches.

For example, GEOMOD is a GIS-based model that simulates the location of deforested cells by using geophysical attributes as well as spatial data of forest cover at different time intervals, and is now commercially available as part of the IDRISI software program. In a review of the current understanding of land-cover change in tropical regions, researchers emphasize the complexity of the issue and highlight the role of market forces, policy interventions, and changes in social organization and attitudes. Relatively simple statistical models are unlikely to capture this complexity, and instead these authors highlight the role of agent-based systems and narrative perspectives to provide an explanation and prediction of land-use changes.

Although important advances have been made in modelling the spatial pattern of deforestation, ecosystem heterogeneity and lack of data (particularly for relevant socioeconomic drivers) continue to constrain analyses and the development of deforestation scenarios. As deforestation cuts across a range of disciplines (sociology, economics, ecology, and geography), some researchers have encouraged a focus on the interactions between these disciplinary approaches rather than on the details of each subcomponent.

MONITORING

National Forest Monitoring Systems

FAO has supported more than 50 countries in their development of robust National Forest Monitoring Systems (NFMS) and assessments, with the goal of developing reliable forest resource information for application in creating national forest policies, planning and sustainable development. Forest monitoring systems include measurement, reporting and verification (MRV) functions and aim to produce high-quality, reliable data on forests, including forest-carbon estimates, that are critical to the battle against climate change caused by among others deforestation and degradation of forests.

NFMS components include:

- Satellite land monitoring systems (SLMS) and other data collection providing information for activity data (AD)
- National Forest Inventories (NFI) or other data collection providing information on emission factors (EF)

Satellite land monitoring systems

FAO helps countries to identify and collect activity data for their relevant land use and its change (LULUC), forestry and REDD+ activities. For example, spatial data on deforestation and afforestation/reforestation is typically collected through satellite data and changes can be monitored through a satellite land monitoring system. FAO supports and facilitates country processes to develop national capacities to conceptualize, design and implement a national SLMS for forestry and LULUC and to monitor area changes. Such information is the basis for data input into national greenhouse gas inventories, Forest Reference (Emission) Level (FREL/FRL) construction, REDD+ results and NDC reporting. Information on forest areas and area changes from SLMS can also support domestic efforts to improve forest management and forest conservation.

Many countries publish their national forest products on public web geoportals, which increases data transparency for the public and for (inter)national reporting mechanisms. FAO supports generation of the activity data by assisting the countries with mapping-related programming, administration and accuracy assessments of existing national and global products in order to generate country estimates. As well, FAO provides technical support to assist with conceptualisation, design and deployment of national portals.

The UN-REDD Programme has deployed 20 portals in countries in three regions, and is in the design phase for a few others, including Ethiopia, Liberia and Sudan. These will likely be launched in 2019, following consultations with government on the data publication and the portals will be maintained in the country by government staff. Training sessions as well as maintenance trainings are also planned. At the country level, technical support for this work includes discussing technical options and practical solutions for SLMS, supporting country decisions on generation and dissemination of data, and supporting domestic momentum toward improved forest monitoring and improved forest management.

National forest inventories

FAO works with countries to strengthen their capacities in designing, collecting, analysing and reporting data from national forest inventories (NFIs) to monitor forest resources through multi-purpose inventories. These involve the collection of bio-physical data that can satisfy national and international information requirements on forests, their management, carbon stock and biodiversity. In addition, in many forest inventories socio-economic information is collected to understand the anthropogenic impact on forests and their role in sustainable livelihoods.

Multipurpose NFI is also one of the key sources of relevant data for national GHG reporting, FREL/FRL construction and reporting. NFIs are a key information source for modern approaches to forest management and associated planning processes.

At the country level, support is provided on all aspects of developing an NFI. Capacity building and training efforts are tailored to each country's specific needs and context, and range from forest-inventory planning, field-data collection, data entry, analysis and reporting, biomass modelling, quality assurance and quality control, integration with remote sensing, communication and dissemination, as well as data archiving, sharing and documentation.

National GHG inventories

FAO supports capacity development for national GHG inventory systems in the context of the UNFCCC. Well-developed national GHG inventory systems will ensure consistency among the forest reference (emission) levels, REDD+ results reports and the national GHG inventory estimates. Work focuses on both technical and functional capacities. Technical capacities for the national GHG inventory include the ability to work with data and carry out calculations. Functional capacities include system-related aspects such as inventory planning and management, data sharing, quality checking and continuous improvement.

At the country level, the capacity development approach includes a range of products and results: mapping of existing institutional arrangements, capacity assessment and gap analysis, carrying out GHG inventory calculations in several software packages, and reviews of existing GHG inventories.

FAO tools for NFMS

The Open Foris (OF) initiative is led by FAO Forestry to support multi-purpose forest inventories, data processing and dissemination of results. OF provides a set of free and open-source software tools to facilitate flexible and efficient data collection, analysis, and reporting for field and satellite data. The initiative is a collaborative effort by numerous public and private institutions and hosted by FAO's Forestry Department. Open Foris provides ideal tools for performing fast, accurate and cost-effective assessments. It is highly customizable for the specific data collection needs and methodologies

Open Foris tools have been applied in forest sector particularly in NFIs in more than 20 countries, including NFIs in Tanzania, Zambia, Bhutan, Sri Lanka, Vietnam, and Argentina. In addition, OF tools are used for example for socio-economic inventory in Mongolia and dairy industry survey in Kazakhstan.

The suite of software tools includes OF Collect, used to define the hierarchical structure of a survey and define the attributes to be collected and their parameters; OF Collect Mobile, a tablet-based tool that enables data entry directly in the field, thus significantly improving the quality of data collected; and OF Calc which supports data analysis and reporting under a variety of sampling design approaches.

Collect Earth, part of Open Foris, is a tool that enables data collection through Google Earth software. The data gathered through Collect Earth is exportable to commonly used formats and can also be exported to Saiku, a tool that facilitates data analysis. In conjunction with Google Earth, Bing Maps and Google Earth Engine, users can analyse high and very-high resolution satellite imagery for a variety of purposes, including:

- Supporting multi-phase nfis,
- Land use, land use change and forestry (lulucf) assessments,
- Monitoring agricultural land and urban areas,
- Validation of existing maps,
- Collection of spatially explicit socio-economic data,
- Quantifying deforestation, reforestation and desertification.

Collect Earth Online developed by NASA-Servir in co-operation with FAO is the next generation of web-based, crowd-sourcing technology for Earth Science analyses. It allows users to collect reference data using high-resolution satellite images and big-data analysis through Google Earth Engine. Multiple users can simultaneously collect information.

Other tools in the area of forest monitoring and reporting include the System for Earth Observation Data Access, Processing and Analysis for Land Monitoring (SEPAL), which FAO has developed with funding from Norway. The cloud-based SEPAL computing platform facilitates countries' access to Earth observation data as well as processing of that data. SEPAL allows users to query and process satellite data quickly and efficiently, tailor their products for local needs, and produce sophisticated and relevant geospatial analyses quickly. An easy-to use computing platform, SEPAL allows countries to overcome processing issues related to poor internet connections or low computing power on local computers.

The following recommendations are made in a critical review of current practice:

- It is essential to clearly define the objectives of monitoring. Many programmes are based on the assumption that collecting any information about a system will be useful, but such an approach is inefficient and can result in large amounts of irrelevant information being collected. Two main types of objective can be differentiated: scientific objectives, focusing on understanding of the behaviour and dynamics of the system, and management objectives, designed to provide information for informing management decisions.
- Monitoring, management and research should be closely integrated, as explicitly required in adaptive management approaches (see above). Research can play a critical role in identifying appropriate management actions, and it can be argued that monitoring is trivial unless it is linked to experimental work to understand the mechanisms underpinning system changes. In practice, monitoring is often limited to the assessment of management policies.
- Decisions regarding what to monitor should be determined by the a priori hypotheses to be addressed by the programme and by the relative values of different components of diversity as specified in management objectives.

- The sampling design will be dependent on the choice of biological diversity measures. For some objectives, it might be adequate to focus on species richness of some groups. For others it will be necessary to estimate the abundances of each species in the community of interest, which requires greater effort. It is also important that monitoring programmes provide data to estimate not only the state variables of interest, but also the rate parameters that determine system dynamics.
- It is important to estimate detection error, which arises because few survey methods permit the detection of all individuals, or even all species, in surveyed areas. Distance sampling methods can be used to estimate detection probabilities associated with count statistics. Monitoring programmes should estimate the detection probabilities associated with the selected survey methods; without these, it is not possible to draw strong inferences about the monitored system.
- Spatial variation is a second source of error because of the inability to survey large areas entirely. Most surveys are not based on an appropriate spatial sampling scheme, and therefore do not provide unbiased estimates of biodiversity at larger spatial scales. Sample locations for monitoring programmes should not be selected arbitrarily, but be selected to permit inference to the larger area of interest. Sampling designs should be chosen with respect to their efficiency, in terms of the precision of resulting estimates.

INDICATORS

The conservation of biological diversity is one of the goals of ecologically sustainable forestry, although the concept encompasses much more than biodiversity conservation alone. Biodiversity includes diversity at the genetic, species, landscape, and ecosystem levels. Given this complexity and the inadequate descriptions of local biodiversity currently available, it is difficult to judge whether forests are being managed in an ecologically sustainable way. Moreover, it is impossible to measure and monitor the effects of various management practices on all species. An array of international and national initiatives have sought to overcome this problem by identifying indicators — a subset of attributes that could serve as surrogates for total biodiversity and be used as indicators to monitor the success or failure of management practices to sustain biodiversity.

The lists of criteria for biodiversity conservation and indicators of biodiversity as part of ecologically sustainable forestry are extensive and have been embraced by many national and international organizations. It is difficult, however, to determine how the criteria and indicators of biodiversity might be identified, measured, interpreted or monitored. In part, this is because ecological knowledge lags behind policy initiatives.

For example, a basic requirement for the use of many of the proposed biodiversity indicators is knowledge of what species are in a forest, where they are, and how they might respond to disturbance. Yet the prospect of complete species inventories, which include location as well as identity, is dim, as is an understanding of population dynamics sufficient to predict levels of abundance and population demographic structures necessary for the assessment of longterm viability.

There are no easy answers to these problems, but they need to be tackled now if conservation biology is to make a contribution to forest policy, planning, and management before the opportunity to do so is gone.

Concepts associated with indicators and indicator species are not well understood, yet there is strong pressure on forest managers to embrace indicators in the conservation of biological diversity (e.g., Commonwealth of Australia 1998). Selection of the wrong or inappropriate indicators could give a false impression of scientific understanding, managerial knowledge, and ecological sustainability. This could have negative effects on biological diversity in forest ecosystems.

We reviewed some of the approaches to indicators of biodiversity. Although biodiversity includes genetic, species, landscape, and ecosystem diversity, we focused on species. Within this more restricted arena, indicators of biodiversity can be divided into two broad groups: (1) biological or taxon-based indicators, particularly indicator species and guilds, and (2) what we call structure based indicators—stand -and landscape-level (spatial) features such as stand structural complexity, plant species composition and connectivity and heterogeneity. We discuss the limitations of these broad groups of indicators and consider what can be done to promote the conservation of biodiversity in forests despite existing limitations.

Taxon-Based Biodiversity Indicators

Indicator Species

The search for indicators of biodiversity has tended to focus on biological entities, such as gene frequencies, populations, species, species assemblages, and communities, that might function as surrogates or proxies for other forms of biodiversity and/or reflect changes in ecosystem patterns or processes. Although indicators are required at a wide range of organizational levels, most efforts to date have focused on particular species or members of species assemblages (e.g., guilds). Researchers defined an indicator species as "an organism whose characteristics (e.g. presence or absence, population density, dispersion, reproductive success) are used as an index of attributes too difficult, inconvenient, or expensive to measure for other species or environmental conditions of interest."

The term *indicator species* can mean many different things, including:

1. A species whose presence indicates the presence of a set of other species and whose absence indicates the lack of that entire set of species;
2. A keystone species, which is a species whose addition to or loss from an ecosystem leads to major changes in abundance or occurrence of at least one other species;

3. A species whose presence indicates humancreated abiotic conditions such as air or water pollution (often called a pollution indicator species);
4. A dominant species that provides much of the biomass or number of individuals in an area;
5. A species that indicates particular environmental conditions such as certain soil a species thought to be sensitive to and therefore to serve as an early warning indicator of environmental changes such as global warming or modified fire regimes (sometimes called a bio-indicator species); and
6. A management indicator species, which is a species that reflects the effects of a disturbance regime or the efficacy of efforts to mitigate disturbance effects.

Types 1, 2, and 4 have been proposed as indicators of biological diversity and types 3, 5, 6, and 7 as indicators of abiotic conditions and/or changes in ecological processes.

The Southern Cassowary (*Casuarius casuarius*) is an example of type 2. Found in the rainforests of Queensland, northern Australia, it is the only disperser of more than 100 plant species that have large fruits. Thus, the loss of the Southern Cassowary could result in the loss of a large set of other species dependent on it for seed dispersal. Type 3 species are typically used to monitor pollution. In fact, the indicator species concept has probably been best developed in the field of pollution monitoring. For example, the absence of lichens that grow on tree trunks in places where they would ordinarily occur indicates the existence of specific air pollutants.

Some plants (type 5 indicators) are virtually restricted to soils derived from serpentine bedrock, a rock low in some essential chemical elements and overabundant in others. For these "almost endemic" species, the major portion of their distribution is on serpentine soil, and they are found elsewhere only rarely. The presence of one of these plants almost certainly indicate that the soil is derived from serpentine rock.

Kirtland's Warbler (*Dendroica kirtlandii*) is an example of type 6. It has been proposed as a potential biological indicator of global warming because it nests only in jack pine (*Pinus banksiana*) trees at the southern edge of their range, where they grow on a soil type suitable for this bird. Examples of type 7 include recovery indicator species such as ants, which are considered useful in assessing the effectiveness of mine site rehabilitation.

Problems with Species as Indicators

Although the concept of indicator species has considerable intuitive appeal, there are many instances where its application would be unsuccessful. The American chestnut (*Castanea dentata*) is an example of the failure of a species as a type 4 indicator. The American chestnut was once one of the dominant trees of the mid-Atlantic forests of the United States. Although a fungal blight essentially eliminated the species as a canopy tree by the 1930s, no other species are known to have become extinct or suffered declines that would leave them threatened as a direct result of the demise of the chestnut. For example, the gray squirrel (*Sciurus carolinensis*), which fed heavily on chestnuts, remains abundant today. Similarly, frogs have been suggested as type 6 indicator species because their decline was thought to indicate global climate change. The decline, however, seems to be a result of many factors that vary with locality. Thus, it remains unclear what environmental or other changes are indicated by declines in frog populations.

There also appears to be problems with the concept of management indicator species (type 7). Researchers argued that the maintenance of suitable habitat for a management indicator species would conserve other taxa with similar requirements. Any species, however, that is the specific target for conservation by particular management actions can no longer be an independent yardstick of those actions and, in turn, be regarded as a suitable indicator species for other taxa. In Australian wood production forests, two species of arboreal marsupials, the greater glider (*Petauroides volans*) and the yellow-bellied glider(*Petaurus australis*), have been proposed as potentially useful management indicator species.

Researchers tested this idea in the forests of central Victoria, southeastern Australia. They found no support for the hypothesis that the presence of greater glider, yellowbellied glider, or any other species of arboreal marsupial was a good surrogate for the presence or abundance of the other taxa in the assemblage, in part because (1) few sites support many different species of arboreal marsupials and (2) there is resource partitioning among the different members of the arboreal marsupial assemblage. There are differences among species in diet, body size, habitat requirements, nest tree use, mating system, and other life-history characteristics. One of the key habitat requirements for virtually all species of arboreal marsupials is hollow trees, which are used as nest and den sites. The abundance of these trees is a factor limiting populations over large areas of forest.

Rather than a management indicator species, the prevalence of hollow trees might serve as a useful indicator. But many other forest taxa are not dependent on hollow trees, and such an indicator would have limited or no relevance to them. Indeed, in another study, the structural complexity of different vegetation types in Western Australia proved to be a poor indicator of species richness in several invertebrates and lizards.

Even in the case of pollution indicator species (type 3 above), for which the indicator species concept has been best developed, the behavior of some indicator species may prove to be contrary to what was first expected. Hence, there are problems stemming simply from choosing the wrong indicator species. The case of the bivalve mollusc (*Velesunio ambiguus*) in Australian river systems is a classic example. Early research suggested that the species was an indicator of the presence of heavy metals. Subsequent work found that the uptake of heavy metals by *V. ambiguus* did not reflect the extent of pollution in the surrounding riverine system, making the mollusc an unreliable and thus unsuitable indicator species. There are other potential problems with the concept of indicator species. One is the lack of taxonomic work within the most diverse groups (e.g., invertebrates).

The indicator species approach is hamstrung and monitoring programs are jeopardized because different invertebrate taxa can have markedly different responses to disturbance. In addition, the population dynamics of many species of invertebrates is episodic, but such marked changes in abundance are rarely synchronous across taxa, limiting the value of the status of any given taxon as a surrogate for the dynamics of another.

A problem also arises from the choice of the appropriate scale over which one taxon could indicate the status of supposedly associated taxa (Weaver 1994). Finally, there are problems associated with the lack of sensitivity of indicators. Indicators with a high threshold response may result in some environmental problems being well advanced (and difficult to reverse) before they are detected. A good example of this is pitch pine (*Pinus rigida*). A second-growth natural forest was subjected to ionizing radiation from cesium - 137 for 15 years at the U.S. Brookhaven National Laboratory, and pitch pine was much more sensitive to cesium- 137 than the other dominant tree species. Radiation levels of 10 roentgens per day killed 100% of the pitch pine but had no effects on other species of trees such as scarlet oak (*Quercus coccinea*). Much higher levels were required to affect scarlet oak, making it a poor indicator species for ionizing radiation.

Problems with Guilds

Several authors have suggested that the response of one member of a guild or other form of assemblage might predict the responses of other members of that guild or assemblage. Members of a guild, however, may not respond in the same way to a given disturbance. Researchers found that of five sympatric, closely related, and morphologically similar rainforest bird species, the population of one declined markedly, another substantially increased, and three were moderately influenced following selective logging.

Researchers have shown that different species of arboreal marsupials respond differently to disturbances such as landscape change and habitat fragmentation and that it was not possible to predict population responses from one species to another, even for species that are closely related. Thus, the patterns of response to change exhibited by different species within the same guild may not be readily predictable, even among groups of closely related taxa.

There are other good reasons to believe that not all members of a guild will respond in the same way to the same phenomena. First, the same group of taxa can be partitioned into different guilds for different purposes, and in some cases guilds so constructed may have only limited ecological meaning. Second, the competitive exclusion principle suggests that each member of a guild must have somewhat different environmental requirements and therefore all the members of the guild cannot be indicators of exactly the same conditions. Finally, if a guild represents species with similar niches, but if the niches differ quantitatively, then conditions that favor one member of the guild may not favor another.

Species Lists as Indicators

Given the problems with the use of species and guilds as indicators, one might attempt to seek a measure of the complete species list or an estimator of that list. A direct measure of this list, however, is not practicable except in certain ecosystems of extreme habitats, such as those in thermal hot springs, where there are few species. Researchers discussed the advantages and problems of a species list as an indicator of ecosystem status. They proposed a foundation for an ecological theory in which the complete species list is the ultimate defining feature of an ecosystem. Researchers discussed how a limited, observed list (a subset of the complete species list) could serve as a practical estimator of the complete list. Simply stated, the knowledge that one species is present tells us that many other species cannot be present; for example, the presence of moose (*Alces alces*) tells us that other species such as whales, mahogany trees, and most of the grasses of the tall grass prairie, are not present. It also tells us that many taxa are likely to be present, including parasites and symbionts of moose. Therefore, each new species observed refines the list.

As the observed list lengthens, an asymptotic curve to total species diversity can be estimated. Although we cannot expect to enumerate the complete species list, we might be able to develop an "indicator species list" that could be used in practice. This approach has potential, but remains undeveloped in practice, and we do not have such estimators at this time.

In summary, the concept of indicator species remains an appealing and potentially important one because of the impossibility of monitoring everything. If indicator species could be identified, they would be a powerful management tool, but taxon-based indicators are not adequate at present. Indeed, although many taxa have been proposed as indicator species, the specific entities or conditions that they are supposed to indicate are often not stated explicitly. In other cases, those entities or conditions are nebulous or are difficult to define rigorously (e.g., ecosystem health).

Moreover, the causal linkages between an indicator and other entities (e.g., species or ecosystem processes) are not well demonstrated. Considerable work is needed before reliable indicator species can be identified and included in ecologically sustainable management programs. Carefully designed experiments are required to test relationships between the presence and abundance of potential indicator species and other taxa and the maintenance of critical forest ecosystem processes such as nutrient cycling.

STRUCTURE-BASED BIODIVERSITY INDICATORS

While studies of indicator species proceed, forests continue to be logged, certification of management practices for ecological sustainability will be required under international trade agreements, and other types of biodiversity indicators will be needed. A potentially powerful and complementary approach is to focus on management practices per se using what we call structure-based indicators that include (1) stand complexity and plant species composition in logged stands, (2) connectivity, and (3) heterogeneity.

Stand Complexity

In many types of forest, timber harvesting operations result in marked, medium-length changes in stand structure and floristic composition. It is possible that the effects of such activities could be mitigated and many species conserved in wood production forests if the natural complexity of stand structure is sustained in the form of features such as large, old trees, logs, and the range of overstory and understory species. For example, a number of Australian studies have highlighted the potential value of retained trees in promoting the recolonization of logged and regenerated forests by vertebrates. Similar results have been obtained in the United States, although not all taxa will respond to such strategies.

The structural and floristic features of uncut stands provide an indication of the attributes that need to be retained and perpetuated in logged forests. Stand complexity would seem most effective for biological diversity if the structural features left after human disturbance closely match those resulting from natural disturbance. Researchers describe how silvicultural systems could be modified to allow the types of structural and floristic features of naturally disturbed forest to be perpetuated in logged forests.

Connectivity

Connectivity among forest stands allows for the exchange of seeds, pollen, and individual animals and it can be important for many processes including, among others, the persistence or re-colonization of cutover areas and the exchange of genes among populations. Natural forests typically consist of a system of stands in different successional stages among which species can migrate. Connectivity in wood production forests is important because timber harvesting has the potential to eliminate species from logged areas and to fragment and isolate populations remaining in uncut areas.

Connectivity may be facilitated by establishing corridors, and these may be useful not only for animals but also plants. What constitutes a suitable corridor varies among species. In the case of animals, corridor suitability is a function of an array of factors, including mode of dispersal, social behavior, diet, status of the surrounding (logged) matrix, and corridor location and dimensions (e.g., width, length, and habitat suitability). Corridors may not be effective for all taxa, such as those for which corridor location is not congruent with movement pathways or those that disperse randomly.

Although corridors such as those located in gullies act as dispersal routes for some terrestrial animals, corridors may be required in other parts of a landscape for species inhabiting forests outside the riparian zone. Thus, the concept of connectivity embodies more than corridors because it relates, in part, to the extent of the hostility or permeability of logged areas for movement.

Hence, connectivity also may involve the retention of some components of the original vegetation on logged areas within managed forest landscapes. For example, the approach to the conservation of the Northern Spotted Owl (*Strix occidentaliscaurina*) has been underpinned by connectivity via stand (green tree) retention regimes on cutover areas throughout logged landscapes.

Heterogeneity

The size and spatial arrangement of habitat patches appear to be important for some taxa, so another objective of forest management should be sustaining spatial complexity or heterogeneity over a range of spatial scales. Habitats within forests include biological features such as the range of forest age classes, the size of patches in each class, and variation in overstory and understory structure and floristics. These, in turn, are related to environmental changes in terrain, aspect, elevation, and soil typ. Human and natural disturbances such as logging and fire alter spatial heterogeneity.

Strategies to perpetuate heterogeneity in wood production forests may include using patch sizes and shapes that fall within the range of those created by natural disturbance regimes (e.g., fire or windstorms) as a template for guiding the spatial location of harvested sites. This would create congruence between natural and human disturbance patterns over a range of spatial scales. This approach is underpinned by the general (although largely untested) concept that human disturbance regimes such as logging will have less impact on biodiversity if they are congruent with ecological processes and within the bounds of natural disturbance regimes.

Studies are required to contrast heterogeneity and landscape composition between cut and uncut landscapes and the response of organisms to such differences. The results of some empirical investigations to date have been equivocal. This is possibly because the metrics employed to characterize heterogeneity are not always particularly meaningful for assessing species response, and other measures that relate better to coincidences of disturbance patterns and animal movement (e.g., home range patterns) may be better predictors. Nevertheless, further empirical investigations are critical because heterogeneity appears to be important for many taxa. Promoting heterogeneity in wood production forests needs careful consideration. Increasing heterogeneity may have negative effects on some species, including those requiring large, intact areas of particular age classes such as old growth.

Limitations of Structure-Based Biodiversity Indicators

Structure-based indicators attempt to ensure that management perpetuates inherent taxonomic, structural, and landscape complexities characteristic of forest ecosystems and in so doing contributes to the conservation of biodiversity. Although these structure-based indicators are intuitively sensible and reflect current knowledge of forest biodiversity, their longterm effectiveness remains unknown. For example, there are few data on the number and spatial configuration of retained trees needed to promote the re-invasion of logged sites by forest biota. Therefore, empirical tests of the value of structure-based indicators are required.

STRATEGIES FOR IMMEDIATE IMPLEMENTATION

The lack of information on taxon- and structure-based indicators leads us to conclude that there are four actions that should be adopted now, with or without adequate information, to enhance the likelihood that forest biodiversity will be protected. (1) Establish an adequate amount of representative biodiversity priority areas (e.g., reserves) managed primarily for the conservation of biological diversity. (2) Within production forests, apply structure-based indicators that include structural complexity, connectivity, and heterogeneity. (3) Employ a risk-spreading approach in wood production forests using multiple conservation strategies at multiple spatial scales. (4) Adopt an adaptive management approach to test the validity of structure-based indices of biological diversity by treating management practices as experiments.

Biodiversity Priority Areas

The goal of a network of biodiversity priority areas should be to represent adequately the biodiversity of a region. Existing protected-area networks contain a biased sample of biodiversity typical of locations in environments less suited to natural resource exploitation. Explicit, efficient, and flexible methods such as reserve selection algorithms and gap analysis can be employed to help resolve some of these problems. The presence of a species within a biodiversity priority area should not imply a lack of need for active management regimes inside that area, such as the restoration of burning regimes that may be required by taxa dependent on particular seral stages or vegetation mosaics.

And, the identification of biodiversity priority areas does not mean that off-reserve management is not needed, because reserves will never sample all biodiversity, may not provide sufficient suitable habitat for some taxa, and may not be large enough to sustain some of the species they do contain without sympathetic management of surrounding areas. Thus, the adoption of the structure-based principles outlined in the previous section will be important for the conservation of biodiversity in wood production areas.

Risk Spreading and Variety of Conservation Strategies at Different Spatial Scales

Because different taxa have different resource, spatial, and other requirements, indicator species often fail as useful measures of biological diversity, and a single conservation strategy (e.g., tree retention on logged sites), although appropriate for a particular species, may not ensure the persistence of all other taxa, even closely related ones. Thus, implementing an array of management strategies, including setting aside biodiversity priority areas or reserves, establishing wildlife corridors, and maintaining elements of original stand structure on cutover sites, may help meet the diverse requirements of different taxa. For example, empirical studies suggest that although the provision of wildlife corridors and retained trees on logged sites will make a major contribution to the conservation of populations of the mountain brushtail possum (*Trichosurus caninus*) in Australian mountain ash forests, uncut biodiversity priority areas containing large, continuous stands dominated by old-growth trees are important for the conservation of the yellow-bellied glider in this same forest type.

Another important advantage of a multifaceted approach to management is that if any one strategy is found to be ineffective (e.g., the establishment of wildlife corridors), others (e.g., tree retention throughout logged areas) will be in place that might better protect those elements of forest biodiversity under threat. This is a form of "risk spreading" in forest management. For example, in the forests of Victoria in southeastern Australia, reserves, corridors, and stand retention strategies each have some potential limitations for the conservation of the endangered Leadbeater's possum (*Gymnobelideus leadbeateri*), and a combination of all three strategies will be needed to ensure its persistence.

Finally, a given management strategy may generate benefits for another strategy implemented at a different spatial scale. For example, increased levels of stand retention on logged sites may limit rates of windthrow and vegetation loss in adjacent wildlife corridors, in particular the attrition of trees with hollows that provide den sites for many cavity-dependent taxa. The integrity of wildlife corridors will, in turn, help maintain their effectiveness as linear strips of habitat for wildlife and may promote their use as movement conduits for dispersing animals.

Instigating Management-by-Experimentand-Monitoring Studies

An adaptive management approach integrating research, monitoring, and management should be adopted to test the validity of principles such as stand complexity, connectivity, and heterogeneity as structure- based indices of biological diversity. The best approach is to treat logging as an experiment and implement monitoring systems and feedback mechanisms to management. This would provide new knowledge to managers and improve the effectiveness of management strategies.

At least five steps underpin effective adaptive logging "experiments." (1) Employ a carefully designed monitoring system, including baseline measurements prior to any new action, to track the response of the forest ecosystem. Indeed, a forest cannot be considered constructively managed if it is not monitored. (2) Use a variety of logging practices within wood production areas. For example, some areas could be clear-cut, some selectively harvested, and others could be left with varying levels of stand retention. Control areas with no timber harvesting are necessary to interpret results. (3) Record and store data on the intensity of disturbance and the extent of vegetation retention on logged sites. (4) Change practices based on the results of the experiments. (5) Continue monitoring and data interpretation as part of any change in management actions, including the development of indicators of the status of biodiversity.

Disturbance experiments must be well designed to ensure that the monitoring data generated can be subjected to rigorous statistical analyses and can identify important trends such as population declines. The design of a monitoring program should include treatment replicates to account for spatial heterogeneity and random variation and to provide error estimates, pretreatment monitoring to establish "natural" trends and pretreatment differences between plots, environmental stratification to detect any interactions between treatments and environmental variables, a period of time sufficient to establish treatment effects and distinguish them from climatic fluctuations or other episodic or stochastic events, and replication at more than one location to avoid location-specific phenomena or geographic bias.

SCENARIO PLANNING FOR SUSTAINABLE DEVELOPMENT

Challenge

Madre de Dios, located in the foothills of Peru's tropical Andes, is one of the most biodiverse places in the world. The region has renowned protected areas and initial land-use plans, which makes sustainable development a challenge. In 2014, a hydrocarbon company began gas exploration activities in Madre de Dios as part of the government's economic development needs. The work overlapped a protected area, called the Amarakaeri Communal Reserve, and included several Indigenous territories.

There were concerns that energy exploration would open the area to illegal mining and deforestation. Local communities feared it would impact their forest and livelihoods. This case study explores how researchers from the Smithsonian Conservation Biology Institute's Center for Conservation and Sustainability used scenario planning to help the company and regional government minimize social and environmental impacts in Madre de Dios and create a strategy for sustainable development.

Background

Madre de Dios is located within the Vilcabamba-Amboró conservation corridor, which connects about 74 million acres (30 million hectares) of wildlife habitat between Peru and Bolivia. The corridor is the most biologically diverse ecosystem in the world. Dividing Madre de Dios and the conservation corridor is the Interoceanic Highway, which connects ports on the Atlantic coast of Brazil to those on the Pacific coast of Peru.

The road, completed in 2010, sped the transition of the regional economy from forest-based extractive industries (logging, Brazil nut harvesting, ecotourism, etc.) to agriculture, cattle production and gold mining. These changes promoted population growth. Today, Madre de Dios is home to about 150,000 people, including 14 native cultures, migrants from neighboring Andean regions, and immigrants from Brazil, Bolivia, and other countries.

Agriculture and gold mining are the main causes of deforestation in the area. The Amazon Forest covers more than 90% of Madre de Dios territory, of which about half is protected. The rest is divided into concessions for logging, hydrocarbon, mining, ecotourism and more.

Evaluation and Analysis

The Center for Conservation and Sustainability suggested a scenario planning approach to minimize the possible impact of the gas development field in Madre de Dios. Scenario planning is a structured way to think about the future of a region and facilitate decision making by exploring a handful of plausible futures, called scenarios. For this purpose, the CCS team needed to compare the economic, social and environmental success of different development scenarios. With more than 30 years of experience in Peru, CCS scientists were able to work with local stakeholders and the government to define a path forward.

When the exploratory drilling well produced no gas, the work ended. However, CCS continued to complete the scenario planning project for Madre de Dios to benefit sustainable development across the entire region (32,935 square miles or 85,301 square kilometers).

Creating sustainable landscapes requires managing roads and cities (gray infrastructure) that provide economic services with rivers and forests (green Infrastructure) that support biodiversity. Recognizing this, CCS developed the Working Landscape Simulator. The simulator combined state-of-the-art modeling with community engagement. It included seven steps:

1. Assess the critical goods and services that people get from ecosystems (called ecosystem services) in Madre de Dios
2. Develop qualitative scenarios about the future of the region with community participation
3. Collect and generate quantitative data (demographic, economic, etc.)
4. Model landscape changes, like deforestation and urbanization, for each scenario
5. Evaluate economic, environmental and social indicators of success
6. Develop lessons learned from the study
7. Share the results

Models of land-use changes, such as trends in deforestation and urbanization, made the best use of the existing data. Engagement with local stakeholders ensured they were interested in the results and more likely to use them for decision making.

Milestone Solutions

This study demonstrates that industrial operations, such as gas exploration, can contribute to sustainable development beyond traditional approaches and legal obligations. In scenario planning, CCS was able to provide:

- Collective recommendations that highlight the need for a land-use plan and provide a roadmap for planning regional land changes and sustainable development

- Maps of critical conservation corridors and other areas that are key to keeping habitats connected within Madre de Dios.
- Essential and previously unavailable data – such as historical land cover maps; ecosystem services; and economic, social, and environmental indicators – needed to guide sustainable development and land-use planning.

RECOMMENDATIONS

The Center for Conservation and Sustainability's study was key to understanding the conservation landscape of Madre de Dios and integrating that knowledge with the region's development needs. Best practices from this approach include:

- Consider scenario planning for large-scale projects to build a shared conservation and development vision
- Engage stakeholders early in the process
- Use high-quality models that track socioeconomic and ecosystem changes
- Develop various indicators of success in the three dimensions of sustainable development (economic prosperity, human wellbeing and environmental integrity)
- Share information by making methods and results publicly available to all

13

INTERNATIONAL CONVENTIONS AND TREATIES DEALING WITH FOREST AND WILDLIFE CONSERVATION

INTRODUCTION

The term "wildlife" encompasses all undomesticated lifeforms, including birds, insects, plants, fungi, and even tiny organisms, in addition to wild animals. Animals, plants, and marine species are just as vital as humans in sustaining a healthy ecological balance on this planet. Each organism on this planet occupies a distinct position in the food chain, contributing to the ecosystem in its unique way.

Unfortunately, many animals and birds have become endangered today. Humans are destroying the natural habitats of animals and plants for land development and farming. Another important fact is that animals are poached and hunted for their fur, jewellery, meat, and leather. Animal poaching and hunting for fur, jewellery, meat, and leather are other major contributors to wildlife extinction. If no immediate action is made to rescue wildlife, it will not be long before they are included in the list of extinct species.

CONVENTION ON INTERNATIONAL TRADE IN ENDANGERED SPECIES OF WILD FAUNA AND FLORA (CITES)

- The Convention on International Trade in Endangered Species of Wild Fauna and Flora (CITES) is an international agreement to which states and regional economic integration organisations voluntarily sign up.
- CITES was drafted as a result of a resolution passed by members of the International Union for Conservation of Nature (IUCN) in 1963.
- The IUCN is a membership organisation that includes both government and non-governmental organisations. It gives knowledge and methods to public,

private, and non-governmental organisations that enable human advancement, economic development, and conservation of nature to take place together.

- CITES came into effect in July 1975. There are now 183 members (including countries or regional economic integration organizations).
- The goal of CITES is to ensure that international trading in wild animal and plant specimens does not endanger their survival.
- The United Nations Environment Programme (UNEP) administers the CITES Secretariat, which is based in Geneva, Switzerland.
- In the operation of the Convention, plays the role of servicing, advisory, and coordinating.
- The Conference of the Parties (CoP) to CITES is the Convention's top decision-making body, and it comprises all of the Convention's Parties.
- In 2016, the 17th CoP was held in Johannesburg, South Africa. In 1981, India hosted the 3rd edition of the CoP.
- Though CITES is legally obligatory on the parties, it does not replace the national legislation of a country. Rather, it establishes a framework that must be followed by each party, which must enact domestic legislation to ensure that CITES is implemented on a national level.

Function of CITES

- The CITES works by imposing restrictions on international trading in specimens of specific species.
- All import, export, re-export, and introduction from the sea of species covered by the Convention must be authorized by a licensing system.
- Each party who is a member of the Convention must appoint one or more Management Authorities to oversee the licencing system, as well as one or more Scientific Authorities to advise them on the impacts of trade on the species' status.
- The Convention's Appendices I, II, and III contain lists of species that have received various levels or forms of protection from over-exploitation.

Appendix I

- It includes a list of the most endangered animals and plants on the CITES list.
- Gorillas, sea turtles, most lady slipper orchids, and giant pandas are examples that are included in this list. There are currently 931 species listed in Appendix 1.
- They are endangered, and CITES forbids international trade in specimens of these species unless the importation is for non-commercial purposes, such as scientific research.

- Trade may take place in extraordinary circumstances if it is authorised by the issuance of both an import permit and an export permit (or re-export certificate).

Appendix II

- It includes species that aren't necessarily endangered right now but might become so if the trade isn't strictly regulated.
- Most CITES species, including American ginseng, paddlefish, lions, American alligators, mahogany, and many corals, are included in this Appendix. There are currently 34,419 species on the list.
- It also contains "look-alike" species, that is species whose trade specimens resemble those species that are classified for conservation purposes.
- The awarding of an export permit or re-export certificate may authorise international trading in specimens of Appendix-II species.
- Under CITES, these species do not require an import permit (although a permit is needed in some countries that have taken stricter measures than CITES requires).
- Permits or certifications should only be issued if the relevant authorities are satisfied that certain conditions are met, most notably that trade will not jeopardise the survival of species in the wild.

Appendix III

- It's a list of species that have been included at the request of a Party that already regulates trade in the species but needs other nations' help to prevent unsustainable or unlawful exploitation.
- Map turtles, walruses, and Cape stag beetles are just a few examples that are included in this Appendix. There are now 147 species on the list.
- International trading in specimens of the species described in this Appendix is only permitted if the appropriate permits or certificates are shown.

Only the Conference of the Parties has the authority to add or remove species from Appendix I and II, or to shift species between them. Species may, however, be unilaterally added to or removed from Appendix III at any time and by any Party.

Contribution of CITES

CITES governs international trade in nearly 35,000 species of plants and animals.

- 3 % of these species are normally barred from international commercial trade,
- The remaining 97 % of international commercial trade is controlled to guarantee that it is lawful, sustainable, and traceable.

For the past 42 years, CITES has been at the forefront of the debate over the sustainable use of biodiversity, and its databases contain records of over 12,000,000 international trade transactions – a trade that has benefited local communities on numerous occasions, such as with the vicuña in South America. Appendix II enables the international trading of wool cloth and other manufactured products derived from the shearing of live vicuña (luxury and knitted handicrafts).

It is estimated that illegal commerce is worth between USD 5 billion and USD 20 billion per year, an illegal activity that is driving numerous species to extinction while also denying local people development options and governments of revenue opportunities.

To combat illegal wildlife trade, the CITES Secretariat, International Criminal Police Organization (INTERPOL), UN Office on Drugs and Crime, the World Bank, and the World Customs Organization formed the International Consortium on Combating Wildlife Crime (ICCWC).

It brings together the whole enforcement chain to aid national and regional enforcement authorities in combating illegal wildlife trading.

India and CITES

India is one of the world's acknowledged mega-diverse countries, home to approximately 7-8 percent of the world's recorded species and four of the 34 globally recognised biodiversity hotspots (Himalaya, Indo-Burma, Western Ghats, and Sundaland).

In addition to biological resources, India has a rich library of traditional knowledge. In the 10 biogeographic areas of the country, about 91,200 animal species and 45,500 plant species have been identified.

Through regular surveys and research, inventories of floral and faunal diversity are being gradually updated with various discoveries.

- India, as a CITES Party, vigorously forbids the international trade in endangered wild species and has put in place several measures to combat invasive alien species concerns (e.g. certificates for exports, permits for imports, etc.).
- India has requested that rosewood (Dalbergia sissoo) be removed from CITES Appendix II. The species spreads quickly and has the potential to become naturalised outside of its native region; it is also invasive in other parts of the world.
- It is not essential to regulate the species' trade to keep it from being eligible for inclusion in Appendix I in the foreseeable future.
- Small-clawed otters (Aonyx cinereus), smooth-coated otters (Lutrogale Perspicillata), and Indian Star Tortoise (Geochelone E#legans) have also been suggested to be moved from Appendix II to Appendix I, offering the species extra protection.
- The plan also calls for the Gekko gecko and Wedgefish (Rhinidae) to be added to CITES' Appendix II. For Chinese traditional medicine, the Gekko gecko is highly traded.

CONVENTION ON THE CONSERVATION OF MIGRATORY SPECIES OF WILD ANIMALS (CMS)

- Convention on the Conservation of Migratory Species of Wild Animals (CMS) is also known as the Bonn Convention. It is the sole treaty that addresses the capture or harvesting of wild species. Currently, it safeguards 173 migratory species from all over the world.
- The Convention was enforced on November 1, 1983. The Secretariat, which administers the Convention, was established in 1984.
- As of November 1, 2019, the Convention had 130 parties, including 129 countries and the European Union. The Maldives is the most recent country to join (November 2019).
- Covered Species: There are two Appendices to the Convention:
    - Appendix I is a list of endangered or threatened migratory species. Parties commit to closely protecting these creatures, conserving or restoring their habitats, removing migration barriers, and limiting other causes that may imperil them.
    - Appendix II of the Convention lists migratory species that require protection and management or would benefit greatly from international cooperation.
- Migratory Species: A migratory species crosses one or more national jurisdictional lines cyclically and reliably due to factors such as food, temperature, and shelter.
- Its purpose is to define States' obligations to conserve species that live within or travel through their national borders/jurisdiction.

India and the CMS

- Since 1983, India has been a signatory to the Convention.
- India and the CMS have signed a non-binding Memorandum of Understanding (MoU) on the protection and management of Siberian Cranes (1998), Marine Turtles (2007), Dugongs (2008), and Raptors (2016).
- With 2.4 percent of the world's land area, India's contribution is about 8% of all known global biodiversity.
- The Indian subcontinent is part of a huge bird flyway network, the Central Asian Flyway, which spans the Arctic and Indian Oceans and includes at least 279 populations of 182 different migratory waterbird species (that includes 29 globally threatened species).
- Several migratory species are temporarily provided shelter in India. Species such as Amur Falcons, Bar-headed Geese, Black-necked Cranes, Marine Turtles, Dugongs, and Humpback Whales, are temporarily housed in India.

- A Convention on Migratory Species (CMS) has been in existence since 1983, under the aegis of the United Nations Environment Programme, to preserve migratory species throughout their range countries.
- A global platform is provided through this convention for the conservation and sustainable use of migratory animals and their habitats. The Convention brings together the States through which migratory animals pass that is the Range States and lays the legal foundation for internationally coordinated conservation measures throughout a migratory range.
- The 13th Conference of Parties (COP) of the Convention on the Conservation of Migratory Species of Wild Animals (CMS) was held in Gandhinagar, Gujarat, from February 17 to 22, 2020.

'Migratory species connect the planet and we welcome them home,' was the theme of CMS COP-13.

'Gibi — The Great Indian Bustard' was the mascot for CMS COP-13. According to the IUCN, it is a critically endangered species that has been given the highest protection classification (listed in Schedule I) under the Wildlife Protection Act, 1972.

CONVENTION ON BIOLOGICAL DIVERSITY (CBD)

Environment leaders agreed on a comprehensive strategy for "sustainable development" during the 1992 Earth Summit in Rio de Janeiro, guaranteeing that we meet our needs while also leaving a healthy and viable world for future generations. The Convention on Biological Diversity (CBD) was one of the most important accords signed in Rio.

The Convention on Biological Diversity is an international legal instrument that has been ratified by 196 countries and is dedicated to "the conservation of biological diversity, the sustainable use of its components, and the fair and equitable sharing of the benefits arising from the utilisation of genetic resources."

Aichi biodiversity targets

The Convention on Biological Diversity (CBD) adopted the "Aichi Biodiversity Targets" during its Nagoya Summit. It is a short-term plan that consists of a series of 20 ambitious yet attainable aims known as the Aichi Targets. They are classified as follows:

- Strategic Goal A: Address the root causes of biodiversity decline by integrating biodiversity into all aspects of government and society.
- Strategic Goal B: Reduce direct impacts on biodiversity and promote long-term use. Strategic Goal B.
- Strategic Goal C: Ecosystems, species, and genetic diversity are all part of this goal, which aims to improve biodiversity.

- Strategic Goal D: Increase the advantages of biodiversity and ecosystem services to all people.
- Strategic Goal E: Enhance implementation through participative planning, knowledge management, and capacity building.

India and CBD

The Sixth National Report on Biological Diversity was recently submitted by India to the Convention on Biological Diversity (CBD).

India is one of the first five countries in the world to submit the report, as well as the first in Asia and among the biodiversity-rich megadiverse countries.

Parties to international treaties, including the CBD, are required to submit national reports as part of their obligations. India, as a responsible nation, has never broken its international commitments and has already filed five national reports to the CBD on time.

Highlights of the report

Highlights of the study include an update on progress toward achieving 12 National Biodiversity Targets (NBTs) created through the convention process and aligned with the 20 global Aichi biodiversity targets.

According to the report, India has exceeded/overachieved two NBTs and is on track to achieve eight NBTs.

According to the report, India has met 17 percent of Aichi goal 11s terrestrial component and 20 percent of the relevant NBT relating to areas under biodiversity management.

In addition, India has been investing a significant amount in biodiversity, either directly or indirectly, through many development programs run by the central and state governments, to the tune of Rs 70,000 crores per year, compared to an estimated yearly requirement of roughly Rs 1,09,000 crore.

India's 12 national biodiversity targets were as follows:

1. By 2020, a large section of the population, particularly the youth, will be aware of the importance of biodiversity and the efforts they may take to protect and use it sustainably.
2. Biodiversity values are included in national and state planning procedures, development programs, and poverty alleviation methods by 2020.
3. By 2020, strategies for lowering the rate of degradation, fragmentation, and loss of all-natural habitats have been finalised, and actions for environmental improvement and human well-being have been implemented.
4. By 2020, the genetic diversity of cultivated plants, farm livestock, and their wild relatives, including other socioeconomically and culturally important species,

will have been preserved, and strategies for minimising genetic erosion and safeguarding their genetic diversity will have been developed and implemented.

5. By 2020, ecosystem services, particularly those related to water, human health, livelihoods, and well-being, will be enumerated, and strategies to protect them will be established, with specific attention paid to the needs of women and local communities, particularly the poor and vulnerable.
6. By 2015, the Nagoya Protocol's access to genetic resources and the fair and equitable sharing of benefits arising from their use will be operational, following national legislation.
7. At all levels of governance, and effective, participative, and updated national biodiversity action plan is put in place by 2020.
8. National initiatives based on communities' traditional biodiversity knowledge are strengthened by 2020, to safeguard this information in compliance with national legislation and international commitments.
9. By 2020, possibilities to expand the availability of financial, human, and technical resources to support the execution of the Strategic Plan for Biodiversity 2011-2020 and national targets will have been identified, and the Resource Mobilization Strategy will have been established.

WORLD HERITAGE CONVENTION

One of the most important worldwide conservation instruments is the World Heritage Convention.

The Convention was established in 1972 with the primary goal of identifying and protecting the world's natural and cultural heritage that is judged to be of Outstanding Universal Value.

It reflects a visionary idea, that some places are so vital that protecting them is not only the responsibility of a single nation, but also the obligation of the entire world community; and not just for this age, but for future generations.

The Operational Guidelines, which specify the procedures for new inscriptions, site protection, danger-listings, and international support under the World Heritage Fund, aid in the implementation of the World Heritage Convention.

The World Heritage Committee oversees the Convention, which is backed by the UNESCO World Heritage Centre, the Convention's administration, and three technical advisory bodies: IUCN, ICOMOS, and ICCROM. The International Union for Conservation of Nature (IUCN) is the Advisory Body on Natural Heritage. It keeps track of listed sites and assesses sites that have been nominated for inclusion on the World Heritage List, using the appropriate natural criteria for selection (vii) – (x):

- (vii) to contain outstanding examples representing major stages of earth's history, such as the record of life, significant on-going geological processes in the development of landforms, or significant geomorphic or physiographic features;

- (viii) to contain superlative natural phenomena or areas of exceptional natural beauty and aesthetic importance;
- (ix) to be outstanding examples representing major stages of earth's history, such as the record of life, significant ongoing geological processes in the development of landforms, or significant geomorphic;
- (x) to constitute exceptional instances of ongoing ecological and biological processes in the growth and development of terrestrial, freshwater, coastal, and marine ecosystems and plant and animal communities;
- (xi) to house the most significant natural habitats for in-situ biological diversity conservation, especially those containing threatened species of outstanding universal importance from a scientific or conservation standpoint.

UNESCO-WHO

The lifting of world heritage sites, which include both cultural and natural sites, is the responsibility of UNESCO's World Heritage Convention. The Ministry of Environment and Forests of India's wildlife department is involved in the conservation of natural world heritage sites. We have also embarked on a wildlife conservation project with the help of external funding.

The project will last ten years and will be divided into two sections. The project will take place in India's Kaziranga National Park, Manas National Park, Nanda Devi National Park, and Keoladeo National Park, which are all world heritage sites.

Natural sites in India

- Great Himalayan National Park Conservation Area (2014)
  - Great Himalayan National Park is well renowned for its high alpine peaks, alpine meadows, and riverine forests, and is located in the western part of the Himalayan highlands in the state of Himachal Pradesh.
  - It also encompasses the catchment area as well as the glacial and snow meltwater sources of multiple rivers.
  - It is a biodiversity hotspot, with 25 different types of forests home to a diverse range of faunal species, some of which are threatened.
- Kaziranga National Park (1985)
  - In 1974 Kaziranga National park was declared as a national park
  - It is located in the State of Assam and covers 42,996 ha. This park is the single largest undisturbed and representative area in the valley of the Brahmaputra floodplain.
  - In 2007 it was declared a tiger reserve. The total tiger reserve area is 1,030 sq km with a core area of 430 sq. km.

- In 1985 it was declared a UNESCO World Heritage Site.
- BirdLife International recognises it as an Important Bird Area.

◉ Keoladeo National Park (1985)

- This wetland, located in Rajasthan, was used as a duck hunting reserve until the end of the nineteenth century. The hunting, however, came to an end soon after, and the region was designated as a national park in 1982.
- Keoladeo National Park is home to 375 bird species as well as a variety of other animals. It also provides a wintering ground for migratory waterfowl from the Palaearctic, the critically endangered Siberian Crane, and the internationally vulnerable Greater Spotted Eagle and Imperial Eagle.
- It is well-known for its resident breeding population of non-migratory birds.

◉ Manas Wildlife Sanctuary (1985)

- Assam's Manas Wildlife Sanctuary is a biodiversity hotspot. It is located alongside the Manas River and is part of the Manas Tiger Reserve.
- The site's spectacular beauty and tranquil environment are due to a variety of forested hills, alluvial grasslands, and tropical evergreen woods.
- It also supports a variety of endangered species, including the tiger, greater one-horned rhino, swamp deer, pygmy hog, and Bengal florican.

◉ Nanda Devi and Valley of Flowers National Parks (1988, 2005)

- Nanda Devi and Valley of Flowers National Parks, both National Parks are high-altitude West Himalayan landscapes with extraordinary beauty, and they are located in the state of Uttarakhand.
- The rough and high-mountain environment of Nanda Devi National Park is dominated by India's second-highest mountain, the peak of Nanda Devi. The Valley of Flowers, on the other hand, boasts aesthetically appealing alpine flower meadows.
- These parks are home to a diverse range of floral and faunal species, as well as a major population of globally vulnerable species like the Snow Leopard and Himalayan Musk Deer.

◉ Sundarban National Park (1987)

- The Sundarbans contain the largest mangrove forests in the world and out of all-natural ecosystems it is one of the most biologically productive.
- It is located at the mouth of the Brahmaputra and the Ganges Rivers between India and Bangladesh. The Sundarbans forest and waterways support a wide range of fauna which includes several species that are threatened with extinction.

  - The mangrove habitat supports the single largest population of tigers in the world which have adapted to an almost amphibious life that are capable of swimming for long distances and feeding on crab, fish, and water monitor lizards. These tigers are also renowned for being "man-eaters", probably due to their relatively high frequency of encounters with the local people.

- Western Ghats (2012)
  - The Western Ghats are a mountain range that runs parallel to India's western coast, going through Kerala, Maharashtra, Goa, Gujarat, Tamil Nadu, and Karnataka.
  - They span for 1600 kilometres and are only interrupted once by a 30-kilometre Palghat gap at roughly 11 degrees North.
  - They also have an impact on the Indian monsoon weather patterns, which help to moderate the region's warm tropical climate and act as a barrier to rainy monsoon winds from the southwest.
  - Tropical evergreen woods, as well as 325 globally endangered species, are found in the Western Ghats.

Mixed sites in India

- Khangchendzonga National Park (2016)
  - Mount Khangchendzonga, the world's third tallest peak, dominates Khangchendzonga National Park in Sikkim.
  - The Park is home to steep-sided valleys, snow-capped mountains, and numerous lakes and glaciers, including the Zemu glacier, which stretches for 26 kilometres around Mount Khangchendzonga's base.
  - It encompasses over a quarter of the state of Sikkim and provides a haven for a variety of indigenous and threatened plant and animal species.

RAMSAR CONVENTION

- The Ramsar Convention is a convention categorising wetlands that were signed in the Iranian city of Ramsar in 1971. The pact was negotiated in the 1960s by many countries and non-governmental organisations for the conservation of migratory waterbirds' wetland habitats.
- The Ramsar Convention was established in 1975 to conserve and properly use all wetlands through local, national, and international collaboration as a contribution to global sustainable development.
- The Ramsar Convention has 171 contracting parties as of October 2020.
- The official name of the treaty is The Convention on Wetlands of International Importance, especially as Waterfowl Habitat.

- It is the only international treaty that specifically addresses a particular ecology (wetland.) Originally, the treaty was focused on the protection of waterbird habitats. With time, the treaty's scope has expanded to include all areas of wetland conservation.
- Three major subjects are covered in the Ramsar Conventions:
  - Under the Ramsar List of Wetlands of International Importance, the 171 contracting parties must designate appropriate wetlands in their territories.
  - The wetlands that have been declared must be carefully managed and cared for.
  - The parties involved must use shared wetland systems over the territory of more than one contracting party sensibly and after adequate consultation.
- There are 2406 wetlands on the list of wetlands of international importance as of OctoberThe Ramsar Convention is not a regulatory regime.
- The Ramsar Convention was modified by the Regina Amendments in 1987 and by the Paris Protocol in 1982.
- The Montreux Record is a mechanism related to the Ramsar Advisory Mission that was established in 1990. It is a list of those Ramsar Sites that require immediate attention.
- The first World Wetlands Day was held in 1997. Every year on February 2nd, it commemorates the Ramsar Convention's 10th anniversary and promotes its objective.
- Every three years, the convention's contracting parties convene for a Conference of the Contracting Parties (COP).
- The Ramsar Convention is supported by six international organisations partners, that are as follows:
  - Birdlife International,
  - IUCN,
  - Wetlands International,
  - WWF,
  - International Water Management Institute,
  - Wildfowl and Wetlands Trust,
  - A six-year strategic plan is included with the convention. The 4th Ramsar Convention Strategic Plan 2016-2024, which was approved during the convention's COP12, is the most recent.

- The Standing Committee of the Ramsar Convention comprises 18 members who are elected at each COP until the following COP elects new members.
- English, Spanish, and French are the three languages used by the Convention.

Purpose of the Ramsar Convention

The Ramsar Convention is founded on three pillars that define its mission:

- Wise Use- Ensure that all wetlands are used wisely.
- List of Internationally Significant Wetlands- Designate appropriate wetlands for effective management under the Ramsar List.
- International Cooperation- To foster international cooperation on transboundary wetlands, shared species, and shared wetland systems.

WETLANDS

Wetlands are defined as "areas of marsh, fen, peatland, or water, whether natural or artificial, permanent or temporary, with water that is static or flowing, fresh, brackish, or salt, including areas of marine water whose depth at low tide does not exceed six metres," according to the broad definition of Ramsar Convention.

Wetland examples include

- Coastal and marine environments
- Estuaries
- Rivers and lakes
- Peatlands and marshes
- Rice paddies, shrimp ponds, and reservoirs are examples of human-made wetlands.
- Groundwater

India and Ramsar Convention

- In India, there are 42 Ramsar Sites designated under the Ramsar Convention.
- India has added ten more wetlands to the Ramsar Convention's list of protected wetlands.

These are the following:

- Uttar Pradesh: Nawabganj, Parvati Agra, Saman, Samaspur, Sandi, and Sarsai Nawar.
- Punjab: Keshopur-Miani, Beas Conservation Reserve, and Nangal.
- Maharashtra: Nandur (state's first).

This expansion will aid India's ambitious 'Nal se Jal' initiative, which aims to supply piped water to every household by 2024.

India and Wetland Conservation

The Ramsar Convention on Wetland Conservation came into force in India on February 1, 1981. The following are India's national wetlands conservation initiatives (4.63 percent of the overall geographical area):

1. Wetlands (Conservation and Management) Rules, 2017.
2. In January 2020 a set of guidelines were released by the Ministry of Environment, Forest & Climate Change for implementation of the Wetland Rules 2017.
3. India regulated the following wetlands:
   - Wetlands are designated under the Ramsar List.
   - Those wetlands are notified under central, state, and UT rules.
4. The following wetlands are not covered under India's Wetlands Rules:
   - River channels.
   - Paddy fields.
   - Human-made water bodies are specifically designed for drinking water, aquaculture, salt production, recreation, and irrigation.
   - Wetlands were protected by the Indian Forest Act of 1927, the Forest (Conservation) Act of 1980, and the State Forest Acts.
   - Wetlands that are protected under the Wildlife (Protection) Act of 1972.
   - Wetlands that are covered under the 2011 Coastal Regulation Zone Notification.

THE WILDLIFE TRADE MONITORING NETWORK (TRAFFIC)

The Wildlife Trade Monitoring Network (TRAFFIC) is a leading non-governmental organisation that works on wildlife trade as part of biodiversity conservation and sustainable development.

It is a joint program of the International Union for Conservation of Nature (IUCN) and the World Wildlife Fund (WWF).

It was founded in 1976 and has grown into a global network dedicated to providing new and practical conservation solutions that are both research-driven and action-oriented.

The headquarters of TRAFFIC is located in Cambridge, United Kingdom.

Its goal is to ensure that trading in wild flora and animals does not threaten environmental protection.

The illegal wildlife trade is one of the primary causes of the extinction of many species. For example in 2011, rhino poaching to feed demand for illegal rhino horn hit an all-time high, with 448 rhinos poached in South Africa alone. This might unravel years of rhino conservation success in Africa.

Governance of TRAFFIC

- The TRAFFIC Committee is a steering group made up of members of TRAFFIC's partner organisations, the WWF and the IUCN.
- In addition, TRAFFIC works closely with the Secretariat of the Convention on International Trade in Endangered Species of Wild Fauna and Flora (CITES)
- Its team of experts comprises biologists, conservationists, academics, researchers, communicators, and investigators.

Functions

- It has aided in the evolution of international wildlife trading conventions since it has been established.
- It focuses on pooling resources, skills, and knowledge about the latest global wildlife trade issues, such as tiger parts, elephant ivory, and rhino horn.
- Large-scale commercial trading in commodities such as lumber and fisheries goods is also addressed and is related to efforts to create quick results and improve policy.

TRAFFIC and India

- Since 1991, TRAFFIC has been a Program Division of WWF-India, situated in New Delhi.
- Since then, it has collaborated closely with the national and state governments, as well as other organisations, to help study, monitor, and influence action to combat illegal wildlife trading.
- Capacity-building programs in India are helping to close the gap in effective wildlife law enforcement: TRAFFIC provides training and feedback to a varied array of officials working on wildlife enforcement and other relevant issues as part of this program.
- Conducting research and providing analysis on wildlife trade and its trends.
- TRAFFIC Study of leopard and tiger poaching and trade in India, peacock feather trade, owl trade, dynamics of the hunting community, trade in medicinal plants, bird trade, and more are among India's ongoing initiatives.

Creating awareness:

- One of TRAFFIC India's earliest consumer awareness campaigns, "Don't Buy Trouble," warns travellers to be cautious about what they buy as souvenirs when travelling. Since 2008, the program has had great success in airports, tiger reserves, national parks, wildlife resorts/hotels, travel agencies, schools, colleges, and other notable areas.

- The WANTED ALIVE series by TRAFFIC focuses on four Asian big cats: the Tiger, Leopard, Snow Leopard, and Clouded Leopard, all of which are threatened by illegal traffic in their body parts.
- Promoting international collaborations in the fight against wildlife crime.
- TRAFFIC was instrumental in bringing South Asian countries together to form the South Asia Wildlife Enforcement Network (SAWEN).
- SAWEN was formally created in January 2011 at an inter-governmental summit hosted by the Royal Government of Bhutan in Paro (a town in Bhutan). The major goal of this SAWEN project was for governments to work together to combat wildlife crime in the region.

UNITED NATIONS FORUM ON FORESTS (UNFF)

International forest policy was one of the most controversial topics during the 1992 United Nations Conference on Environment and Development (the Earth Summit). Since then, significant progress has been achieved in building a global policy framework for sustainable forest management. The Intergovernmental Panel on Forests (IPF) was the major platform for international forest policy development from 1995 to 1997, and the Intergovernmental Forum on Forests (IFF) was the key forum for international forest policy development from 1997 to 2000. A key outcome from these organisations was the IPF/IFF Proposals for Action towards sustainable forest management. The UNFF was created to build on the processes and outcomes of the IPF and IFF.

United Nations Forum on forests mandate

The United Nations Forum on Forests (UNFF) was established by the Economic and Social Council of the United Nations (ECOSOC) in October 2000 as a subsidiary body with the main goal of promoting "the management, conservation, and sustainable development of all types of forests and to strengthen long-term political commitment to this end..." The UNFF's roles and functions evolved from the Rio Earth Summit.

The UNFF has universal membership and is made up of all United Nations Member States as well as specialised agencies.

UNITED NATIONS FOREST INSTRUMENT (UNFI)

The landmark Non-Legally Binding Instrument on All Types of Forests (NLBI) was accepted by the Forum's Seventh Session on April 28, 2007, after intensive discussions. The instrument is regarded as a watershed moment.

It was the first time that all of the UN's Member States agreed on a global framework for sustainable forest management. The instrument is projected to have a significant impact on international collaboration as well as national action to reduce deforestation, forest degradation, promote sustainable livelihoods, and alleviate poverty among forest-dependent populations.

The instrument was renamed the United Nations Forest Instrument at the 11th UNFF session in 2015. (UNFI). The UNFI has received support from several countries. The UNFI has been endorsed by the United Nations General Assembly. The UNFF website provides information about the organization's operations.

INTERNATIONAL WHALING COMMISSION (IWC)

- International Whaling Commission (IWC) is a non-governmental organisation established under the International Convention for the Regulation of Whaling (ICRW).
- The International Whaling Commission regulates the commercial, scientific, and aboriginal subsistence whaling practices of fifty-nine member countries. In 1946, it was signed in Washington, D.C., United States.
- The headquarters of ICRW is situated in Impington, near Cambridge, England.
- It adopted a moratorium on commercial whaling in 1986. This prohibition is still in effect.
- The International Whaling Commission is a non-governmental organisation dedicated to whale conservation and whaling management.
- The International Convention for the Regulation of Whaling serves as the IWC's legal basis. This Convention was one of the earliest pieces of international environmental legislation, having been formed in 1946. This Convention has been signed by all the IWC member countries. Currently, 88 governments from all over the world are members of the IWC.
- The International Whaling Commission (IWC) was established by the International Convention for the Regulation of Whaling, which was signed in Washington, DC on December 2, 1946.
- The goal of the Convention, according to the preamble, is to ensure the correct conservation of whale stocks. As a result, the whaling industry can develop in a more organised manner.

Whale sanctuary

- It established the Southern Ocean Whale Sanctuary, which surrounds the continent of Antarctica, in 1994. The IWC has outlawed all forms of commercial whaling in this area.

- So yet, the IWC has recognised only two such sanctuaries. Another is Seychelles' Indian Ocean Whale Sanctuary, which is located on a small island nation in the Indian Ocean.

Aims

- Since its inception in 1946, the Commission's responsibility has grown.
- Bycatch and entanglement, ocean noise, pollution, and debris, collisions between whales and ships, and sustainable whale watching are among the challenges that the IWC tries to address.

Membership

- Any country in the globe that formally ratifies the 1946 Convention is eligible to join the IWC.
- Each member country is represented by a Commissioner, who is aided by specialists and advisers.
- The Chair and Vice-Chair are elected from among the Commissioners themselves.
- They normally serve for four years, first as Vice-Chair for two years and then as Chair for the remaining two years.

Organisation and structure

- The Commission's work is divided into six committees, each of which is made up of several sub-groups.
- Some of these sub-groups are long-term permanent committees, while others were formed to finish a single project.
- Commissioners, other members of national delegations, or subject matter experts from the IWC community chair the groups.
- The IWC has a full-time Secretariat with headquarters near Cambridge, United Kingdom.

IWC FUNDING

- The IWC is devoted to financial management that is both sound and transparent.
- All IWC member governments' annual financial contributions are calculated.
- The size of the donation varies by government and is influenced by three factors:
  - The size of their delegation at the most recent biennial Commission meeting.
  - Any whaling action they may have participated in;
  - The capacity of the government to pay.

INTERNATIONAL UNION FOR CONSERVATION OF NATURE (IUCN)

- International Union for Conservation of Nature (IUCN) was founded in the French town of Fontainebleau on October 5, 1948. It was the first worldwide environmental union, bringing governments and civil society organisations together with a common purpose of protecting the environment. Its goal was to promote worldwide cooperation while also providing scientific knowledge and tools to help conservationists make better decisions.
- IUCN has become the global authority on the status of the natural world and the steps required to safeguard it since its establishment in 1948. The knowledge and resources provided by the IUCN are essential for ensuring that human advancement, economic development, and environmental conservation all occur together.
- The IUCN's principal objective throughout its first decade was to investigate the impact of human activities on nature. It raised awareness of pesticides' negative impacts on biodiversity and encouraged the adoption of environmental impact assessments, which have subsequently become standard practice across all sectors and businesses.
- In the 1960s and 1970s, much of the IUCN's work focused on the protection of species and the ecosystems that are crucial for their survival. The IUCN Red List of Threatened Species™ was founded in 1964, and it has since evolved into the world's most comprehensive data source on the global extinction risk of species.
- IUCN also played a fundamental role in the creation of key international conventions. These conventions include the Ramsar Convention on Wetlands (1971), the World Heritage Convention (1972), the Convention on International Trade in Endangered Species (1974), and the Convention on Biological Diversity.
- The World Conservation Strategy, published in 1980 by the IUCN in collaboration with the UN Environment Programme (UNEP) and the World Wildlife Fund (WWF), was a groundbreaking document that helped define the concept of "sustainable development" and shaped the global conservation and sustainable development agenda.
- In the run-up to the 1992 Earth Summit, the three organisations released a revised version of the strategy called Caring for the Earth. It guided the formation of the Rio Conventions on Biodiversity (CBD), Climate Change (UNFCCC), and Desertification (UNFCCC) and served as the foundation for worldwide environmental policy (UNCCD).
- The UN General Assembly gave IUCN formal observer status in 1999, as environmental issues continued to acquire relevance on the international scene.

- IUCN launched its corporate engagement approach in the early 2000s. Its goal is to ensure that any use of natural resources is fair and ecologically sustainable, with a focus on industries that have a substantial impact on nature and livelihoods, such as mining and oil and gas.
- IUCN pioneered 'nature-based solutions' later in the 2000s, which are efforts to conserve nature that simultaneously address global concerns like food and water security, climate change, and poverty reduction.
- IUCN is the world's largest and most diverse environmental network, with more than 1300 Members – including States, government agencies, NGOs, and Indigenous Peoples' Organizations – and over 15,000 international specialists. It continues to promote nature-based solutions as critical to implementing international agreements like the Paris Climate Change Agreement and the 2030 Sustainable Development Goals.

GLOBAL TIGER FORUM (GTF)

- The Global Tiger Forum (GTF) is an inter-governmental and international organisation founded by members from willing countries to launch a global campaign, common approach, and promotion of appropriate programs and controls to save the world's remaining five subspecies of tigers, which are distributed across 14 tiger range countries.
- GTF, which was founded in 1994 and has its headquarters in New Delhi, is the sole intergovernmental and multinational organisation working to rescue the TIGER all over the world.
- The GTF General Assembly meets every three years.

Objectives

- To support a global movement to rescue the tiger, its prey, and its habitat;
- To promote a legislative framework for biodiversity conservation in the nations concerned;
- To expand the tiger's habitat protected area network and make inter-tiger migration easier in the range countries;
- To encourage communities living in and around protected areas to participate in eco-development programs;
- To encourage governments to sign necessary accords for tiger protection and the abolition of illegal trading;
- To encourage and carry out a scientific study to develop information helpful to tigers, their prey, and their environment, and to communicate that information in a user-friendly manner;

- To urge a range of countries to establish and implement their action plans for the protection and growth of the tiger population and its prey base by encouraging the development and exchange of relevant technology and training programs for scientific wildlife management. Habitat improvement and a joint preservation program can be undertaken bilaterally by range countries with adjacent habitats, but their execution should be done individually by each range country.
- Involve intergovernmental organisations in tiger conservation; establish a participatory fund of sufficient size to raise awareness in all areas where people eat tiger derivatives, reduce such consumption, and develop substitutes in the interests of conservation.
- The Global Tiger Forum (GTF) is conducting a situation analysis research to assess tiger habitat conditions in high altitude ecosystems in collaboration with the governments of Bhutan, India, and Nepal, as well as the World Wide Fund for Nature (WWF).
- The study was funded by the International Union for Conservation of Nature's (IUCN) Integrated Tiger Habitat Conservation Program (ITHCP) and KfW. (German Development Bank).

Findings

- The study identified possible viable habitats, corridor links, anthropogenic pressures, and induced landscape-level changes in its most recent report, which will be used to develop an in-situ conservation strategy.
- It also included an action plan for a high-altitude tiger master plan, as well as a profitable portfolio for local communities and a commitment to tiger conservation as a development priority.
- The study's findings will be used by the Indian government to establish a high-altitude tiger master plan. Because they are a high-value ecosystem with numerous hydrological and ecological processes providing ecosystem services, tiger habitats at high altitudes will require protection through sustainable land use. They will also need adaptation techniques to counteract the negative consequences of climate change.

Integrated Tiger Habitat Conservation Program (ITHCP)

- The ITHCP, which was established in 2014, is a strategic funding instrument that aims to save wild tigers, their habitats, and human populations in important Asian locales.
- It has already aided 12 programs to better manage Tiger Conservation Landscapes in six countries (Bangladesh, Bhutan, India, Indonesia, Nepal, and Myanmar).

- It is helping to fund the Global Tiger Recovery Programme (GTRP), which aims to increase the number of wild tigers by 2022.
- The GTF is the only intergovernmental international organisation with members from willing countries to launch a global campaign to protect the tiger.
- The GTF is the only intergovernmental international organisation with members from willing countries to launch a global campaign to protect the tiger.
- It was founded in 1993 as a result of proposals made at an international conference on tiger conservation held in New Delhi, India.

COALITION AGAINST WILDLIFE TRAFFICKING (CAWT)

- The Coalition Against Wildlife Trafficking (CAWT) was established by the US Department of State in 2005 as a "voluntary public-private [international] coalition of like-minded governments and organisations that share the goal of eradicating the illegal trade in wildlife and wildlife products." The CAWT's three objectives are to lower consumer demand, limit supply through increased enforcement, and mobilize high-level political support.
- The founding members of the CAWT, according to a 2005 State Department press release were Conservation International, Save the Tiger Fund, Smithsonian Institution, Traffic International, WildAid, Wildlife Conservation Society, and American Forest and Paper Association.
- The Association of Southeast Asian Nations – Wildlife Enforcement Network (ASEAN-WEN) received nearly $3 million in funding from the United States Agency for International Development (USAID) and $1 million from the U.S. State Department under the umbrella of the CAWT. This money was used to support capacity-building activities in nations where wildlife goods are in high demand.
- In 2008, the CAWT released its public awareness videos, which included Harrison Ford, an actor and wildlife conservationist. Consumers are encouraged not to buy anything made using illegally traded wildlife or wildlife products after watching the movies. You may see the PSAs here.
- Since its inception, the Coalition has raised awareness of the issue of wildlife trafficking online and the role of the private sector in combating it at key global fora such as the London Conference on Illegal Wildlife Trade (2018), the INTERPOL/IFAW Cyber-Enabled Wildlife Crime Workshop (2018), the CITES Conference of the Parties 18 (2019), the INTERPOL Wildlife Crime Working Group (2019), and the First High-Level Convention on Wildlife Trafficking (2019). These gatherings aimed to encourage multi-sector collaboration between NGOs, policymakers, and law enforcement while also accelerating major policy changes.

- Technology and communication advancements, together with expanding power and demand for illegal wildlife items, have made it easier to transfer illegal wildlife products from poachers to customers across continents. As a result, criminals can sell unlawfully stolen wildlife and its products on mainly uncontrolled online markets. Live tiger cubs, reptiles, primates, and birds for the exotic pet trade, as well as products produced from elephants, pangolins, and sea turtles, are readily available for purchase online. In collaboration with WWF, TRAFFIC, and IFAW, the Coalition to End Wildlife Trafficking Online brings together e-commerce, search, and social media organisations from around the world to help keep wildlife safe online and in the wild.
- The Coalition works with businesses to improve prohibited wildlife regulations, train employees to detect prohibited wildlife content, educate users about the issue and how to report suspicious listings, improve automated detection, and more. The Coalition collaborates with businesses to improve forbidden wildlife regulations, train employees to detect prohibited wildlife content, educate users about the issue and how to report suspicious listings, improve automated detection, and share learning across platforms.
- Coalition companies reported removing or disabling over 3 million endangered species listings from their platforms in their most recent progress report (dated March 2020). Strengthened wildlife policies, increased staff ability to detect illegal wildlife products, regular monitoring and data sharing from wildlife experts, reports sent in by volunteers through a Cyber Spotter program, enhanced algorithms as a result of key search word monitoring, and collation and shared learning all contributed to this.

Conclusion

Almost all wildlife authorities in different countries are currently dealing with illegal poaching and animal killing by smugglers for animal skins, tusks, and horns. What these people don't comprehend is that while they may be making large sums of money from these activities today, in 10 or 20 years there will be no animals remaining to provide them with income. There would be no forests if there were no animals, and no forests would imply soil erosion, warm temperatures, the drying up of lakes and rivers, no rains, and hence no crops and plants.

This is a cascading consequence that every human being should be aware of right now. And saving wildlife isn't a one-man effort. It is a team effort that requires everyone's participation. The wildlife authorities should develop severe strategies to reduce human intrusion in the core forest areas, as well as appropriate wildlife tourist policies, to successfully manage unlawful operations.

Further, as individuals, we should begin by taking tiny steps from our homes, such as reducing the use of electronic equipment such as air conditioners, which contribute to global warming, and pooling vehicles when travelling to reduce pollution and carbon emissions.

GLOSSARY

Acre: an area of land measuring 43,560 square feet. A square 1-acre plot measures 209 feet by 209 feet; a circular acre has a radius of 117.75 feet.

Aerial release spraying: an aerial application of herbicide to reduce hardwood competition in young pine stands or to eliminate exotic species. This will generally be done on natural stands that have had a pre-commercial thinning.

All-aged stand: all, or almost all, age classes of trees represented.

Allegany hardwood forest type: a portion of maryland's northern hardwood forest in which black cherry, white ash, and red oak are dominant species.

Amphibian: any of a class of vertebrates that regulate their body temperature externally; lay shell-less eggs in wet areas; live in water during early development and live both in water and on land as adults; and use lungs, gills and their skin for breathing. Most have four legs and smooth, moist skin without scales.

Angiosperm: a plant that has true flowers and bears its seeds in fruits. In temperate zones, many angiosperms are deciduous trees, while in tropical zones, many are evergreen trees. Examples include oaks, willows, maples and birches.

Annual ring: the combination of one earlywood layer (light colored) and one latewood layer (dark colored) seen in a cross-section of a tree. One annual ring usually represents one year of growth.

Annual rings: see growth rings.

Artificial regeneration: the growth of new trees through seeding and planting.

Aspect: the compass direction toward which a slope faces.

Association: a collection of plants with ecologically similar requirements, including one or more dominant species from which the group derives a definite character.

Bark: the tough exterior covering of a woody root or stem that protects the tree from injury caused by insects and other animals, by other plants, by disease and by fire.

Basal area (of a tree): the cross-sectional area of the trunk 4 1/2 feet above the ground; **(per acre)** the sum of the basal areas of the trees on an acre; used as a measure of forest density.

Best management practices: procedures employed during harvesting and/or timber stand improvement activities that reduce erosion and prevent or control water pollution.

Biltmore stick: a stick similar to a yardstick in appearance, but usually about 25 inches long. One side is scaled to read a tree's diameter by holding the stick horizontally at arm's length and against the tree at breast height. A merritt hypsometer runs along one edge of the stick and is scaled to read a tree's height from 66 feet away from the tree's base. These two measurements are then used to find the tree's volume according to the volume table printed on one face of the stick.

Biodiversity: the variety of life forms in a given area; can be categorized in terms of number of species, variety of plant and animal communities, genetic variability or some combination of these categories.

Biological diversity: the variety of life in all its forms and all its levels of organization. Biodiversity refers to diversity of genetics, species, ecosystems, and landscapes.

Bird: any of a class of vertebrates that regulate their body temperature internally, have bodies that are covered almost entirely with feathers and have forelimbs modified as wings that enable most to fly.

Blaze: to mark a tree, usually by painting or cutting the bark. Forest properties often are delineated by blazing trees along the boundary lines.

Blowdown: see windthrow.

Board foot: a unit for measuring wood volume in a tree, log, or board. A board foot is commonly 1 foot by 1 foot by 1 inch, but any shape containing 144 cubic inches of wood equals one board foot.

Bole: the trunk of a tree.

Boots: firefighters wear all leather boots to protect their feet on the fireline and rugged terrain.

Breast height: 4½ feet above ground level. See diameter at breast height.

Broadleaf: a class of trees that have broad, flat leaves of many different shapes; most are deciduous; also called hardwood because most broad-leaved trees have harder wood than do conifers. Examples include oak, hickory, maple and ash.

Browse: parts of woody plants, including twigs, shoots, and leaves, eaten by forest animals.

Buck: to cut trees into shorter lengths, such as logs or cordwood.

Buffer strip: a narrow zone or strip of land, trees or vegetation bordering an area. Common examples include visual buffers, which screen the view along roads, and streamside buffers, which are used to protect water quality. Buffers may also be used to prevent the spread of forest pests.

Butt log: a log cut from the bole immediately above the stump.

Caliper: a tool to measure the diameter of a tree.

Cambium: a thin layer of specialized cells within a tree's trunk that divide to produce new inner bark cells to the outside and new sapwood cells to the inside. The narrow band of cells that is responsible for the tree's growth in circumference.

Canopy: the continuous cover formed by tree crowns in a forest.

Carrying capacity: the maximum number of healthy wildlife that a given habitat or area can support without degradation of the habitat.

Cellulose: the scientific name for wood fiber.

Chain: a distance of 66 feet.

Clearcut: the harvest of all the trees in an area. Clearcutting is used to aid species whose seedlings require full sunlight to grow well.

Clinometer: an instrument used to determine the height of a tree.

Codominant tree: a tree that extends its crown into the canopy and receives direct sunlight from above but limited sunlight from the sides. One or more sides of a codominant tree are crowded by the crowns of dominant trees.

Commercial clearcut: a harvest cut that removes all merchantable timber from the area.

Commercial forestland: any area capable of producing 20 cubic feet of timber per acre per year that has not been protected from such use by law or statute.

Commercial treatments: timber stand improvements, such as thinning, that generate income from the sale of the trees removed.

Community: a collection of living organisms thriving in an organized system through which water, energy, and nutrients cycle.

Competition: the struggle between trees to obtain sunlight, nutrients, water and growing space. Every part of the tree, from the roots to the crown, competes for space and food.

Conifer: any tree that produces seeds in cones. See softwood.

Conservation: planned management and wise use of natural resources for present and future generations.

Consulting forester: an independent professional who manages forests and markets forest products for private woodland owners. Consulting foresters do not have direct connections with firms that buy wood products, but are retained by woodland owners as their agents.

Cooperative extension service (ces): the educational arm of the usda that links university research to people who can benefit from it.

Cord: a unit of wood cut for fuel that is equal to a stack 4 x 4 by 8 feet or 128 cubic feet. A cord is the legal measure of fuelwood volume in maryland.

Cordwood: small diameter or low quality wood suitable for firewood, pulp, or chips. Cordwood is not suitable for sawlogs.

Cover: (a) any plant that intercepts rain drops before they reach the soil or that holds soil in place; (b) a hiding place or vegetative shelter for wildlife from predators or inclement weather.

Critical area:land in maryland that lies within 1,000 feet of the chesapeake bay and its tributaries and is subject to forestry and other land use regulations.

Crook: a tree defect characterized by a sharp bend in the main stem.

Crop tree: a young tree of a desirable species with certain characteristics desired for timber value, water quality enhancement, or wildlife or aesthetic uses.

Crown: the uppermost branches and foliage of a tree.

Crown classes: see codominant, dominant, intermediate, overtopped, and suppressed.

Crown cover or crown closure: the percentage of a given area covered by tree crowns.

Crown ratio or live-crown ratio: the ratio of the leaved portion of a tree's height to its total height.

Cruise: a forest survey used to obtain inventory information and develop a management plan.

Cull: a sawtimber sized tree that has no timber value as a result of poor shape or damage from injury, insects or disease.

Cutting cycle: the period of time between major harvests in a stand.

Deciduous: shedding or losing leaves annually; the opposite of evergreen. Trees such as maple, ash, cherry, and larch are deciduous.

Decomposition: the process by which organic material such as leaves and branches are broken down by bacteria, fungi, protozoans and the many different kinds of animals that live in the soil.

Defects: characteristics of an individual tree that reduce its quality and utility.

Deferment harvest: the residual stand resembles a seed tree cut when the regenerated stand becomes established. Instead the residual trees remain until the regenerated stand is at the end of a rotation. Similar to a seed tree harvest, but the residual trees are left indefinitely for seed production.

Den tree: a tree with cavities suitable for birds or mammals to nest in.

Dendrology: the study of trees; tree identification.

Diameter at breast height (dbh): standard measurement of a tree's diameter, usually taken at 4 1/2 feet above the ground.

Diameter-limit scale: a timber sale in which all trees over a specified dbh may be cut. Diameter-limit sales often result in high grading.

Diameter tape: a steel measuring tape that has a scale calibrated to read a tree's diameter when wrapped around the tree's circumference.

Dimension lumber: hardwood dimension lumber is processed to be used whole in the manufacture of furniture or other products. Softwood dimension lumber consists of boards more than 2 inches thick but less than 5 inches thick. This wood is used in construction and is sold as 2 by 4s, 4 by 8s, or 2 by 10s.

Dominant trees: trees that extend above surrounding individuals and capture sunlight from above and around the crown.

Drip torch: firefighters use these torches, that drip a flaming liquid mixture of diesel fuel and gasoline, to ignite fires in burnout operations.

Earlywood: wood cells produced at the beginning of a tree's growing season that are generally light in color. Also called springwood.

Ecological succession: the gradual change of plant and animal communities over time.

Ecology: the science or study of the relationships between organisms and their environment.

Ecosystem: organisms and the physical factors that make up their environment.

Ecotone: a transition area between two distinct, but adjoining, communities.

Edge: the boundary between two ecological communities, for example, field and woodland. Edges provide wildlife habitat. Consideration of an edge can reduce the impact of a timber harvest.

Endangered species: any species that has been classified by the u.s. Fish and wildlife service or a state wildlife agency as being in danger of extinction throughout all or a significant portion of its range. A species is endangered when the total number of remaining members may not be sufficient to reproduce enough offspring to ensure survival of the species.

Environment: the sum of all external living and non-living conditions and influences that affect the development and survival of an organism.

Epicormic branching: branches that grow out of the main stem of a tree from buds produced under the bark. Severe epicormic branching increases knottiness and reduces lumber quality.

Erosion: the wearing away or removal of land or soil by the action of wind, water, ice or gravity.

Even-aged management: a forest management method used to produce stands that are all the same age or nearly the same age by harvesting all trees in an area at one time or in several cuttings over a short time. This management method is commonly applied to shade-intolerant conifers and hardwoods.

Evergreen: a group of trees that do not lose all of their leaves every year but go through a gradual replacement by dropping only their oldest leaves each year. Instead of being bare in winter, these trees have leaves all year.

Extension forester: a cooperative extension service professional who educates woodland owners on how they can effectively manage their forests.

Farm services agency (FSA): the branch of the U.S. Department of agriculture (Usda) that administers cost-sharing programs for such forestry practices as tree planting and timber stand improvement.

Felling: the cutting of standing trees.

Fertilization: the addition of nutrient elements to increase growth rate or overcome a nutrient deficiency in the soil.

Final harvest: the removal of the remaining crop trees in an even-aged stand.

Fire pack: firefighters use these to carry tools, equipment, and supplies on their backs.

Fire resistant clothing: these yellow shirts and green pants are the trademarks of wildland firefighters.

Fire shelter: firefighters use this personal protection as a last resort if a wildfire traps them and they cannot escape. Firefighters can get into the tent-like shelter, made of heat reflective material, in about 25 seconds.

Fire shovel: these shovels, specifically designed for constructing a fireline, feature a tapered blade with both edges sharpened for scraping, digging, grubbing, cutting, and throwing dirt.

First commercial thinning – this will occur on plantations at age 12-20 years old to facilitate forest health and promote development of larger trees over a shorter amount of time. This is accomplished in plantations by removing every 5th row of trees and selectively thinning between rows. In naturally regenerated stands, thinning corridors will be established every 50 feet and the stand will be selectively thinned along both sides of the corridor. Approximately 30-35% of the total stand volume will be removed in this process.

Forage: vegetation such as leaves, stems, buds and some types of bark, that can be eaten for food and energy.

Forb: any herb other than grass.

Forest: a biological community dominated by trees and other woody plants.

Forest floor: the lowest level of the forest that is made up of tree seedlings, dead leaves and needles, grasses, ferns, flowers, fungi, and decaying plants and logs.

Forest fragmentation: the subdivision of large natural landscapes into smaller, more isolated fragments. Fragmentation affects the viability of wildlife populations and ecosystems.

Forest management: caring for a forest so that it stays healthy and vigorous and provides the products and values the landowner desires.

Forest stewardship plan: a written document listing activities that enhance or improve forest resources (wildlife, timber, soil, water, recreation and aesthetics) on private land over a five-year period.

Forest type: a designation or name given to a forest based on the most abundant tree type or types in the stand; groups of tree species commonly growing in the same stand because their environmental requirements are similar. Examples of north carolina forest types include (a) pine; (b) mixed hardwood; (c) cypress, tupelo and black gum; and (d) oak and hickory.

Forested wetland: an area characterized by woody vegetation taller than 20 feet where soil is at least periodically saturated or covered by water.

Forestry incentives program (F.I.P.): a federal cost-sharing program that reimburse part of the costs landowners incur in completing certain forestry practices. The F.I.P. is administered by the NRCS.

Forestry: the art and science of managing forests to produce various products and benefits including timber, wildlife habitat, clean water, biodiversity and recreation.

Fork: a tree defect characterized by the division of a bole or main stem into two or more stems.

Frilling: the method of killing trees by inflicting a series of cuts around the bole and applying an herbicide to the wounds. Frilling or girdling can be used to reduce the density of a stand or to kill individual undesirable trees.

Fuel loading: a buildup of easily ignited leaves, pine straw, branches and trees on the forest floor.

Fusee: firefighters also use these colored flares to ignite fires in burnout operations.

Girdling: a method of killing trees by cutting through the stem, thus interrupting the flow of water and nutrients.

Ground spraying: a ground application of herbicides to discourage unwanted hardwood competition. It is almost always associated with fertilization. This combined treatment is typically done a year or two after a successful thinning to achieve larger trees faster.

Group selection: (a) the removal of small groups of trees to regenerate shade-intolerant trees in the opening (usually at least 1/4 acre); (b) a specific type of selective cutting.

Group selection: a process of harvesting patches of trees to open the forest canopy and encourage the reproduction of unevenaged stands.

Growth rings: the layers of wood a tree adds each season; also called annual rings. These rings frequently are visible when a tree is cut and can be used to estimate its age and growth rate.

Gymnosperm: a plant whose seeds are not enclosed in flowers. Most gymnosperms produce their seeds on the surface of the scales of female cones and are pollinated by wind. Conifers are the most common type of gymnosperm.

Habitat: an area in which a specific plant or animal naturally lives, grows and reproduces; the area that provides a plant or animal with adequate food, water, shelter and living space.

Hardwoods: trees with broad, flat leaves as opposed to coniferous or needled trees. Wood hardness varies among the hardwood species, and some are actually softer than some softwoods.

Harvest: the cutting, felling, and gathering of forest timber.

Headlamp: firefighters who work on the fireline at night wear these flashlights on the front of their helmets.

Heartwood: the central core of a tree, which is made up of dense, dead wood and provides strength to the tree.

Herbaceous vegetation: low-growing, non-woody plants, including wildflowers and ferns, in a forest understory.

High grading: to remove all mature, good quality trees from a stand and leave inferior species and individuals. High grading should be distinguished from even-aged management in which mature and immature trees are removed to aid regeneration.

Hypsometer: any of several tools or instruments designed to measure the height of trees. The clinometer is such a tool.

Improvement cut: a weeding done in stands of pole-size or larger trees.

Increment borer: a hollow auger-like tool with a screw bit used to remove core samples from trees.

Industrial forester: a professional employed by a wood-using industry, usually a sawmill, who purchases timber from private woodland owners. Many industrial foresters offer free forest management or marketing services to the landowners who sell timber to the forester's employer.

Intermediate crown class: trees with crowns that extend into the canopy with dominant and codominate trees. These trees receive little direct sunlight from above and none from the sides. Crowns generally are small and crowded on all sides.

Intermediate tolerance: a characteristic of certain tree species that allows them to survive, though not necessarily thrive, in relatively low light conditions.

Intolerance: a characteristic of certain tree species that does not permit them to survive in the shade of other trees.

Introduced species: a non-native species that was intentionally or unintentionally brought into an area by humans.

Landing: a cleared area within a timber harvest where harvested logs are processed, piled, and loaded for transport to a sawmill or other facility.

Latewood: wood cells produced at the end of the growing season that make up the darker section of an annual ring. Also called summerwood.

Limiting factor: any requirement for wildlife survival that is in limited supply.

Loblolly-shortleaf forest type: an association of tree species common to the southeastern united states that includes loblolly and shortleaf pines and oaks.

Log rule: a method for calculating wood volume in a tree or log by using its diameter and length. The international ¼-inch rule is the legal rule in maryland.

Logger: an individual who harvests timber for a living.

Lopping: cutting tree tops to a maximum specified height above the ground after a tree is felled.

Mammal: any of a class of higher vertebrates whose bodies are covered with hair, who give birth to live young, nourishes their young with milk from mammary glands, regulate their body temperature internally, have four types of well-developed teeth and typically have four well-developed legs with toes that have nails, claws or hoofs.

Marking timber: indicating by paint or other means which trees are to be cut or otherwise treated. It is advisable to mark trees to be harvested twice-once at eye level and once on the stump.

Mast: fruits or nuts used as a food source by wildlife. Soft mast includes most fruits with fleshy coverings, such as persimmon, dogwood seed or black gum seed. Hard mast refers to nuts such as acorns and beech, pecan and hickory nuts.

Mcleod: firefighters also use this combination hoe, rake and scraping tool and rake to remove plants and shrubs when building a fireline.

Merchantable height: the point on a tree stem to which the stem is salable. Limits are: the point at which a sawlog tree is less than 8 inches in diameter, measured inside the bark (DIB); the point at which a pulpwood tree is less than 4 inches dib; or the point on any tree where a defect is found that cannot be processed out.

Merritt hypsometer: a scale that measures the number of 16-foot logs in a tree.

Multiple-use management: the management of land or forest for more than one purpose, such as wood production, water quality, wildlife, recreation, aesthetics and clean air.

Natural regeneration: the growth of new trees in one of the following ways without human assistance: (a) from seeds carried by wind or animals, (b) from seeds stored on the forest floor, or (c) from stumps that sprout.

Natural resources conservation service (NRCS): the branch of the USDA that coordinates and implements soil conservation practices on private lands. The NRCS can provide woodland owners with detailed information on his or her soil.

Naval stores: products such as turpentine, pitch and rosin that come from pine trees and are used in the construction and maintenance of wooden sailing vessels.

Niche: the physical and functional "address" of an organism within an ecosystem; or, where a living thing is found and what it does there.

Nongame wildlife: wildlife species that are protected by state wildlife laws and can not be hunted. Examples include songbirds, eagles, etc.

Nonindustrial private forestland (nipf): forestland owned by a private individual, group, or corporation not involved in wood processing. Eighty-five percent of maryland's forests are in this category.

Nontidal wetlands: wetlands not affected by ocean tides. Nontidal wetlands are subject to special regulations.

Northern hardwood forest type: an association of tree species common to the northeastern united states that includes sugar maple, red maple, yellow birch, hemlock, and american beech.

Oak-gum-cypress forest type: an association of tree species common to the bottom lands of the southeastern united states.

Oak-hickory forest type: an association of tree species common to the northeastern united states that includes oak, hickory, yellow poplar, and red maple.

Oak-pine forest type: an association of tree species common to the southeastern united states that includes loblolly pine, virginia pine, northern red oak, and white ash.

Old-growth forest: a wooded area, usually greater than 200 years of age, that has never been altered or harvested by humans. An old-growth forest often has large individual trees, a multi-layered crown canopy, and a significant accumulation of coarse woody debris including snags and fallen logs.

Overmature: a quality exhibited by trees that have declined in growth rate because of old age and loss of vigor.

Overstocked: the situation in which trees are so closely spaced that they compete for resources and do not reach full growth potential.

Overstory removal: a silvicultural technique where the trees to be removed are all in the dominant or codominant crown class or position. This basically is performed to harvest mature trees and to remove competition from preferred understory trees.

Overstory: the level of forest canopy that includes the crowns of dominant, codominant, and intermediate trees.

Overtopped: the situation in which a tree cannot sufficiently extend its crown into the overstory and receive direct sunlight. Overtopped trees that lack shade tolerance lose vigor and die.

Patch cut: a clearcut on a small area.

Phloem: the part of a tree that carries sap from the leaves to the rest of the tree. Also called inner bark.

Photosynthesis: the process by which a plant or tree combines water and carbon dioxide with energy from the sun to make glucose and oxygen.

Plant succession: the progression of plants from bare ground to mature forest.

Pole stand: a stand of trees whose average dbh is between 4 and 10 inches.

Pole timber: trees 4 to 10 inches dbh.

Precommercial operations: cutting in forest stands to remove wood too small to be marketed. Precommercial operations improve species composition and increase the quality, growth, and vigor of remaining trees.

Precommercial thinning: a harvest made purely as investments in the future growth of stands so young so young that none of the felled trees are extracted and utilized. This treatment is usually completed by hand on trees 5-10 years of age.

Precommercial treatments: forestry operations that require landowner investment, such as cleaning or weeding stands to remove trees that have little or no cash value. See commercial treatments.

Prescribed burning: the practice of using regulated fires to reduce or eliminate material on the forest floor, for seedbed preparation or to control competing vegetation. Prescribed burning simulates one of the most common natural disturbances. Also called controlled burning.

Prescribed fire: fires set deliberately, under proper supervision and certain conditions, to achieve a specific management goal such as enhancing wildlife habitat, encouraging fire-dependant plant species, reducing fuel loads that feed wildfires, and preparing sites for planting. Sometimes referred to prescribed burning.

Prospectus: a document that describes the location of a property, indicates trees marked for cutting, and states that the timber will be sold in accordance with a suitable contact. A prospectus includes the number of trees marked, their diameter classes, and a volume estimate for each species.

Pruning: the act of sawing or cutting branches from a living tree. In forest management, pruning is done to promote the growth of clear, valuable wood on the tree bole.

Pulaski: firefighters use this tool, which combines a single-bitted axe blade for chopping with a narrow blade for trenching, to clear vegetation when constructing a fireline.

Pulpwood harvest: a harvest where the trees are to utilized for paper pulp. This type of harvest usually is preformed as a commercial thinning where the trees are all pole sized (4" to 11" d.b.h.), but definitely less than sawtimber sized (11" d.b.h. Or greater).

Pulpwood: wood used in the manufacture of paper, fiberboard or other wood fiber products. Pulpwood-sized trees are usually a minimum of 4 inches in diameter.

Reforestation – the reestablishment of forest cover either naturally (by natural seeding, coppice, or root suckers) or artificially (by direct seeding or planting).

Regeneration: the process by which a forest is reseeded and renewed. Advanced regeneration refers to regeneration that is established before the existing forest stand is removed.

Regeneration cut: a timber harvest designed to promote natural establishment of trees.

Release: to free a tree from competition with its immediate neighbors by removing the surrounding trees. This occurs naturally and artificially.

Release harvest: the act of freeing a young stand of desirable trees, not past the sapling stage, from the competition of undesirable trees that threaten to suppress them.

Renewable resource: a naturally occurring raw material or form of energy that has the capacity to replenish itself through ecological cycles and sound management practices.

Reptile: any of a class of vertebrates that regulates its body temperature externally, has dry, glandless skin covered with scales, breathes through lungs and lays large eggs that develop on land.

Residual stand: the trees remaining intact following any cutting operation.

Resin: a group of sticky liquid substances secreted by plants that appear on the plant's external surface after a wound.

Riparian buffer zone: vegetated areas adjacent to or influenced by a perennial or intermittent bodies of water. These buffers are established and managed to protect aquatic, wetland, shoreline, and/or terrestrial environments.

Roots: the underground portion of a tree that helps anchor the tree in the ground and absorbs water and nutrients from the soil.

Rot: a tree defect characterized by woody decay in a standing tree or log.

Rotation: the number of years required to establish and grow trees to a specified size, product or condition of maturity. A pine rotation may range from as short as 20 years for pulpwood to more than 60 years for sawtimber.

Salvage cut: the harvesting of dead or damaged trees, or the harvesting of trees in danger of being killed by insects, disease, flooding or other factors in order to save their economic value.

Sapling stand: a stand of trees whose average DBH is between 1 and 4 inches.

Sapling: a tree at least 4½ feet tall and up to 4 inches in diameter.

Sawlog: a log large enough to be sawed economically on a sawmill. Sawlogs are usually at least 8 inches in diameter at the small end.

Sawlog tree: a tree at least 11 inches dbh and suitable for conversion to lumber. Sometimes, trees 11 to 14 inches dbh are called small sawlog trees, and trees larger than 18 inches dbh are called large sawlog trees.

Sawtimber: wood of large enough size to be used to produce lumber for construction and furniture.

Scale stick: a calibrated stick used to estimate wood volume in a log.

Sealed-bid sale: a timber sale, usually offered through a consulting forester, in which buyers submit secret bids.

Second commercial thinning: usually performed on stands 20-28 years old, the objective is to lengthen the rotation age of the stand and produce larger trees faster. In some cases, this technique is used to improve habitat for the delmarva fox squirrel (dfs). Approximately 30-35% of the total stand volume is be removed in this process.

Sedimentation: the deposition or settling of soil particles suspended in water.

Seed tree cut: a harvesting method in which a few scattered trees are left in the area to provide seeds for a new forest stand. Selection of seed trees is based on growth rate, form, seeding ability, wind firmness and future marketability. This harvesting method produces an even-aged forest.

Seed tree: a mature tree left uncut to provide seed for regeneration of a harvested stand.

Seed tree harvest: the felling of all the trees in an area except for a few desirable individuals that provide seed for the next forest.

Selection harvest: the harvest of all individual trees or small groups at regular intervals to maintain an uneven-aged forest. Selection harvests are used to manage species that do not need sunlight to survive.

Selective cutting: the periodic removal of individual trees or groups of trees to improve or regenerate a stand.

Service forester: a professional forester employed by the maryland department of natural resources forest service. Service foresters help private woodland owners develop and implement forest management plans. There is at least one service forester in each maryland county.

Shade-intolerant species: trees that require full sunlight to thrive and cannot grow in the shade of larger trees.

Shade-tolerant species: trees that have the ability to grow in the shade of other trees and in competition with them.

Shelterwood cut: removing trees in the harvest area in a series of two or more cuttings so that new seedlings can grow from the seeds of older trees. This method produces an even-aged forest.

Shelterwood harvest: the harvest of all mature trees in an area in a series of two or more cuts, leaving enough trees of other sizes to provide shade and protection for forest seedlings.

Silviculture: the art, science and practice of establishing, tending and reproducing forest stands of desired characteristics. It is based on knowledge of species' characteristics and environmental requirements.

Site: the combination of biotic, climatic, topographic, and soil conditions of an area.

Site index: a relative measure of forest site quality based on the height (in feet) of the dominant trees at a specific age (usually 25 or 50 years, depending on rotation length). Site index information helps estimate future returns and land productivity for timber and wildlife.

Site preparation: treatment of an area prior to reestablishment of a forest stand. Site preparation can include mechanical clearing, burning, or chemical (herbicide) vegetation control.

Skidder: a rubber-tired machine with a cable winch or grapple used to drag logs out of the forest.

Skidding: the act of moving trees from the site of felling to a leading area or landing. Tractors, horses, or specialized logging equipment can be used for skidding. Skidding methods vary in their impact on soils and the remaining stands.

Slash: branches and other woody material left on a site after logging.

Snag: a dead tree that is still standing. Snags provide important food and cover for a wide variety of wildlife species.

Softwood: any tree in the gymnosperm group, including pines, hemlocks, larches, spruces, firs, and junipers. Softwoods often are called conifers although some, such as junipers and yews do not produce cones.

Species: a group of related organisms having common characteristics and capable of interbreeding. Loblolly and virginia pine are common tree species that can interbreed.

Sprout: a tree growing from a cut stump or previously established root system.

Stand: a group of forest trees of sufficiently uniform species composition, age, and condition to be considered a homogeneous unit for management purposes.

Stand density: the quantity of trees per unit area, usually evaluated in terms of basal area, crown cover and stocking.

Stocking: the number and density of trees in a forest stand. Stands are often classified as understocked, well-stocked or overstocked.

Stratification: division of a forest, or any ecosystem, into separate layers of vegetation that provide distinct niches for wildlife. See canopy, understory, and herbaceous vegetation.

Streamside management zone (SMZ) : an area adjacent to a stream in which vegetation is maintained or managed to protect water quality.

Stump height: the distance from the ground to the top of the stump. Good logging practice dictates that stumps be as low as possible (preferably as low as 12 inches) to reduce waste, to minimize visual impact on the logging site, and to promote resprouting of trees.

Stumpage: the value of standing trees in a forest.

Stumpage price: the price paid for standing forest trees.

Succession: the natural replacement of one plant (or animal) community by another over time in the absence of disturbance.

Suppressed: a tree condition characterized by low growth rate and low vigor as a result of competition with overtopping trees. See overtopped.

Suppression: the process by which a tree loses its vigor due to inadequate light, water and nutrients.

Sustained yield: an ideal forest management objective in which the volume of wood removed equals growth within the total forest.

Sweep: a tree defect characterized by a gradual curve in the main stem.

Thinning: a tree removal practice that reduces tree density and competition between trees in a stand. Thinning concentrates growth on fewer, high-quality trees, provides periodic income and generally enhances tree vigor. Heavy thinning can benefit wildlife through the increased growth of ground vegetation.

Threatened species: any species that has been classified by the u.s. Fish and wildlife service or a state wildlife agency as likely to become endangered within the foreseeable future throughout all or a significant portion of its range. A threatened species has declining or dangerously low populations but still has enough members to maintain or increase numbers.

Timber stand improvement (T.S.I.): any practice that increases the value or rate of value growth in a stand of potential sawtimber trees. Pruning and thinning are considered T.S.I.

Tolerance: a tree species' capacity to grow in shade

Transpiration: the loss of water through leaves.

Tree caliper: a metal or wooden device consisting of an arm and two prongs, one of which is free to slide along a graduated scale on the arm. The prongs are placed against opposite sides of a tree to read its diameter on the scale.

Tree farm: a privately owned forest managed on a multiple use basis with timber production as an important management goal.

Trim allowance: the extra 2 or 3 inched left on a bucked log to allow logs with end checks, pulls, or slanting buck cuts to be trimmed to standard lumber lengths.

Turpentine: a distilled chemical produced from tapping into a living pine and harvesting the sap.

Understocked: a stand of trees so widely spaced, that even with full growth potential realized, crown closure will not occur.

Understory: the area below the forest canopy that comprises shrubs, snags and small tree. Because the understory receives little light, many of the plants at this level tolerate shade and will remain part of the understory. Others will grow and replace older trees that fall.

Uneven-aged stand: three or more age classes of trees represented.

Unit sale: a timber sale in which the buyer makes regular (weekly, monthly) payments based on mill receipts. Unit sales are useful when the amount of timber sold is so large that a preharvest, lump-sum payment would be prohibitive.

Variable retention harvest system: an approach to harvesting based on the retention of structural elements or biological legacies (trees, snags, logs, etc.) From the harvested stand for integration into the new stand to achieve various ecological objectives.

Vegetation: low-growing, non-woody plants, including wildflowers and ferns, in a forest understory.

Veneer log: a high-quality log of a desirable species suitable for conversion to veneer. Veneer logs must be large, straight, of minimum taper, and free from defects.

Virgin forest: an area of old-growth trees that never has been harvested by humans.

Watershed: a region defined by patterns of stream drainage. A watershed includes all the land that contributes water to a particular stream or river.

Well-stocked: the situation in which a forest stand contains trees spaced widely enough to prevent competition yet closely enough to utilize the entire site.

Wildlife: a broad term that includes non-domesticated vertebrates, especially mammals, birds and fish.

Wildlife habitat: the native environment of an animal. Habitats ideally provide all the elements needed for life and growth: food, water, cover and space.

Windthrow: a tree felled by wind. Windthrows, also known as blowdowns, are common among shallow-rooted species and in areas where cutting has reduced stand density.

Wolf tree: a large older tree with a spreading crown and little or no timber value, but often great value for wildlife.

Wood: the solid interior of a tree.

Wood chemicals: chemicals that are found naturally in the various parts of a tree.

Woodland: see forest.

Xylem: the part of a tree that transports water and nutrients up from the roots to the leaves. Older xylem cells become part of the heartwood. Also called sapwood.

BIBLIOGRAPHY

Akçakaya, H.R.; Radeloff, V.C.; Mladenoff, D.J.; He, H.S. Integrating landscape and metapopulation modeling approaches: Viability of the sharp-tailed grouse in a dynamic landscape. *Conserv. Biol.* 2004, *18*, 526–537.

Albert, C.H.; Rayfield, B.; Dumitru, M.; Gonzalez, A. Applying network theory to prioritize multispecies habitat networks that are robust to climate and land-use change. *Conserv. Biol.* 2017, *31*, 1383–1396.

Alderman, D.R. Jr., Bumgardener, M.S., Baumgras, J.E. (September 1, 2005). An Assessment of Red Maple Resource in the Northeastern United States. N*orthern Journal of Applied Forestry*. 22(3).

Alexander, H. M., Foster, B. L., Ballantyne, F., Collins, C. D., Antonovics, J., & Holt, R. D. (2012). Metapopulations and metacommunities: combining spatial and temporal perspectives in plant ecology. *Journal of Ecology, 100*(1), 88-103.

America's Longleaf Restoration Initiative (ALRI). (n.d.). *American's Longleaf.* http://www.americaslongleaf.org/home/ µ!

American Conifer Society (2020, September 4). *10 Types of Cedar Everyone Should Know.* https://conifersociety.org/conifers/articles/types-of-cedar-trees/ µ!

American Forest. (n.d.). *Champion Trees National Register.* http://www.american forests.org/explore-forests/americas-biggest-trees/champion-trees-national-register/ µ!

Anderson, T. M., Schütz, M., & Risch, A. C. (2012). Seed germination cues and the importance of the soil seed bank across an environmental gradient in the Serengeti. *Oikos, 121*(2), 306-312.

Angelstam, P.; Persson, R.; Schlaepfer, R. The sustainable forest management vision and biodiversity: Barriers and bridges for implementation in actual landscapes. *Ecol. Bull.* 2004, *51*, 29–49.

Angelstam, P.; Törnblom, J.; Degerman, E.; Henrikson, L.; Jougda, L.; Lazdinis, M.; Malmgren, J.C.; Myhrman, L. From forest patches to functional habitat. In *Farming, Forestry and the Natural Heritage: Towards a More Integrated Future*; Davison, R., Garblraith, C.A., Eds.; Scottish Natural Heritage: Edinburgh, UK, 2006; pp. 193–209.

Armstrong, J. A., and W. G. H. Ives, eds. 1995. Forest insect pests in Canada. Natural Resources Canada, Canadian Forest Service, Science and Sustainable Development Directorate, Ottawa, Ont.

Asner, G.P.; Mascaro, J. 2014. Mapping tropical forest carbon: calibrating plot estimates to a simple LiDAR metric. Remote Sensing of Environment. 140: 614–624.

Auffret, A. G., & Cousins, S. A. O. (2011). Past and present management influences the seed bank and seed rain in a rural landscape mosaic. *Journal of Applied Ecology, 48*, 1278-1285.

Ayre, D. J., Ottewell, K. M., Krauss, S. L., & Whelan, R. J. (2009). Genetic structure of seedling cohorts following repeated wildfires in the fire-sensitive shrub *Persoonia mollis* ssp. *nectens*. *Journal of Ecology, 97*(4), 752-760. 227

Basic Biology. (May 21, 2015). *Basal Angiosperms*. https://basicbiology.net/plants/angiosperms/basal-angiosperms µ!

Baskin, C. C., & Baskin, J. M. (2014). *Seeds. Ecology, Biogeography and Evolution of Dormancy and Germination* (2nd ed.). Academic Press.

Beazley, K.; Cardinal, N. A systematic approach for selecting focal species for conservation in the forests of Nova Scotia and Maine. *Environ. Conserv.* 2004, *31*, 91–101.

Bell, F.W.; Dacosta, J.; Larocque, G.R. Considering forest biodiversity indicators within a pressure, state, benefit, and response framework. In *Ecological Forest Management Handbook*; Larocque, G.R., Ed.; CRC Press: Boca Raton, FL, USA, 2016; pp. 337–359.

Beresford, A.E.; Donald, P.F.; Buchanan, G.M. Repeatable and standardised monitoring of threats to Key Biodiversity Areas in Africa using Google Earth Engine. *Ecol. Indic.* 2020, *109*, 105763.

Bernes, C.; Jonsson, B.G.; Junninen, K.; Lõhmus, A.; MacDonald, E.; Müller, J.; Sandström, J. What is the impact of active management on biodiversity in boreal and temperate forests set aside for conservation or restoration? A systematic map. *Environ. Evid.* 2015, *4*, 25.

Bertuol-Garcia, D.; Morsello, C.; El-Hani, C.N.; Pardini, R. A conceptual framework for understanding the perspectives on the causes of the science–practice gap in ecology and conservation. *Biol. Rev.* 2018, *93*, 1032–1055.

Bias, M.A.; Gutiérrez, R. 1992. Habitat associations of California spotted owls in the central Sierra Nevada. The Journal of Wildlife Management: 584–595.

Blakesley, J.A.; Franklin, A.B.; Gutiérrez, R. 1992. Spotted owl roost and nest site selection in northwestern California. The Journal of Wildlife Management: 388–392.

Blakesley, J.A.; Noon, B.R.; Anderson, D.R. 2005. Site occupancy, apparent survival, and reproduction of California spotted owls in relation to forest stand characteristics. Journal of Wildlife Management. 69(4): 1554–1564.

Blanchard, S.; Jakubowski, M.; Kelly, M. 2011. Object-based image analysis of downed logs in a disturbed forest landscape using lidar. Remote Sensing. 3(11): 2420–2439.

Blattert, C.; Lemm, R.; Thees, O.; Hansen, J.; Lexer, M.J.; Hanewinkel, M. Segregated versus integrated biodiversity conservation: Value-based ecosystem service assessment under varying forest management strategies in a Swiss case study. *Ecol. Indic.* 2018, *95*, 751–764.

Bond, M.L.; Gutierrez, R.; Franklin, A.B.; LaHaye, W.S.; May, C.A.; Seamans, M.E. 2002. Short-term effects of wildfires on spotted owl survival, site fidelity, mate fidelity, and reproductive success. Wildlife Society Bulletin: 1022–1028.

Bond, M.L.; Lee, D.E.; Siegel, R.B.; Ward, J.P. 2009. Habitat use and selection by California spotted owls in a postfire landscape. The Journal of Wildlife Management. 73(7): 1116–1124.

Bonnot, T.W.; Thompson, F.R., III; Millspaugh, J.J.; Jones-Farrand, D.T. Landscape-based population viability models demonstrate importance of strategic conservation planning for birds. *Biol. Conserv.* 2013, *165*, 104–114.

Bouget, C.; Parmain, G.; Gilg, O.; Noblecourt, T.; Nusillard, B.; Paillet, Y.; Pernot, C.; Larrieu, L.; Gosselin, F. Does a set-aside conservation strategy help the restoration of old-growth forest attributes and recolonization by saproxylic beetles? *Anim. Conserv.* 2014, *17*, 342–353.

Brooks, R.P. Improving habitat suitability index models. *Wildl. Soc. Bull.* 1997, *25*, 163–167.

Brooks, T.M.; Pimm, S.L.; Akçakaya, H.R.; Buchanan, G.M.; Butchart, S.H.; Foden, W.; Hilton-Taylor, C.; Hoffmann, M.; Jenkins, C.N.; Joppa, L.; et al. Measuring terrestrial area of habitat (AOH) and its utility for the IUCN red list. *Trends Ecol. Evol.* 2019, *34*, 977–986.

Brown, D. (1992). Estimating the composition of a forest seed bank: a comparison of the seed extraction and seedling emergencemethods. *Canadian Journal of Botany, 70*, 1063-1612.

Brown, D.J.; Ribic, C.A.; Donner, D.M.; Nelson, M.D.; Bocetti, C.I.; Deloria-Sheffield, C.M. Using a full annual cycle model to evaluate long-term population viability of the conservation-reliant Kirtland's warbler after successful recovery. *J. Appl. Ecol.* 2017, *54*, 439–449.

Brown, J. S., & Venable, D. L. (1986). Evolutionary ecology of seed-bank annuals in temporally varying environments. *The American Naturalist*, 127, 31-47.

Burley, J.B. Habitat suitability models: A tool for designing landscape for wildlife. *Landsc. Res.* 1989, *14*, 23–26.

Burnett, K.M.; Reeves, G.H.; Miller, D.J.; Clarke, S.; Vance-Borland, K.; Christiansen, K. Distribution of salmon-habitat potential relative to landscape characteristics and implications for conservation. *Ecol. Appl.* 2007, *17*, 66–80.

Cabin, R. J., Marshall, D. L., & Mitchell, R. J. (2000). The demographic role of soil seed banks. II. Investigations of the fate of experimental seeds of the desert mustard *Lesquerella fendleri*. *Journal of Ecology, 88*(2), 293-302.

Caro, T. *Conservation by Proxy: Indicator, Umbrella, Keystone, Flagship, and Other Surrogate Species*; Island Press: Washington, DC, USA, 2010.

Carroll, C. Interacting effects of climate change, landscape conversion, and harvest on carnivore populations at the range margin: Marten and lynx in the northern Appalachians. *Conserv. Biol.* 2007, *21*, 1092–1104.

Carroll, C. Role of climatic niche models in focal-species-based conservation planning: Assessing potential effects of climate change on northern spotted owl in the pacific northwest, USA. *Biol. Conserv.* 2010, *143*, 1432–1437.

Carroll, C.; Dunk, J.R.; Moilanen, A. Optimizing resiliency of reserve networks to climate change: Multispecies conservation planning in the pacific northwest, USA. *Glob. Chang. Biol.* 2010, *16*, 891–904.

Carroll, C.; Johnson, D.S. The imp ortance of being spatial (and reserved): Assessing northern spotted owl habitat relationships with hierarchical bayesian models. *Conserv. Biol.* 2008, *22*, 1026–1036.

Cash, D.W.; Clark, W.C.; Alcock, F.; Dickson, N.M.; Eckley, N.; Guston, D.H.; Jäger, J.; Mitchell, R.B. Knowledge systems for sustainable development. *Proc. Natl. Acad. Sci. USA* 2003, *100*, 8086–8091.

Castillo, L. S., & Stevenson, P. R. (2010). Relative importance of seed-bank and post-disturbance seed dispersal on early gap regeneration in a Colombian Amazon forest. *Biotropica, 42*(4), 488-492.

Chazdon, R.L.; Guariguata, M.R. *Decision Support Tools for Forest Landscape Restoration: Current Status and Future Outlook*; Center for International Forest Research: Bogor, Indonesia, 2018.

Chesson, P., & Huntly, N. (1997). The roles of harsh and fluctuating conditions in the dynamics of ecological communities. *The American Naturalist*, 150, 519-553.

Ciesla, W. M. 2000. Remote sensing in forest health protection. United States Department of Agriculture, Forest Service, Forest Health Technology Enterprise Team, Remote Sensing Applications Center, Salt Lake City, Utah. Report, No. 00–03.

Clark, C. J., Poulsen, J. R., Levey, D. J., & Osenberg, C. W. (2007). Are plant populations seed limited? A critique and meta-analysis of seed addition experiments. *The American Naturalist*, *170*(1), 128-142.

Cohen, D. (1968). A general model of optimal reproduction in a randomly varying environment. *Journal of Ecology*, 56, 219-228.

Colyn, R.B.; Ehlers Smith, D.A.; Ehlers Smith, Y.C.; Smit-Robinson, H.; Downs, C.T. Predicted distributions of avian specialists: A framework for conservation of endangered forests under future climates. *Divers. Distrib.* 2020, *26*, 652–667.

Corona, P.; Chirici, G.; McRoberts, R.E.; Winter, S.; Barbati, A. Contribution of large-scale forest inventories to biodiversity assessment and monitoring. *For. Ecol. Manag.* 2011, *262*, 2061–2069.

Corona, P.; Franceschi, S.; Pisani, C.; Portoghesi, L.; Mattioli, W.; Fattorini, L. Inference on diversity from forest inventories: A review. *Biodivers. Conserv.* 2017, *26*, 3037–3049.

Cressie, N.; Calder, C.A.; Clark, J.S.; Hoef, J.M.V.; Wikle, C.K. Accounting for uncertainty in ecological analysis: The strengths and limitations of hierarchical statistical modeling. *Ecol. Appl.* 2009, *19*, 553–570.

Csontos, P., & Tamás, J. (2003). Comparisons of soil seed bank classification systems. *Seed Science Research*, 13(2), 101-111.

Culvenor, D. 2003. Extracting individual tree information: A survey of techniques for high spatial resolution imagery. In Remote Sensing of Forest Environments: Concepts and Case Studies, ed. M. Wulder, and S. Franklin, 255–277. Dordrecht, Boston, London: Kluwer Academic Publishers.

De Jaegere, T.; Hein, S.; Claessens, H. A review of the characteristics of small-leaved lime (*Tilia cordata* Mill.) and their implications for silviculture in a changing climate. *Forests* 2016, *7*, 56.

Debinski, D. M., K. Kindscher, and M. E. Jakubauskas. 1999. A remote sensing and GIS based model of habitat and biodiversity in the Greater Yellowstone Ecosystem. International Journal of Remote Sensing 20: 3281–3291.

del Cacho, M., & Lloret, F. (2012). Resilience of Mediterranean shrubland to a severe drought episode: the role of seed bank and seedling emergence. *Plant Biology, 14*, 458-466.

Dobler, A.H.; Geist, J.; Stoeckl, K.; Inoue, K. A spatially explicit approach to prioritize protection areas for endangered freshwater mussels. *Aquat. Conserv. Mar. Freshw. Ecosyst.* 2019, *29*, 12–23.

Dostal, P. (2005). Is the population turnover of patchy-distributed annuals determined by dormancy dynamics or dispersal processes? *Ecography, 28*, 745-756.

Edenius, L.; Mikusiñski, G. Utility of habitat suitability models as biodiversity assessment tools in forest management. *Scand. J. For. Res.* 2006, *S7*, 62–72.

Elpel, T.J. (2013). *Botany in a Day: The Patterns Method of Plant Identification* (6th ed.). Hops Press, LLC. µ!

Enright, N. J., Mosner, E., Miller, B. P., Johnson, N., & Lamont, B. B. (2007). Soil vs. canopy seed storage and plant species coexistence in species-rich Australian shrublands. *Ecology, 88*, 2292-2304.

Eriksson, O. (1996). Regional dynamics of plants: a review of evidence for remnant, source–sink and metapopulations. *Oikos, 77*, 248-258.

Esaete, J., Eycott, A. E., Reiniö, J., Telford, R. J., & Vandvik, V. (2014). The seed and fern spore bank of a recovering African tropical forest. *Biotropica*, 46(6), 677-686.

European Environment Agency. Halting the loss of biodiversity by 2010: Proposal for a first set of indicators to monitor progress in Europe. *EEA Tech. Rep.* 2007, *11*, 1–182.

Evans, A. S., & Cabin, R. J. (1995). Can dormancy affect the evolution of post-germination traits – the case of *Lesquerella fendleri*. *Ecology, 76*, 344-356.

Evans, M. E. K., & Dennehy, J. J. (2005). Germ banking: bet-hedging and variable release from egg and seed dormancy. *Quarterly Review of Biology, 80*, 431-451.

Eycott, A. E., Watkinson, A. R., & Dolman, P. M. (2006). The soil seedbank of a lowland conifer forest: the impacts of clear-fell management and implications for heathland restoration. *Forest Ecology and Management, 237*(1-3), 280-289.

Eyvindson, K.; Repo, A.; Mönkkönen, M. Mitigating forest biodiversity and ecosystem service losses in the era of bio-based economy. *For. Policy Econ.* 2018, *92*, 119–127.

Failing, L.; Gregory, R. Ten common mistakes in designing biodiversity indicators for forest policy. *J. Environ. Manag.* 2003, *68*, 121–132.

Faist, A. M., Ferrenberg, S., & Collinge, S. K. (2013). Banking on the past: seed banks as a reservoir for rare and native species in restored vernal pools. *AoB Plants, 5*, plt043.

Fenner, M., & Thompson, K. (2005). *The Ecology of Seeds*. Cambridge: Cambridge University Press.

Franklin, S. E. 2001. Remote sensing for sustainable forest management. Boca Raton, Florida: CRC Press.

Fraser, R. H., Z. Li, and J. Cihlar. 2000. Hotspot and NDVI differencing synergy (HANDS): A new technique for burned area mapping over boreal forest. Remote Sensing of Environment 74: 362–376.

Garibaldi, A.; Turner, N. Cultural keystone species: Implications for ecological conservation and restoration. *Ecol. Soc.* 2004, *9*, 1.

Garwood, N. C. (1989). Tropical soil seed banks: a review. In M. A. Leck, V. T. Parker, & R. L. Simpson (Eds.), *Ecology of Soil Seed Banks* (pp. 149-209). San Diego, CA: Academic Press Inc.

Garwood, N. C., & Tebbs, M. C. (2009). *Seedlings of Barro Colorado Island and the Neotropics.* Comstock Publishing Associates/Cornell University Press.

Gaston, K.J.; Jackson, S.F.; Cantu-Salazar, L.; Cruz-Piñón, G. The ecological performance of protected areas. *Annu. Rev. Ecol. Evol. Syst.* 2008, *39*, 93–113.

Graul, W.D.; Miller, G.C. Strengthening ecosystem management approaches. *Wildl. Soc. Bull.* 1984, *12*, 282–289.

Graul, W.D.; Torres, J.; Denney, R. A species-ecosystem approach for nongame programs. *Wildl. Soc. Bull.* 1976, *4*, 79–80.

Gremer, J. R., & Venable, D. L. (2014). Bet hedging in desert winter annual plants: optimal germination strategies in a variable environment. *Ecology Letters, 17*, 380-387.

Guisan, A.; Thuiller, W. Predicting species distribution: Offering more than simple habitat models. *Ecol. Lett.* 2005, *8*, 993–1009.

Guisan, A.; Thuiller, W.; Zimmermann, N.E. *Habitat Suitability and Distribution Models with Applications in R*; Cambridge University Press: Cambridge, UK, 2017.

Guisan, A.; Tingley, R.; Baumgartner, J.B.; Naujokaitis-Lewis, I.; Sutcliffe, P.R.; Tulloch, A.I.; Regan, T.J.; Brotons, L.; McDonald-Madden, E.; Mantyka-Pringle, C.; et al. Predicting species distributions for conservation decisions. *Ecol. Lett.* 2013, *16*, 1424–1435.

Haavik, A.; Dale, S. Are reserves enough? Value of protected areas for boreal forest birds in southeastern Norway. *Ann. Zool. Fenn.* 2012, *49*, 69–80.

Habel, J.C.; Gossner, M.M.; Meyer, S.T.; Eggermont, H.; Lens, L.; Dengler, J.; Weisser, W.W. Mind the gaps when using science to address conservation concerns. *Biodivers. Conserv.* 2013, *22*, 2413–2427.

Haché, S.; Cameron, R.; Villard, M.-A.; Bayne, E.M.; MacLean, D.A. Demographic response of a neotropical migrant songbird to forest management and climate change scenarios. *For. Ecol. Manag.* 2016, *359*, 309–320.

Hahn, W.A.; Knoke, T. Sustainable development and sustainable forestry: Analogies, differences, and the role of flexibility. *Eur. J. For. Res.* 2010, *129*, 787–801.

Hall, R. J., D. R. Peddle, D. L. and Klita. 2000. Mapping conifer under story within boreal mixed woods from Landsat TM satellite imagery and forest inventory information. The Forestry Chronicle 76: 887–902.

Halme, P.; Holec, J.; Heilmann-Clausen, J. The history and future of fungi as biodiversity surrogates in forests. *Fungal Ecol.* 2017, *27*, 193–201.

Hanski, I. Extinction debt and species credit in boreal forests: Modelling the consequences of different approaches to biodiversity conservation. *Ann. Zool. Fenn.* 2000, *37*, 271–280.

Hanski, I. Habitat loss, the dynamics of biodiversity, and a perspective on conservation. *Ambio* 2011, *40*, 248–255.

Henrikson, L.; Arvidsson, B.; Österling, M. *Aquatic Conservation with Focus on Margaritifera Margaritifera*; Karlstads University: Karlstad, Sweden, 2012.

Hill, D. A., and Leckie, D. G., eds. 1999. International forum: Automated interpretation of high spatial resolution digital imagery for forestry. Proceedings of a workshop held February 10–12, 1998. Natural Resources Canada, Canadian Forest Service, Victoria, B.C.

Hines, A.; Bengston, D.N.; Dockry, M.J.; Cowart, A. Setting up a horizon scanning system: A US federal agency example. *World Futures Rev.* 2018, *10*, 136–151.

Humphrey, J.; Ray, D.; Brown, T.; Stone, D.; Watts, K.; Anderson, R. Using focal species modelling to evaluate the impact of land use change on forest and other habitat networks in western oceanic landscapes. *Forestry* 2009, *82*, 119–134.

Hunter, M., Jr.; Westgate, M.; Barton, P.; Calhoun, A.; Pierson, J.; Tulloch, A.; Beger, M.; Branquinho, C.; Caro, T.; Gross, J.; et al. Two roles for ecological surrogacy: Indicator surrogates and management surrogates. *Ecol. Indic.* 2016, *63*, 121–125.

IUCN Standards and Petitions Committee. *Guidelines for Using the IUCN Red List Categories and Criteria. Version 14*; IUCN: Gland, Switzerland, 2019; Available online: http://www.iucnredlist.org/documents/RedListGuidelines. pdf (accessed on 19 May 2020).

James Hutton Institute (2011). *Handbook of Field Sampling Protocols for Biodiversity Indicator Monitoring.* http://www.hutton.ac.uk/sites/default/files/files/A%20 Handbook%20of%20biodiversity%20monitoring%20protocols(1).pdf

Johansson, V.; Felton, A.; Ranius, T. Long-term landscape scale effects of bioenergy extraction on dead wood-dependent species. *For. Ecol. Manag.* 2016, *371*, 103–113.

Johansson, V.; Ranius, T.; Snäll, T. Epiphyte metapopulation persistence after drastic habitat decline and low tree regeneration: Time-lags and effects of conservation actions. *J. Appl. Ecol.* 2013, *50*, 414–422.

Johnson, E. A. 1992. Fire and vegetation dynamics: Studies from the North American Boreal forest. Cambridge, United Kingdom: Cambridge University Press.

Johnson, K.N.; Duncan, S.; Spies, T.A. Regional policy models for forest bio-diversity analysis: Lessons from coastal Oregon. *Ecol. Appl.* 2017, *17*, 81–90.

Jones, H.P.; Schmitz, O.J. Rapid recovery of damaged ecosystems. *PLoS ONE* 2009, *4*, e5653.

Kadmon, R., & Shmida, A. (1990). Spatiotemporal demographic processes in plant populations: an approach and a case study. *The American Naturalist*, *135*(3), 382-397.

Kalamees, R., & Zobel, M. (2002). The role of the seed bank in gap regeneration in a calcareous grassland community. *Ecology*, *83*, 1017-1025.

Kasischke, E. S., and B. J. Stocks, eds. 2000. Fire, climate change, and carbon cycling in the boreal forest. New York: Springer-Verlag.

Kliskey, A.D.; Lofroth, E.C.; Thompson, W.A.; Brown, S.; Schreier, H. Simulating and evaluating alternative resource-use strategies using GIS-based habitat suitability indices. *Landsc. Urban Plan.* 1999, *45*, 163–175.

Kotiaho, J.S.; Mönkkönen, M. From a crisis discipline towards prognostic conservation practise: An argument for setting aside degraded habitats. *Ann. Zool. Fenn.* 2017, *54*, 27–37.

Kröger, M.; Raitio, K. Finnish forest policy in the era of bioeconomy: A pathway to sustainability? *For. Policy Econ.* 2017, *77*, 6–15.

Kuuluvainen, T. Forest management and biodiversity conservation based on natural ecosystem dynamics in northern Europe: The complexity challenge. *Ambio* 2009, *38*, 309–315.

Laasimer, L. *Vegetation of the Estonian S.S.R.*; Valgus: Tallinn, Estonian, 1965.

Lambeck, R.J. Focal species and restoration ecology: Response to Lindenmayer et al. *Conserv. Biol.* 2002, *16*, 549–551.

Lambeck, R.J. Focal species: A multi-species umbrella for nature conservation. *Conserv. Biol.* 1997, *11*, 849–856.

Lamont, B.B., & Enright, N. J. (2000). Adaptive advantages of aerial seed banks. *Plant Species Biology*, *15*(2), 157-166.

Landuyt, D.; Perring, M.P.; Seidl, R.; Taubert, F.; Verbeeck, H.; Verheyen, K. Modelling understorey dynamics in temperate forests under global change – Challenges and perspectives. *Perspect. Plant Ecol. Evol. Syst.* 2018, *31*, 44–54.

Lee, B. S., M. E. Alexander, B. C. Hawkes, T. J. Lynham, B. J. Stocks, and P. Englefi eld. 2002. Information systems in support of wildland fi re management decision making in Canada. Computers and Electronics in Agriculture 37: 185–198.

Leslie, H.M.; Goldman, E.; Mcleod, K.L.; Sievanen, L.; Balasubramanian, H.; Cudney-Bueno, R.; Feuerstein, A.; Knowlton, N.; Lee, K.; Pollnac, R.; et al. How good science and stories can go hand-in-hand. *Conserv. Biol.* 2013, *27*, 1126–1129.

Leverett, R., Bertolette, D. (2015). *American Forests Champion Trees Measuring Guidelines Handbook.* https://www.americanforests.org/wp-content/uploads/2014/12/AF-Tree-Measuring-Guidelines_LR.pdf µ!

Levrel, H.; Kerbiriou, C.; Couvet, D.; Weber, J. OECD pressure–state–response indicators for managing biodiversity: A realistic perspective for a French biosphere reserve. *Biodivers. Conserv.* 2009, *18*, 1719–1732.

Lim, K., P. Treitz, M. Wulder, B. St-Onge, and M. Flood. 2003. LiDAR remote sensing of forest structure. Progress in Physical Geography 27(1): 88–106.

Lindenmayer, D.B.; Cunningham, S.A. Six principles for managing forests as ecologically sustainable ecosystems. *Landsc. Ecol.* 2013, *28*, 1099–1110.

Lindenmayer, D.B.; Manning, A.D.; Smith, P.L.; Possingham, H.P.; Fischer, J.; Oliver, I.; McCarthy, M.A. The focal-species approach and landscape restoration: A critique. *Conserv. Biol.* 2002, *16*, 338–345.

Lindenmayer, D.B.; Margules, C.R.; Botkin, D.B. Indicators of biodiversity for ecologically sustainable forest management. *Conserv. Biol.* 2000, *14*, 941–950.

Lõhmus, A.; Fridolin, H.; Leivits, A.; Tõnisson, K.; Rannap, R. Prioritizing research gaps for national conservation management and policy: The managers' perspective in Estonia. *Biodivers. Conserv.* 2019, *28*, 2565–2579.

Lõhmus, A.; Kohv, K.; Palo, A.; Viilma, K. Loss of old-growth, and the minimum need for strictly protected forests in Estonia. *Ecol. Bull.* 2004, *51*, 401–411.

Lõhmus, A.; Kraut, A.; Rosenvald, R. Dead wood in clearcuts of semi-natural forests in Estonia: Site-type variation, degradation, and the influences of tree retention and slash harvest. *Eur. J. For. Res.* 2013, *132*, 335–349.

Lõhmus, A.; Leivits, M.; Pçterhofs, E.; Zizas, R.; Hofmanis, H.; Ojaste, I.; Kurlavièius, P. The Capercaillie (*Tetrao urogallus*): An iconic focal species for knowledge-based integrative management and conservation of Baltic forests. *Biodivers. Conserv.* 2017, *26*, 1–21.

Lõhmus, A.; Lõhmus, P. Coarse woody debris in mid-aged stands: Abandoned agricultural versus long-term forest land. *Can. J. For. Res.* 2005, *35*, 1502–1506.

Lõhmus, A.; Lõhmus, P. Old-forest species: The importance of specific substrata vs. stand continuity in the case of calicioid fungi. *Silva Fenn.* 2011, *45*, 1015–1039.

Lõhmus, A.; Vunk, E.; Runnel, K. Conservation management for forest fungi in Estonia: The case of polypores. *Folia Cryptog. Estonica* 2018, *55*, 79–89.

Lõhmus, K.; Paal, T.; Liira, J. Long-term colonization ecology of forest-dwelling species in a fragmented rural landscape–dispersal versus establishment. *Ecol. Evol.* 2014, *4*, 3113–3126.

Lõhmus, P.; Lõhmus, A. The potential of production forests for sustaining lichen diversity: A perspective on sustainable forest management. *Forests* 2019, *10*, 1063.

Loveridge, A.J.; Sousa, L.L.; Seymour-Smith, J.; Hunt, J.; Coals, P.; O'Donnell, H.; Lindsey, P.A.; Mandisodza-Chikerema, R.; Macdonald, D.W. Evaluating the spatial intensity and demographic impacts of wire-snare bush-meat poaching on large carnivores. *Biol. Conserv.* 2020, *244*, 108504.

Lucas, N. S., and P. J. Curran. 1999. Forest ecosystem simulation models: The role of remote sensing. Progress in Physical Geography 23: 391–423.

Lundemo, S., Falahati-Anbaran, M., & Stenøien, H. K. (2009). Seed banks cause elevated generation times and effective population sizes of *Arabidopsis thaliana* in northern Europe. *Molecular Ecology, 18*, 2798–2811.

Mandak, B., Zákravský, P., Mahelka, V., & Plaèková, I. (2012). Can soil seed banks serve as genetic memory? A study of three species with contrasting life history strategies. *PLoS ONE, 7*, e49471.

Map of Biodiversity Importance. Available online: https://www.natureserve.org/conservation-tools/projects/map-biodiversity-importance (accessed on 20 June 2020).

Måren, I. E., & Vandvik, V. (2009). Fire and regeneration: the role of seed banks in the dynamics of northern heathlands. *Journal of Vegetation Science, 20*, 871-888.

Marzluff, J.M.; Millspaugh, J.J.; Ceder, K.R.; Oliver, C.D.; Withey, J.; McCarter, J.B.; Mason, C.L.; Comnick, J. Modeling changes in wildlife habitat and timber revenues in response to forest management. *For. Sci.* 2002, *48*, 191–202.

Mattsson, B.J.; Cooper, R.J. Louisiana waterthrushes (*Seiurus motacilla*) and habitat assessments as cost-effective indicators of instream biotic integrity. *Freshw. Biol.* 2006, *51*, 1941–1958.

Mazziotta, A.; Pouzols, F.M.; Mönkkönen, M.; Kotiaho, J.S.; Strandman, H.; Moilanen, A. Optimal conservation resource allocation under variable economic and ecological time discounting rates in boreal forest. *J. Environ. Manag.* 2016, *180*, 366–374.

McComb, B.C.; Spies, T.A.; Olsen, K.A. Sustaining biodiversity in the Oregon Coast Range: Potential effects of forest policies in a multi-ownership province. *Ecol. Soc.* 2007, *12*, 29.

McInerny, G.J.; Chen, M.; Freeman, R.; Gavaghan, D.; Meyer, M.; Rowland, F.; Spiegelhalter, D.J.; Stefaner, M.; Tessarolo, G.; Hortal, J. Information visualisation for science and policy: Engaging users and avoiding bias. *Trends Ecol. Evol.* 2014, *29*, 148–157.

McLachlan, S.M.; Bazely, D.R. Recovery patterns of understory herbs and their use as indicators of deciduous forest regeneration. *Conserv. Biol.* 2001, *15*, 98–110.

Mealy, S.P.; Horn, J.R. Integrating wildlife habitat objectives into the forest plan. In *Transactions of the 46th North Americal Wildlife and Natural Resources Conference*; Wildlife Management Institute: Washington, DC, USA, 1981; pp. 488–500.

Meli, P.; Holl, K.D.; Benayas, J.M.R.; Jones, H.P.; Jones, P.C.; Montoya, D.; Mateos, D.M. A global review of past land use, climate, and active vs. passive restoration effects on forest recovery. *PLoS ONE* 2017, *12*, e0171368.

Michanek, G.; Bostedt, G.; Ekvall, H.; Forsberg, M.; Hof, A.R.; De Jong, J.; Rudolphi, J.; Zabel, A. Landscape planning—Paving the way for effective conservation of forest biodiversity and a diverse forestry? *Forests* 2018, *9*, 523.

Mladenoff, D.J.; Host, G.E.; Boeder, J.; Crow, T.R. LANDIS: A spatial model of forest landscape disturbance, succession, and management. In *GIS and Environmental Modeling: Progress and Research Ideas*; Goodchild, M.F., Steyaert, L.T., Parks, B.O., Johnston, C., Maidment, D., Crane, M., Glendinning, S., Eds.; John Wiley & Sons: New York, NY, USA, 1996; pp. 175–179.

Moilanen, A. Landscape zonation, benefit functions and target-based planning: Unifying reserve selection strategies. *Biol. Conserv.* 2007, *134*, 571–579.

Molau, U., & Mølgaard, P. (1996). *ITEX Manual – International Tundra Experiment* (2nd ed.) Copenhagen: Danish Polar Centre. http://ibis.geog.ubc.ca/itex/PDFs/ITEXmanual.pdf

Mönkkönen, M.; Reunanen, P.; Kotiaho, J.S.; Juutinen, A.; Tikkanen, O.P.; Kouki, J. Cost-effective strategies to conserve boreal forest biodiversity and long-term landscape-level maintenance of habitats. *Eur. J. For. Res.* 2011, *130*, 717–727.

Muller, F. M. (1978). *Seedlings of the North-Western European Lowland.* Dordrecht: Springer.

Nel, J.L.; Roux, D.J.; Driver, A.; Hill, L.; Maherry, A.C.; Snaddon, K.; Petersen, C.R.; Smith-Adao, L.B.; Van Deventer, H.; Reyers, B. Knowledge co-production and boundary work to promote implementation of conservation plans. *Conserv. Biol.* 2016, *30*, 176–188.

Nelson, G., Earle, C., Spellenberg, R. (2014). *Trees of Eastern North America, Princeton Field Guides.* Princeton University Press. µ!

Noon, B.R.; McKelvey, K.S. Management of the spotted owl: A case history in conservation biology. *Annu. Rev. Ecol. Syst.* 1996, *27*, 135–162.

Noreika, N.; Helm, A.; Öpik, M.; Jairus, T.; Vasar, M.; Reier, Ü.; Kook, E.; Riibak, K.; Kasari, L.; Tullus, H.; et al. Forest biomass, soil and biodiversity relationships originate from biogeographic affinity and direct ecological effects. *Oikos* 2019, *128*, 1653–1665.

Noss, R.F. High-risk ecosystems as foci for considering biodiversity and ecological integrity in ecological risk assessments. *Environ. Sci. Policy* 2000, *3*, 321–332.

Ochoa-Hueso, R., & Manrique, E. (2010). Nitrogen fertilization and water supply affect germination and plant establishment of the soil seed bank present in a semi-arid Mediterranean scrubland. *Plant Ecology, 210*(2), 263-273.

Orlemann, A., M. Saurer, A. Parsons, and B. Jarvis. 2002. Rapid delivery of satellite imagery for burned area emergency response (Baer). In Proceedings of the 9th Biennial Forest Service Remote Sensing Applications Conference, ed. J. D. Greer. Bethesda, Md.: American Society of Photogrammetry and Remote Sensing.

Orozco, L., Brumér, C. (eds). Inventarios forestales para bosques latifoliados en América Central. Serie Técnica Manual Técnico no 50. Centro Agronómico Tropical para la Investigación y Educación (CATIE), 2002. Turrialba, Costa Rica.

Ortega Adarme, M.; Feitosa, R.Q.; Happ, N.P.; De Almeida, C.A.; Rodrigues Gomes, A. Evaluation of deep learning techniques for deforestation detection in the Brazilian Amazon and cerrado biomes from remote sensing imagery. *Remote Sens.* 2020, *12*, 910.

Parsons, A., and A. Orlemann. 2002. Mapping post-wildfi re burn severity using remote sensing and GIS. Annette Parsons, Andrew In Proceedings 2002 ESRI User Conference. ESRI. Redlands, Calif. gis.esri.com/library/userconf/proc02/pap0431/p0431.htm

Paschen, J.A.; Ison, R. Narrative research in climate change adaptation—Exploring a complementary paradigm for research and governance. *Res. Policy* 2014, *43*, 1083–1092.

Pebesma, E. Simple features for R: Standardized support for spatial vector data. *R J.* 2018, *10*, 439–446.

Peterson, A.T. Ecological niche conservatism: A time-structured review of evidence. *J. Biogeogr.* 2011, *38*, 817–827.

Pigott, C.D. *Tilia cordata* Miller. *J. Ecol.* 1991, *79*, 1147–1207.

Plassmann, K., Brown, N., Jones, M. L. M., & Edwards-Jones, G. (2008). Can atmospheric input of nitrogen affect seed bank dynamics in habitats of conservation interest? The case of dune slacks. *Applied Vegetation Science, 11*(3), 413-420.

Plue, J., & Cousins, S. A. O. (2013). Temporal dispersal in fragmented landscapes. *Biological Conservation, 160*, 250-262.

Plue, J., & Hermy, M. (2012). Consistent seed bank spatial structure across semi-natural habitats determines plot sampling. *Journal of Vegetation Science,* 23, 505-516.

Plue, J., Colas, F., Auffret, A. G., & Cousins, S. A. O. (2017). Methodological bias in the seed bank flora holds significant implications for understanding seed bank community functions. *Plant Biology, 19*(2), 201-210.

Plue, J., Hermy, M., Verheyen, K., Thuillier, P., Saguez, R., & Decocq, G. (2008). Persistent changes in forest vegetation and seed bank 1600 years after human occupation. *Landscape Ecology, 23*(6), 673-688.

Plue, J., Thompson, K., Verheyen, K., & Hermy, M. (2012). Seed banking in ancient forest species: why total sampled area really matters. *Seed Science Research, 22*(2), 123-133.

Poiani, K. A., & Johnson, W. C. (1988). Evaluation of the emergence method in estimating seed bank composition of prairie wetlands. *Aquatic Botany, 32*(1-2), 91-97.

Prodan, M., Peters, R., Cox, F., Real. Pedro. Mensura Forestal. Serie investigación y educación en desarrollo sostenible no.1. Deutsche Gesellschaft für Technische Zusammenarbeit (GTZ) GmbH, Instituto Interamericano de Cooperación para la Agricultura (IICA), 1997. San José, Costa Rica. 586 p.

QGIS Development Team. *QGIS Geographic Information System*. Open Source Geospatial Foundation Project. 2020. Available online: http://qgis.osgeo.org (accessed on 18 May 2020).

Quarshie, A.; Salmi, A.; Wu, Z. From equivocality to reflexivity in biodiversity protection. *Organ. Environ.* in press.

Raphael, M.G.; Wisdom, M.J.; Rowland, M.M.; Holthausen, R.S.; Wales, B.C.; Marcot, B.G.; Rich, T.D. Status and trends of habitats of terrestrial vertebrates in relation to land management in the interior Columbia river basin. *For. Ecol. Manag.* 2001, *153*, 63–87.

Reier, Ü.; Tuvi, E.L.; Pärtel, M.; Kalamees, R.; Zobel, M. Threatened herbaceous species dependent on moderate forest disturbances: A neglected target for ecosystem-based silviculture. *Scand. J. For. Res.* 2005, *20*, 145–152.

Rempel, R.S.; Naylor, B.J.; Elkie, P.C.; Baker, J.; Churcher, J.; Gluck, M.J. An indicator system to assess ecological integrity of managed forests. *Ecol. Indic.* 2016, *60*, 860–869.

Reza, M.I.H.; Abdullah, S.A.; Nor, S.B.M.; Ismail, M.H. Integrating GIS and expert judgment in a multi-criteria analysis to map and develop a habitat suitability index: A case study of large mammals on the Malayan Peninsula. *Ecol. Indic.* 2013, *34*, 149–158.

Roberge, J.M.; Öhman, K.; Lämås, T.; Felton, A.; Ranius, T.; Lundmark, T.; Nordin, A. Modified forest rotation lengths: Long-term effects on landscape-scale habitat availability for specialized species. *J. Environ. Manag.* 2018, *210*, 1–9.

Rondeux, J.; Sanchez, C. Review of indicators and field methods for monitoring biodiversity within national forest inventories. Core variable: Deadwood. *Environ. Monit. Assess.* 2010, *164*, 617–630.

Rookwood, P. Landscape planning for biodiversity. *Landsc. Urban Plan.* 1995, *31*, 379–385.

Runnel, K.; Lõhmus, A. Deadwood-rich managed forests provide insights into the old-forest association of wood-inhabiting fungi. *Fungal Ecol.* 2017, *27*, 155–167.

Runnel, K.; Sell, I.; Lõhmus, A. Recovery of the Critically Endangered bracket fungus *Amylocystis lapponica* in the Estonian network of strictly protected forests. *Oryx* 2019, in press.

Sætersdal, M.; Gjerde, I. Prioritising conservation areas using species surrogate measures: Consistent with ecological theory? *J. Appl. Ecol.* 2011, *48*, 1236–1240.

Schamberger, M.; Krohn, W.B. Status of the habitat evaluation procedures. In *Transactions of the 47th North Americal Wildlife and Natural Resources Conference*; Sabol, K., Ed.; Wildlife Management Institute: Washington, DC, USA, 1982; pp. 154–164.

Schreuder, H.T., Gregoire, T.G., Wood, G.B. Sampling methods for multiresource forest inventory. John Wiley and Sons Inc. 1993. New York. 446 p.

Schumaker, N.H.; Brookes, A. HexSim: A modeling environment for ecology and conservation. *Landsc. Ecol.* 2018, *33*, 197–211.

Sepp, T.; Liira, J. Factors influencing the species composition and richness of herb layer in old boreo-nemoral forests. *For. Stud.* 2009, *50*, 23–41.

Shahin, A.; Mahbod, M.A. Prioritization of key performance indicators: An integration of analytical hierarchy process and goal setting. *Int. J. Oper.* 2007, *56*, 226–240.

Shorohova, E.; Kuuluvainen, T.; Kangur, A.; Jõgiste, K. Natural stand structures, disturbance regimes and successional dynamics in the Eurasian boreal forests: A review with special reference to Russian studies. *Ann. For. Sci.* 2009, *66*, 1–20.

Siewert, W., & Tielbörger, K. (2010). Dispersal-dormancy relationships in annual plants: putting model predictions to the test. *The American Naturalist*, *176*(4), 490-500.

Sinclair, S.J.; White, M.D.; Newell, G.R. How useful are species distribution models for managing biodiversity under future climates? *Ecol. Soc.* 2010, *15*, 8.

Sotirov, M.; Blum, M.; Storch, S.; Selter, A.; Schraml, U. Do forest policy actors learn through forward-thinking? Conflict and cooperation relating to the past, present and futures of sustainable forest management in Germany. *For. Policy Econ.* 2017, *85*, 256–268.

Sparks, T.H.; Butchart, S.H.; Balmford, A.; Bennun, L.; Stanwell-Smith, D.; Walpole, M.; Bates, N.R.; Bomhard, B.; Buchanan, G.M.; Chenery, A.M.; et al. Linked indicator sets for addressing biodiversity loss. *Oryx* 2011, *45*, 411–419.

Spies, T.A.; Johnson, K.N.; Burnett, K.M.; Ohmann, J.L.; McComb, B.C.; Reeves, G.H.; Bettinger, P.; Kline, J.D.; Garber-Yonts, B. Cumulative ecological and socioeconomic effects of forest policies in coastal Oregon. *Ecol. Appl.* 2007, *17*, 5–17.

Spies, T.A.; McComb, B.C.; Kennedy, R.S.H.; McGrath, M.T.; Olsen, K.; Pabst, R.J. Potential effects of forest policies on terrestrial biodiversity in a multi-ownership province. *Ecol. Appl.* 2007, *17*, 48–65.

Suislepp, K.; Rannap, R.; Lõhmus, A. Impacts of artificial drainage on amphibian breeding sites in hemiboreal forests. *For. Ecol. Manag.* 2011, *262*, 1078–1083.

Sutherland, W.J.; Bailey, M.J.; Bainbridge, I.P.; Brereton, T.; Dick, J.T.; Drewitt, J.; Dulvy, N.K.; Dusic, N.R.; Freckleton, R.P.; Gaston, K.J.; et al. Future novel threats and opportunities facing UK biodiversity identified by horizon scanning. *J. Appl. Ecol.* 2008, *45*, 821–833.

Tarr, N.M.; Rubino, M.J.; Costanza, J.K.; McKerrow, A.J.; Collazo, J.A.; Abt, R.C. Projected gains and losses of wildlife habitat from bioenergy-induced landscape change. *GCB Bioenergy* 2017, *9*, 909–923.

Telewski, F. W., & Zeevaart, J. A. (2002). The 120 yr period for Dr. Beal's seed viability experiment. *American Journal of Botany, 89*(8), 1285-1288.

Ter Heerdt, G. N. J., Verweij, G. L., Bekker, R. M., & Bakker, J. P. (1996). An improved method for seed-bank analysis: seedling emergence after removing the soil by sieving. *Functional Ecology, 10*, 144-151.

Theobald, D.M.; Hobbs, N.T.; Bearly, T.; Zack, J.A.; Shenk, T.; Riebsame, W.E. Incorporating biological information in local land-use decision making: Designing a system for conservation planning. *Landsc. Ecol.* 2000, *15*, 35–45.

Thompson, J.R.; Simons-Legaard, E.; Legaard, K.; Domingo, J.B. A LANDIS-II extension for incorporating land use and other disturbances. *Environ. Model. Softw.* 2016, *75*, 202–205.

Thompson, K., & Grime, J. P. (1979). Seasonal variation in the seed banks of herbaceous species in ten contrasting habitats. *Journal of Ecology, 67*(3), 893-921.

Thompson, K., Bakker, J. P., & Bekker, R. M. (1997). *The Soil Seed Banks of North West Europe: Methodology, Density and Longevity* (Vol. 1). Cambridge: Cambridge University Press.

Tikkanen, O.P.; Heinonen, T.; Kouki, J.; Matero, J. Habitat suitability models of saproxylic red-listed boreal forest species in long-term matrix management: Cost-effective measures for multi-species conservation. *Biol. Conserv.* 2007, *140*, 359–372.

Tremblay, J.A.; Boulanger, Y.; Cyr, D.; Taylor, A.R.; Price, D.T.; St-Laurent, M. Harvesting interacts with climate change to affect future habitat quality of a focal species in eastern Canada's boreal forest. *PLoS ONE* 2018, *13*, e0191645.

Trisurat, Y.; Kanchanasaka, B.; Kreft, H. Assessing potential effects of land use and climate change on mammal distributions in northern Thailand. *Wildl. Res.* 2014, *41*, 522–536.

Tulloch, V.J.D.; Tulloch, A.I.T.; Visconti, P.; Halpern, B.S.; Watson, J.E.M.; Evans, M.C.; Auerbach, N.A.; Barnes, M.; Beger, M.; Chadès, I.; et al. Why do we map threats? Linking threat mapping with actions to make better conservation decisions. *Front. Ecol. Environ.* 2015, *13*, 91–99.

Uliczka, H.; Angelstam, P.; Roberge, J.-M. Indicator species and biodiversity monitoring systems for non-industrial private forest owners - is there a communication problem? *Ecol. Bull.* 2004, *51*, 379–384.

United States Department of Agriculture (USDA) (n.d.). *PLANTS Database.* USDA Natural Resource Conservation Service. https://plants.usda.gov/java/ µ!

Van der Biest, K.; Meire, P.; Schellekens, T.; D'hondt, B.; Bonte, D.; Vanagt, T.; Ysebaert, T. Aligning biodiversity conservation and ecosystem services in spatial planning: Focus on ecosystem processes. *Sci. Total Environ.* 2020, *712*, 136350.

Vandekerkhove, K.; De Keersmaeker, L.; Walleyn, R.; Köhler, F.; Crevecoeur, L.; Govaere, L.; Thomaes, A.; Verheyen, K. Reappearance of old growth elements in lowland woodlands in northern Belgium: Do the associated species follow? *Silva Fenn.* 2012, *45*, 909–935.

Vandvik, V., & Goldberg, D. E. (2006). Sources of diversity in a grassland metacommunity: quantifying the contribution of dispersal to species richness. *The American Naturalist*, *168*, 157-167.

Vandvik, V., Klanderud, K., Meineri, E., Måren, I. E., & Töpper, J. P. (2016). Seed banks are biodiversity reservoirs: species–area relationships above versus below ground. *Oikos*, *125*, 218-228.

Vellend, M.; Verheyen, K.; Jacquemyn, H.; Kolb, A.; Van Calster, H.; Peterken, G.; Hermy, M. Extinction debt of forest plants persists for more than a century following habitat fragmentation. *Ecology* 2006, *87*, 542–548.

Virginia Botanical Association. (2017). *Digital Atlas of Virginia Flora.* http://www.vaplantatlas.org µ!

Virginia Department of Forestry (2016). *Common Native Trees of Virginia Tree Identification Guide.* Virginia Department of Forestry Publication P00026. www.dof.virginia.gov µ!

Ward, M.; Rhodes, J.R.; Watson, J.E.; Lefevre, J.; Atkinson, S.; Possingham, H.P. Use of surrogate species to cost-effectively prioritize conservation actions. *Conserv. Biol.* 2020, *34*, 600–610.

Watts, M.T. (1998). *A Manual for the Identification of Trees by Their Leaves.* Nature Study Guide Publications. µ!

Wenger, K.F. (ed). Forestry Handbook. Second edition. John Wiley and Sons, Inc. 1984. New York. 1335 p.

Wickham, H.; François, R.; Henry, L.; Müller, K. *dplyr: A Grammar of Data Manipulation*, R package version 0.8.4; 2020. Available online: https://CRAN.R-project.org/package=dplyr (accessed on 18 May 2020).

Wilhere, G.F.; Linders, M.J.; Cosentino, B.L. Defining alternative futures and projecting their effects on the spatial distribution of wildlife habitats. *Landsc. Urban Plan.* 2007, *79*, 385–400.

Yashina, S., Gubin, S., Maksimovich, S., Yashina, A., Gakhova, E., & Gilichinsky, D. (2012). Regeneration of whole fertile plants from 30,000-y-old fruit tissue buried in Siberian permafrost. *Proceedings of the National Academy of Sciences USA*, *109*(10), 4008-4013.

Zhu, M.J.; Hoctor, T.S.; Volk, M.; Frank, K.I.; Zwick, P.D.; Carr, M.H.; Linhoss, A.C. Spatial conservation prioritization to conserve biodiversity in response to sea level rise and land use change in the Matanzas River Basin, Northeast Florida. *Landsc. Urban Plann.* 2015, *144*, 103–118.

INDEX